Concept Mapping in Mathematics

Karoline Afamasaga-Fuata'i
Editor

Concept Mapping in Mathematics

Research into Practice

 Springer

Editor
Karoline Afamasaga-Fuata'i
University of New England
Australia
kafamasa@une.edu.au

ISBN 978-0-387-89193-4   e-ISBN 978-0-387-89194-1
DOI 10.1007/978-0-387-89194-1

Library of Congress Control Number: 2008942431

Printed on acid-free paper

springer.com

Foreword

It is a real pleasure to write a foreword to this first book that seeks to illustrate how concept mapping can be used to facilitate meaningful learning in mathematics. I believe the authors succeed in showing that mathematics can be more than the memorization of procedures to get answers to textbook types of problems. Through all of my elementary and high school studies of mathematics, I thought that what was required was to learn the procedures for getting answers, and I recall thinking that after doing 2 or 3 textbook problems of a given type, mathematics was rather tedious and relatively boring. By contrast, I saw the study of science as the search for understanding of fundamental concepts, such as the nature of matter, energy, and evolution. This I found to be exciting and I was always eager to seek deeper understanding of basic science concepts. It came as a somewhat shocking surprise to me when I was studying calculus at the University of Minnesota that there were fundamental mathematics concepts also, such as limit, slope, proportionality, etc. I recall feeling cheated that none of my teachers had helped me to gain a *conceptual* understanding of mathematics!

Another fact that I pondered during my youth was that there were child prodigies who could do unusually difficult mathematics at ages 10 or 12, but there were very few such prodigies in sciences, literature, or history. With the invention of the concept mapping tool, it became clear to me why the latter was the case in some disciplines. To achieve relative mastery in a field of science, there were many concepts that had to be learned and understood. In contrast, in music or mathematics, if one gains an early understanding of a few dozen fundamental concepts, such as those discussed in the chapters of this book, you can move on to understanding of major domains of mathematics, and perhaps even do some creative mathematics. History has shown this to be the case. Of course, we are all familiar with the musical early genius of Wolfgang Amadeus Mozart. My hypothesis is that if we can transform the teaching of mathematics to a field that is *conceptually transparent* to students from pre-school through high school, this might become one of the easiest fields of study instead of the opaque and often-dreaded study mathematics has been for so many students.

As a first pioneering effort, this book omits discussion of many issues that hinder the kind of school instruction in mathematics that would make the subject conceptually transparent and meaningful to all students. Nevertheless, the authors of the following chapters provide examples of how to teach mathematics better, and data to support the validity of the idea that teaching mathematics for understanding of mathematics concepts is better than what we are now doing in most school classrooms.

This book can also serve as a primer for mathematics teachers at all levels, who wish to make the study of mathematics a meaningful learning experience for all students. It can also provide guidance for future research using concept mapping tools that can expand our understanding of meaningful teaching and learning in mathematics. I see a bright future for the improvement of mathematics education.

Pensacola, FL   Joseph D. Novak

Contents

Contributors

Karoline Afamasaga-Fuata'i School of Education, University of New England, Armidale, Australia, kafamasa@une.edu.au

Mario Aspée Universidad, Nacional Experimental Del Tachira, San Cristóbal, Estado Táchira, Venezuela, maspee@unet.edu.ve

William Caldwell University of North Florida, Jacksonville, FL, USA, wcaldwel@unf.edu

Alberto J. Cañas Florida Institute for Human & Machine Cognition, Pensacola, FL, USA, acanas@ihmc.us

Fermín M. González Public University of Navarra, Pamplona, Spain, fermin@unavarra.es

Greg McPhan SiMERR National Centre, University of New England, Armidale, Australia, gmcphan2@une.edu.au

Joseph D. Novak Cornell University, Ithaca, NY, USA; Florida Institute for Human & Machine Cognition, Pensacola, FL, USA, jnovak@ihmc.us

Rafael Pérez Flores Universidad Autónoma Metropolitana, Mexico City, México, pfr@correo.azc.uam.mx

Edurne Pozueta Public University of Navarra, Pamplona, Spain, epozueta@unavarra.es

Maria S. Ramirez De Mantilla Universidad, Nacional Experimental Del Tachira, San Cristóbal, Estado Táchira, Venezuela, marimant@unet.edu.ve

Irma Sanabria Universidad, Nacional Experimental Del Tachira, San Cristóbal, Estado Táchira, Venezuela, irmasa66@hotmail.com

Jean Schmittau State University of New York, Binghamton, Vestal, NY, USA, jschmitt@binghamton.edu

Neyra Tellez Universidad, Nacional Experimental Del Tachira, San Cristóbal, Estado Táchira, Venezuela, ntellez@unet.edu.ve

James J. Vagliardo State University of New York at Binghamton, Vestal, NY, USA, jjvags@gmail.com

Introduction

Karoline Afamasaga-Fuata'i

This book is the first comprehensive book on concept mapping in mathematics. It provides the reader with an understanding of how the meta-cognitive tool, namely, hierarchical concept maps, and the process of concept mapping can be used innovatively and strategically to improve planning, teaching, learning, and assessment at different educational levels. The book consists of a collection of articles on research conducted critically to examine the usefulness of concept maps in the educational setting ranging from primary grade classrooms through secondary mathematics to preservice teacher education, undergraduate mathematics and post-graduate mathematics education. A second meta-cognitive tool, called vee diagrams, was also critically examined by some authors particularly its value in improving mathematical problem solving and mathematical modeling in physics.

Thematically, the book flows from a historical development overview of concept mapping in the sciences to applications of concept mapping in mathematics by teachers and preservice teachers as a means of analyzing mathematics topics, planning for instruction and designing assessment tasks including applications by school and university students as learning and review tools. The book provides case studies and resources that have been field tested with school and university students alike. The findings presented have implications for enriching mathematics learning and making problem solving more accessible and meaningful for students.

The theoretical underpinnings of concept mapping and of the studies in the book include Ausubel's cognitive theory of meaningful learning, constructivist and Vygotskian psychology to name a few. There is evidence particularly from international studies such as the Program for International Student Assessment (PISA) and Trends in International Mathematics and Science Study (TIMSS) and mathematics education research, which suggest that students' mathematical literacy and problem solving skills should be enhanced through students collaborating and interacting as they work, discuss and communicate mathematically. This book proposes the meta-cognitive strategy of concept mapping as one viable means of promoting, communicating and explicating students' mathematical thinking and reasoning publicly

K. Afamasaga-Fuata'i (✉)
University of New England, Armidale, Australia

in a social setting (e.g., mathematics classrooms) as they engage in mathematical dialogues and discussions.

Whilst a number of books have been published on concept mapping in the sciences and science education, none is dedicated to mathematics and mathematics education.

Shared commitments to develop and promote meaningful learning of mathematics and a more conceptual understanding of problem solving beyond simply knowing the formulas and procedures subsequently led the contributing authors to investigate the value of concepts maps as an educational tool in a variety of settings with different types of students from primary to university levels. The book, organised into four main parts, presents a diversity of applications as researched by the contributing authors.

Part I provides a historical overview of the development of concept mapping by Joseph Novak, the inventor of hierarchical concept maps. While Part II focuses on research conducted in primary mathematics with primary student teachers, teachers and students, Part III focuses on research conducted in secondary mathematics with secondary student teachers, teachers and students. Research conducted in university mathematics with university teachers and students are presented in Part IV whereas Part V poses questions about potential directions for future research in concept mapping in mathematics.

Individual chapters within each of Parts I–V are briefly summarised below:

Part I: A Historical Overview of Concept Mapping

In Chapter 1, Joseph Novak and Alberto Cañas describe how his research team, in response to a need for a tool to describe explicit changes to children's conceptual understanding, invented *concept mapping* in 1972. Underlying the research program and the development of the concept mapping tool was an explicit cognitive psychology of learning and an explicit constructivist epistemology. As well as describing the various applications of concept maps since its creation, leading up to the development of a concept mapping software at the Institute for Human and Machine Cognition to facilitate concept mapping, Novak and Cañas further propose *A New Model for Education*.

Part II: Primary Mathematics Teaching and Learning

In Chapter 2, Karoline Afamasaga-Fuata'i presents the case study of a primary student teacher who, over one semester, applied concept maps and vee diagrams as tools to conceptually analyse the *Measurement* strand of a primary mathematics syllabus, to communicate her subsequent interpretations and understanding of syllabus outcomes, and to pedagogically plan learning activities to ensure the development of students' conceptual understanding of *length, volume, surface area,* and *capacity.*

In Chapter 3, Jean Schmittau and James Vagliardo use a case study to illustrate the power of concept mapping to reveal both the centrality of the *Positional* concept within elementary mathematics and the pedagogical content knowledge required to teach the concept of positional system and other mathematics concepts to which it is related.

Chapter 4 by Karoline Afamasaga-Fuata'i explores a post-graduate student's use of concept maps and vee diagrams as tools to analyse the *Fraction* strand of a primary mathematics curriculum and related problems and to record his developing pedagogical understanding over the semester as a consequence of social critiques and further revision.

Karoline Afamasaga-Fuata'i and Greg McPhan, in Chapter 5, presents a case study to highlight the kinds of concerns and issues that can impede the introduction of concept mapping to real classrooms and to demonstrate ways in which concept maps can be used to reinforce and review learning of mathematics and science topics in two primary classrooms. The ultimate highlight was the initiative by the two primary teachers and their students to come together for peer tutoring and peer collaborations as the older students mentored and assisted the younger ones in using a computer software *Inspiration*^{TM} to collaboratively construct concept maps.

Part III: Secondary Mathematics Teaching and Learning

In Chapter 6, Edurne Pozueta Mendia and Fermín González present a study to illustrate how concept maps can be used to monitor and identify the extent of secondary students' meaningful learning of *Proportionality* after teaching an innovative instructional module. Comparing individually-constructed concept maps to an expert map enabled them to distinguish students who had learnt proportionality more meaningfully from those who learnt by rote learning or had misconceptions.

In Chapter 7, Jean Schmittau examines how concept mapping can be used to assess whether secondary teachers possess the requisite knowledge to teach both concepts and procedures with understanding premised on the view that mathematical algorithms are not merely mechanical procedures to be learned by rote, but as fully conceptual cultural historical products.

Karoline Afamasaga-Fuata'i in Chapter 8 describes the case of a secondary student teacher, to demonstrate how concept maps can be used to provide a macro view of a two-year mathematics curriculum and to innovatively develop a teaching sequence and lesson plan on *Derivatives*.

James Vagliardo in Chapter 9 explores how concept maps may be used to highlight the importance of mediating a deeper meaning of *Logarithms* and its connections to other mathematical ideas by locating its conceptual essence from a cultural-historical context.

In Chapter 10, Maria Ramirez, Mario Aspee, Irma Sanabria & Neyra Tellez provide practical guidelines that are theoretically driven to assist mathematics students to construct concept maps and vee diagrams to illustrate their understanding of mathematical functions used to model physical phenomena.

William Caldwell explores in Chapter 11 the potential of concept mapping for increasing meaningful learning in mathematics at middle school level for both teachers and students and examines the results of mathematical professional development and student learning activities using concept mapping.

Part IV: University Mathematics Teaching and Learning

In Chapter 12, Karoline Afamasaga-Fuata'i presents a study that investigated the use of concept maps and vee diagrams to learn about new advanced mathematics topics students had not encountered before. Students displayed their developing understanding and knowledge on concept maps and vee diagrams for public scrutiny.

Rafael Pérez Flores in Chapter 13 deals with a particular way of using concept maps to contribute to engineering students' meaningful learning of mathematics by implementing a didactic strategy that is guided by the professor's concept maps, to facilitate the development of students' critical thinking and understanding and the application of these process in solving mathematics problems.

In Chapter 14, Karoline Afamasaga-Fuata'i presents the case of a student who used concept maps to illustrate and communicate his evolving understanding of *Differential Equations* over the semester as a result of his own research and revisions subsequent to social critiques during seminar presentations and individual consultations.

In Chapter 15, Karoline Afamasaga-Fuata'i explores a group of students' use of concept maps and vee diagrams as tools to visually display their developing and growing understanding of the conceptual structure of selected topics and the connections between this structure and procedures for solving problems.

Part V: Future Directions

In Chapter 16, Karoline Afamasaga-Fuata'i provides a synopsis of chapter findings and implications for incorporating concept mapping in real classrooms. Also included are suggestions of potential directions for future research in concept mapping in mathematics education.

Part I
A Historical Overview of Concept Mapping

Chapter 1
The Development and Evolution of the Concept Mapping Tool Leading to a New Model for Mathematics Education

Joseph D. Novak and Alberto J. Cañas

A research program at Cornell University that sought to study the ability of first and second grade children to acquire basic science concepts and the affect of this learning on later schooling led to the need for a new tool to describe explicit changes in children's conceptual understanding. *Concept mapping* was invented in 1972 to meet this need, and subsequently numerous other uses have been found for this tool. Underlying the research program and the development of the concept mapping tool was an explicit cognitive psychology of learning and an explicit constructivist epistemology, described briefly in this paper.

In 1987, collaboration began between Novak and Cañas and others at the Institute for Human and Machine Cognition, then part of the University of West Florida. This led to the development of software to facilitate concept mapping, evolving into the current version of CmapTools, now widely used in schools, universities, corporations, and governmental and non-governmental agencies.

CmapTools allows for selective use of Internet and other digital resources that can be attached to concept nodes and accessed via icons on a concept, providing a kind of knowledge portfolio or knowledge model. This capability permits a new kind of learning environment wherein learners build their own knowledge models, individually or collaboratively, and these can serve as a basis for life-long meaningful learning. Combined with other educational practices, use of CmapTools permits a New Model for Education, described briefly. Preliminary studies are underway to assess the possibilities of this New Model.

J.D. Novak (✉)
Cornell University, Ithaca, New York, USA; Florida Institute for Human & Machine Cognition, Pensacola, FL, USA
e-mail: jnovak@ihmc.us

A.J. Cañas (✉)
Florida Institute for Human & Machine Cognition, Pensacola, FL, USA
e-mail: acanas@ihmc.us

K. Afamasaga-Fuata'i (ed.), *Concept Mapping in Mathematics*,
DOI 10.1007/978-0-387-89194-1_1, © Springer Science+Business Media, LLC 2009

Introduction: The Invention of Concept Mapping

During the 1960s, Novak's research group first at Purdue University and then at Cornell University sought to develop a coherent theory of learning and theory of knowledge that would form a basis for more systematic research in education and a scientific basis for school curriculum design. We found that Ausubel's assimilation theory of learning presented in his *Psychology of Meaningful Verbal Learning* (1963) spoke to what we were most interested in, namely, how do learners grasp the meanings of concepts in a way that permits them to use these concepts to facilitate future learning and creative problem solving? Ausubel stressed the distinction between learning by rote and learning meaningfully. Rote learning or memorizing information may permit short-term recall of this information, but since it does not involve the learner in actively integrating new knowledge with concepts and propositions already known, it does not lead to an improved organization of knowledge in the learner's cognitive structure. In contrast, meaningful learning requires that the learner chooses actively to seek integration of new concept and propositional meanings into her/his cognitive structure, thereby enhancing and enriching her/his cognitive structure. Since the 1960s, many studies have shown that what distinguishes the naive or novice learner from the expert is the extent to which the person has a highly organized cognitive structure and metacognitive strategies to employ this knowledge in new learning or novel problems solving (Bransford, Brown, & Cocking, 1999). In short, one builds expertise in any discipline by building powerful knowledge structures that characterize the key intellectual achievements in that discipline, as well as strategies to use this knowledge.

Also occurring in the 1960s was a philosophical movement away from *positivism* where knowledge creation was seen as a search for "truths" unfettered by prior ideas or emotion. Kuhn's (1962) book, *The Structure of Scientific Revolutions* marked a turning point toward *constructivist* ideas that saw knowledge creation as a human endeavor that involved changing methodologies and paradigms and an evolving set of ideas and methodologies leading to useful but evolving paradigms and ideas. We saw that this constructivist epistemology and cognitive psychology was equally applicable to mathematics and mathematics teaching. The challenge was: "how do we get educators and the school contexts to change to enhance the utilization of these new insights" (Novak, 1986, p. 184). As we proceeded in our mathematics education studies, we found we could work with a theory of learning that explained how new concepts are acquired and used that complemented a theory of knowledge that focused on the evolving creation of new concepts and problem solving approaches.

Working with elementary school children, we sought to design new instruction in such a way that meaningful learning would be enhanced, and to demonstrate that such learning could facilitate future learning and problem solving. To do this we found that we needed an assessment method that could monitor the evolving knowledge frameworks of our learners. Moreover, we were interested in demonstrating that young children (ages 6–8) could acquire significant science concepts and that this learning would facilitate later learning. The advances in learning theory and

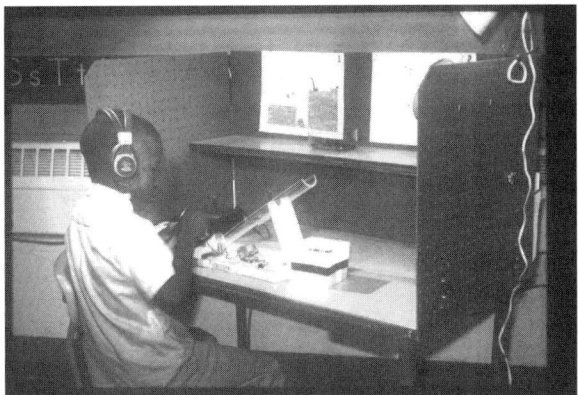

Fig. 1.1 An audio-tutorial carrel unit showing a 7 year old student learning about energy transformations

epistemology permitted Novak to construct a theory of education, first presented in 1977 (Novak, 1977) and modified and elaborated in 1998 (Novak, 1998). This theory has been guiding our work for some three decades.

Given our interest in teaching young children basic science concepts such as the nature of matter, energy, and energy transformations and related ideas, we were faced with the reality that most primary grade teachers did not posses the knowledge to teach these ideas. Therefore, we developed a series of audio-tutorial lessons where children were guided by audio instruction in the manipulation of pertinent materials and presented the vocabulary needed to code the concepts they were learning. Figure 1.1 shows an example of one of these audio-tutorial lessons and the carrel unit in which they were presented.

Lessons proceeded on a schedule of one new lesson every two weeks placed in classrooms in Ithaca Public schools. Earlier studies had shown that a variety of paper and pencil tests were inadequate for monitoring the growth in the children's understanding of concept meanings, so we used interviews to probe the children's knowledge. This, however created another problem in that it was difficult to see in the interview transcripts just how the children's cognitive structure was changing and how new concepts were being integrated into the child's cognitive structure. After struggling with the problem for some weeks, Novak's research group came up with the idea of transforming the interview transcripts into a hierarchically arranged picture showing the concepts and proposition revealed in the interview. We called the resulting drawing a *concept map*. We found that we could now see explicitly what concept and propositions were being integrated into a child's mind as they progressed through the audio-tutorial lessons, and also in later years as they encountered school science studies.

Figure 1.2 shows an example of a concept map drawn from an interview with one child at the end of grade two and another for the same child in grade 12.

This child was obviously a meaningful learner and not only were some misconceptions remediated, but he developed an excellent knowledge structure for this area of science. The results from this 12-year longitudinal study demonstrated several

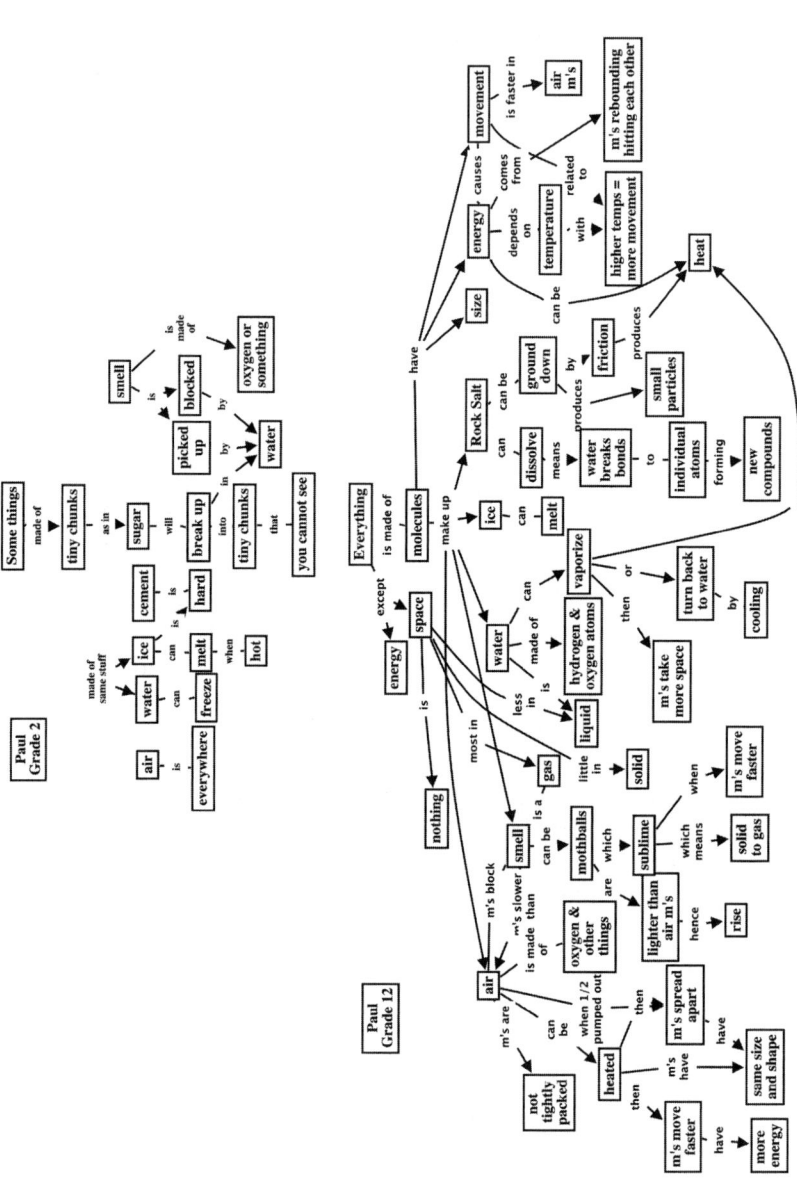

Fig. 1.2 Two concept maps drawn from interviews with children in grade 2, (*top*) and grade 12 (*bottom*). These show the enormous growth in conceptual understanding for this child over 10 years of schooling

things: concept maps could be a powerful knowledge representation and assessment tool; young children can acquire significant understanding of basic science concepts (widely disputed in the 1960s and 70s); technology can be used to deliver meaningful instruction to students; early meaningful learning of science concepts highly influenced learning in later science studies (Novak & Musonda, 1991).

Shortly after we developed the concept map tool for the assessing of changes in learner's cognitive structure, we found that our staff and others were reporting that concept maps were very helpful as a study tool for virtually any subject matter. This led Novak to develop a course called "Learning To Learn", and he taught this course at Cornell for some 20 years. One of the outcomes from the course was the book *Learning How to Learn* (Novak & Gowin, 1984). The book has been published in 9 languages and remains as popular as when it was first published. Concept maps are also powerful metacognitive tools helping students to understand the nature of knowledge and the nature of meaningful learning (Novak, 1990). More recently we have found concept maps to be an excellent tool for capturing expert knowledge for archiving and training in schools and corporations, and also for team problem solving. Beginning some 10 years ago, the Institute for Human and Machine Cognition has developed *CmapTools*, software that not only facilitates building concept maps but also offers new opportunities for learning, creating, and using knowledge, as will be discussed further.

The Use of Concept Maps in Mathematics

Our early work using concept maps in mathematics was focused on demonstrating how mathematical ideas could be represented in this form. Cardemone (1975) showed how the key ideas in a remedial college math course could be represented using concept maps. He found that the use of concept maps could help teachers design a better sequence of topics and helped students see relationships between topics. Minemier (1983) found that when students made concept maps for the topics they studied they not only performed better on problems solving tests but they also gained increased confidence in their ability to do mathematics. Fuata'i (1985, 1998) used concept maps along with vee diagrams with Form Five students in Western Samoa. She found that students became more autonomous learners and better at solving novel problems as compared with students not using these tools. Figure 1.3 shows an example of a concept map produced by one of her students.

CmapTools and the Internet

CmapTools goes beyond facilitating the construction of concept maps through an easy-to-use map editor, leveraging on the power of technology and particularly the Internet and WWW to enable students to collaborate locally or remotely in the construction of their maps, search for information that is relevant to their maps, link all types of resources to their maps, and publish their concept maps,

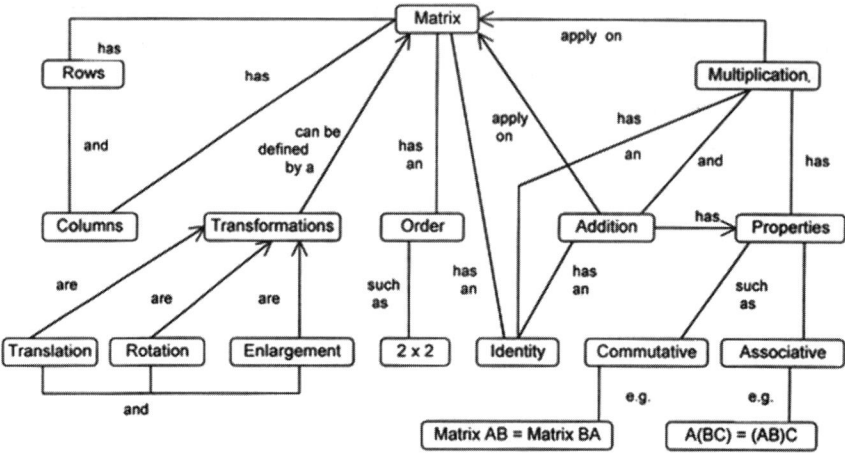

Fig. 1.3 A concept map made by a Western Samoan student in a study by Fuata'i (1998, p. 65)

(Cañas et al., 2004)[1]. CmapTools facilitates the linking of digital resources (e.g. images, videos, text, Web pages, or other digital concept maps) to further explain concepts through a simple drag-and-drop operation. The linked resources are depicted by an icon under the concept that represents the type of resource linked. By storing the concept maps and linked resources on a CmapServer (Cañas, Hill, Granados, C. Pérez, & J. D. Pérez, 2003), the concept maps are converted to Web pages and can be browsed through a Web browser. The search feature in CmapTools takes advantage of the context provided by a concept map to perform Web searches related to the map, producing more relevant results that may include Web pages and resources, or concept maps stored in the CmapTools network (Carvalho, Hewett, & Cañas, 2001). Thus, a small initial map can be used to search for relevant information which the student can investigate, which leads to an improved map, to another search, and so on. The student can link relevant resources found to the map, create other related maps, and organize these into what we call a *knowledge model* (Cañas, Hill, & Lott, 2003). Figure 1.4 illustrates a concept map made with CmapTools and the insets show some of the resources attached to this map that can be opened by clicking on the icon for the resource. Other examples will be given in subsequent chapters of this book.

A New Model for Education

Given the new technological capabilities available now, and combined with new ideas for applying the latest thinking about teaching and learning, it is possible to propose a New Model for Education (Novak & Cañas, 2004; Cañas & Novak, 2005). The New Model involves these activities that will be further elaborated:

---

[1]CmapTools can be downloaded and used at no cost from: http://cmap.ihmc.us

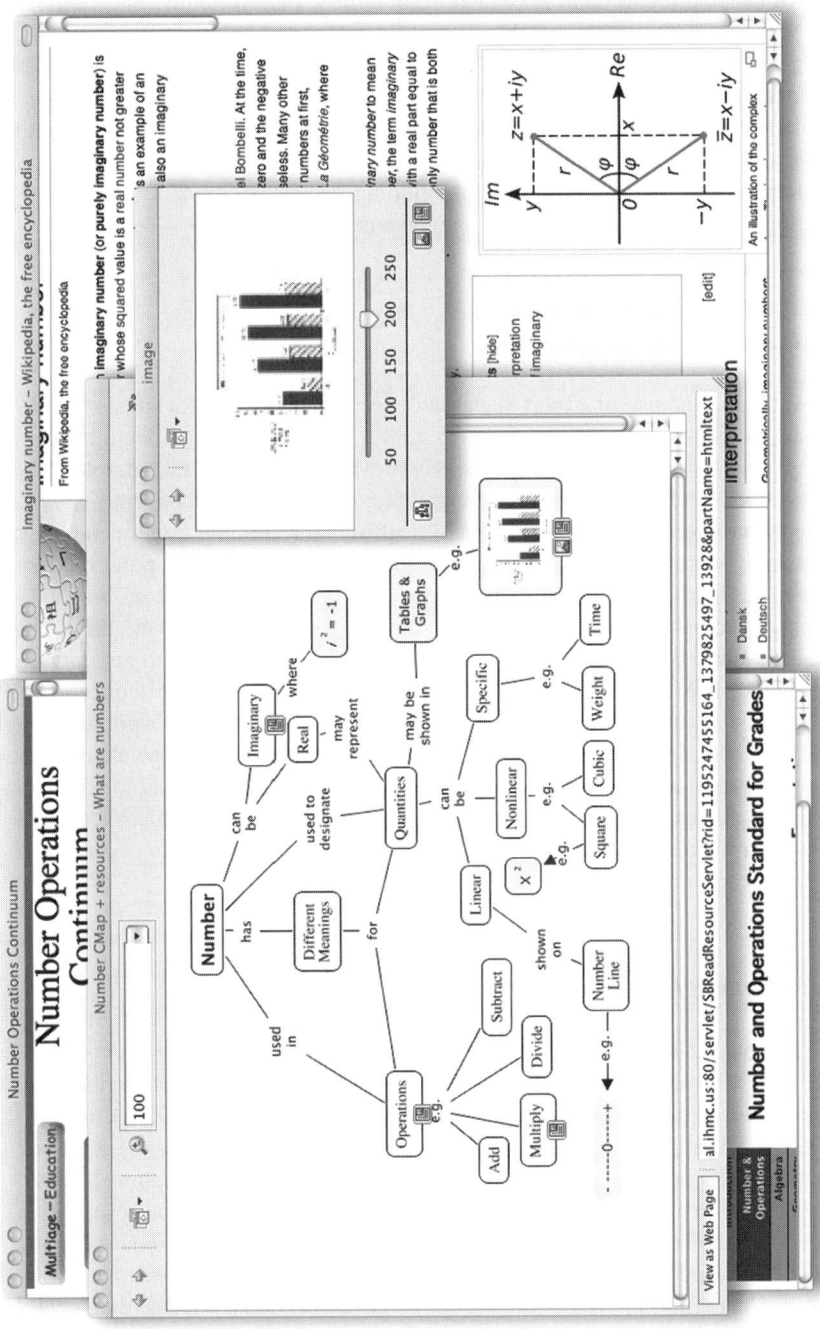

Fig. 1.4 A concept map about numbers showing some resources that have been opened by clicking on the various icons under the concepts

1. Use of "expert skeleton concept maps" to *scaffold* learning.
2. Use of CmapTools to build upon and expand the expert skeleton concept maps by drawing on resources available on the CmapServers, on the Web, plus texts, image, videos and other resources.
3. Collaboration among students to build "knowledge models".
4. Explorations with real world problems providing data and other information to add to developing knowledge models.
5. Written, oral, and video reports and developing knowledge models.
6. Sharing and assessing team knowledge models.

Expert Skeleton Concept Maps

The idea behind the use of expert skeleton concept maps is that for most students (and many teachers) it is difficult to begin with a "blank sheet" and begin to build a concept map for some topic of interest. By providing a concept map prepared by an expert with 10–15 concepts on a given topic, this "skeleton" concept map can help the learner get started by providing a "scaffold" for building a more elaborate concept map. Vygotsky (1934), Berk and Winsler (1995) and others point out that the apprentice learner is often very insecure in their knowledge and needs both cognitive and affective encouragement. While the teacher can best provide the latter, the skeleton concept map can provide the cognitive encouragement to get on with the learning task. Moreover, students (and often teachers) may have misconceptions or faulty ideas about a topic that would impede their learning if they were to begin with a "blank sheet". The scaffolding provided by the expert map can get the learner off to a good start, and as they begin to research relevant resources and to add concepts and resources to their map, there is a good chance that their misconceptions will also be remediated (Novak, 2002). Figure 1.5 shows an example of an expert skeleton concept map.

Adding Concepts and Resources Using CmapTools

It is well known today that meaningful learning requires that the learner chooses to interact with the learning materials in an active way and that he/she seeks to integrate new knowledge into her/his existing knowledge frameworks (Novak, 1998; Bransford et al., 1999). Through the drag-and-drop feature of CmapTools that allows for linking supplemental resources to a concept map by simply dragging and dropping the icon for an URL, an image, video, another concept map or any other digital resource on a concept, a learner can build an increasingly complex knowledge model for any domain of knowledge, which can serve as a starting point for later related learning. Moreover, CmapTools provide for easy collaboration between learners either locally or remotely, and either synchronously or asynchronously. When the recorder option of CmapTools is turned on, it will record step-by-step the *history* of the creation of a concept map, indicating the sequence of building steps and who did what at each step.

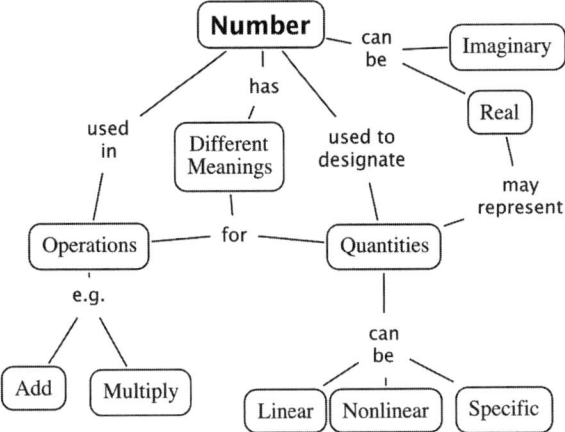

Fig. 1.5 An example of an "expert skeleton concept map" that can serve as a starting point for building a knowledge portfolio about number ideas. Figure 1.4 above shows an example of how this skeleton map could be elaborated using CmapTools and resources drawn from the Web. In general, the number of concepts expected to be added by the student is proportional to (e.g. two or three times) the number of concepts originally in the skeleton map

Obviously this also provides a new tool for cognitive learning studies, but such work is just beginning. For example, Miller, Cañas, & Novak (2008) using the Recorder tool has shown that in the process of learning how to construct concept maps, patterns in map changes for teachers in Proyecto Conéctate al Conocimiento in Panama (Tarté, 2006) were similar for teachers who had no previous experience using computers when compared with those who reported previous experience.

A major problem in mathematics studies is that there is rarely clear focus on the concepts underlying the mathematical operations learners are asked to do. It is even more rare to have an explicit record of the *conceptual thinking* of the learners as they progress in their studies, and a record the learner can turn to when related materials are studied. Another important advantage is that concept maps can be easily related to one another, for example as sub-concept maps in a more general, more encompassing concept map. Examples of this are shown in other chapters.

Collaboration Among Students

With the rediscovery of the studies of Vygotsky (1934) in the past 20 years, educators are increasingly recognizing the importance of social exchange in the building of cognitive structure, as well as for motivation for meaningful learning. Although the work of the Johnson brothers (1988) and others have shown some of the merits of "cooperative learning", most of these studies could not take advantage of the facilitation offered by CmapTools for cooperative learning. Often the advantages of cooperative learning were found to be small at best. We need new research studies showing the effect of collaboration on learning using CmapTools.

Exploration with Real World Problems

One of the important conclusions from many recent studies on "situated cognition" is the importance of placing learning into a meaningful, real world context. We recognize this value in our New Model and urge that whenever possible, new ideas in mathematics should be introduced within the framework of some real world problem. While math teachers have been using for decades the idea that ratio and proportion problems, for example, should be introduced with tangible activities, such as using objects of different densities for comparison, most of these activities have not made explicit the *mathematics concepts* involved in the problems given, and the focus has been mostly on procedures to get the "correct" answer. Of course, this has also carried over into physics teaching and teaching in other subjects.

When real world activities are tied in with the use of CmapTools and the creation of knowledge models for the domains studied, research is beginning to show the resulting improvement in learning (Cañas, Novak, & González, 2004; Cañas & Novak, 2006; Cañas et al., 2008).

In the Proyecto Conéctate al Conocimiento (Tarté, 2006) effort in Panama, we are finding that the collaboration possibilities available with CmapTools are leading not only to sharing in knowledge building but also to a variety of social exchanges. During the training programs for teachers, teachers are invited to prepare a concept map in the form of biography referred to as "Who am I?". This has led to teachers and principals also building concept maps about their schools, communities and a variety of related exchanges (Sánchez et al., 2008). This personal engagement has had strong motivating effects for pursuing other collaborations and we expect this will increase over time as the social network grows. Figure 1.6 shows a sample montage of some of the work done by teachers and school principals in Panama. With thousands of teachers and students involved in this project, we are learning many new possibilities for ways to use concept maps to facilitate meaningful learning.

Written, Oral, and Video Reports and Developing Knowledge Models

While we will continue to see an increase in the use of electronic communications in the future, there will always be an important place for written and oral reports. Whether in schools or corporate settings, our students need to become effective written and oral communicators. In one of our early studies, we found that fifth grade children who prepared concept maps prior to attempting to write out their ideas not only wrote better stories but they were also better able to tell their stories (Ben-Amar, 1990). In fact, they wrote a play derived from their stories and it was so well received they were invited to present it at other elementary schools!

The full range of capabilities for organizing knowledge available using Cmap-Tools is too recent to have an empirical research base to document the value of the possibilities created, including what we call A New Model for Education. Hopefully,

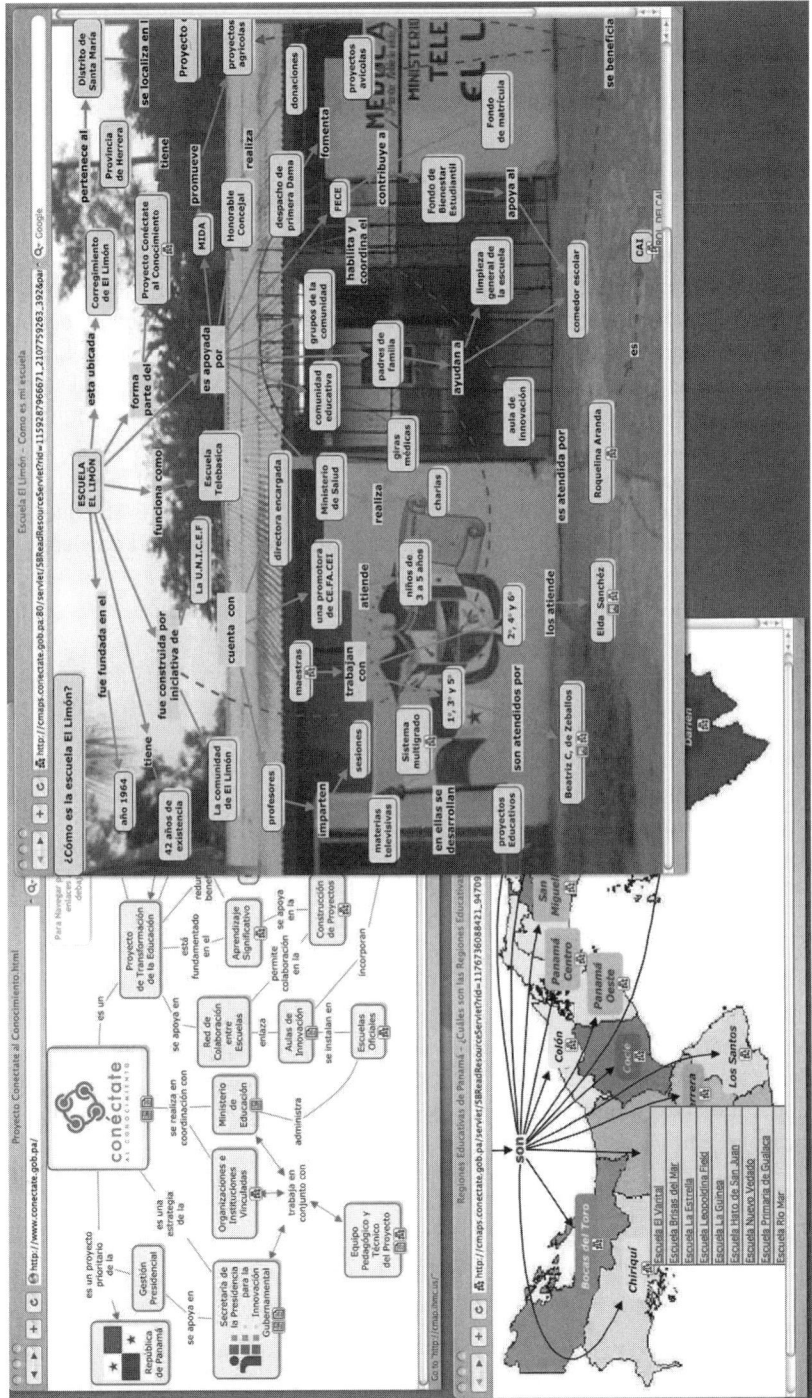

Fig. 1.6 Concept maps drawn by Panamanian teachers illustrating "Who am I?" concept maps posted on Web sites and serving to provide motivation and recognition, as well as a basis for collaborations

after the publication of this book, many empirical studies will be done to assess the value of CmapTools not only for improving instruction in mathematics, but also in improving students' ability to communicate their mathematics ideas and a new excitement for learning mathematics.

Sharing and Assessing Team Knowledge Models

Already indicated above are ways in which the sharing of individual and team knowledge models can be facilitated using the collaboration tools of CmapTools. However, many teachers want to know how they can *evaluate* knowledge models. In our own teaching, we have used a variety of strategies including having teams post their knowledge models anonymously and then asking students to rank the models from lowest to highest, including criteria for their rankings. Using digital knowledge models created with CmapTools, these can be posted on the class server and provide easy access for assessment. Students can be very insightful, and often brutally honest, in their assessments. Furthermore, serious assessment is an educational experience, and students learn how they can improve their own knowledge models.

In Conclusion

Our objective in this chapter was to provide a brief history of the development of the concept mapping tool, including the development of the computer software, Cmap-Tools, designed to facilitate concept map making and to provide new opportunities for individual and collaborative learning. Although the research to date supports the value of concept mapping to facilitate meaningful learning (Coffey et al., 2003), very little research has been done in the field of mathematics education. It is our hope this book will encourage such research. We also hope to see studies in mathematics learning that will utilize what we call a New Model for Education, and that libraries of "expert skeleton concept maps" in mathematics will be posted on web sites. We observed an increase in the number of papers dealing with mathematics education presented at international conferences on concept mapping from 2004 to 2008 (Cañas, Novak, et al., 2004; Cañas & Novak, 2006; Cañas et al., 2008) and we are hopeful that even more and improved studies will be presented at following conferences (see http://cmc.ihmc.us).

References

Ausubel, D. P. (1963). *The psychology of meaningful verbal learning.* New York: Grune and Stratton.
Ben-Amar Baranga, C. (1990). *Meaningful learning of creative writing in fourth grade with a word processing program integrated in the whole language curriculum.* Unpublished M.S. thesis, Cornell University.

Berk, L. E., & Winsler, A. (1995). *Scaffolding children's learning: Vygotsky and early child-hood education.* Washington, DC: National Assn. for Education for the Education of Young Children.

Bransford, J., Brown, A. L., & Cocking, R. R. (Eds.). (1999). *How people learn: Brain, mind, experience, and school.* Washington, DC: National Academy Press.

Cardemone, P. F. (1975). *Concept maps: A technique of analyzing a discipline and its use in the curriculum and instruction in a portion of a college level mathematics skills course.* Unpublished M.S. thesis, Cornell University.

Cañas, A. J., Hill, G., & Lott, J. (2003). *Support for constructing knowledge models in CmapTools* (Technical Report No. IHMC CmapTools 2003-02). Pensacola, FL: Institute for Human and Machine Cognition.

Cañas, A. J., Hill, G., Granados, A., Pérez, C., & Pérez, J. D. (2003). *The network architecture of CmapTools* (Technical Report IHMC CmapTools 2003-01). Pensacola, FL: Institute for Human and Machine Cognition.

Cañas, A. J., Hill, G., Carff, R., Suri, N., Lott, J., Eskridge, T., et al. (2004). CmapTools: A knowledge modelling and sharing environment. In A. J. Cañas, J. D. Novak, & F. Gonázales (Eds.), *Concept maps: Theory, methodology, technology. Proceedings of the first international conference on concept mapping* (Vol. I, pp. 125–133). Pamplona, Spain: Universidad Pública de Navarra.

Cañas, A. J., Novak, J. D., & González, F. (Eds.) (2004). *Concept maps: Theory, methodology, technology. Proceedings of the first international conference on concept mapping.* Pamplona, Spain: Universidad Publica de Navarra.

Cañas, A. J., & Novak, J. D. (2005). *A concept map-centered learning environment.* Symposium at the 11th Biennial Conference of the European Association for Research in Learning and Instruction (EARLI), Cyprus.

Cañas, A. J., & Novak, J. D. (2006). Re-examining the foundations for effective use of concept maps. In A. Canãs & J. D. Novak (Eds.), *Concept maps: theory, methodology, technology. Proceedings of the second international conference on concept mapping* (Vol. 1, pp. 494–502). San Jose, Costa Rica: Universidad de Costa Rica.

Cañas, A. J., Reiska, P., Åhlberg, M. K., & Novak, J. D. (2008). *Concept mapping: Connecting educators. Third international conference on concept mapping.* Tallinn, Estonia: Tallinn University.

Carvalho, M., Hewett, R. R., & Canãs, A. J. (2001). Enhancing web searches from concept map-based knowledge models. In N. Callaos, F. G. Tinetti, J. M. Champarnaud, & J. K. Lee (Eds.), *Proceedings of SCI 2001: Fifth multiconference on systems, cybernetics and informatics* (pp. 69–73). Orlando, FL: International Institute of Informatics and Systemics.

Coffey, J. W., Carnot, M. J., Feltovich, P. J., Hoffman, R. R., Canãs, A. J., & Novak, J. D. (2003). *A summary of literature pertaining to the use of concept mapping techniques and technologies for education and performance support.* Technical Report submitted to the US Navy Chief of Naval Education and Training, Institute for Human and Machine Cognition, Pensacola, FL.

Fuata'i, K. A. (1985). *The use of Gowin's Vee and concept maps in the learning of form five mathematics in Samoa College, Western Samoa.* Unpublished M.S Thesis, Cornell University.

Fuata'i, K. A. (1998). *Learning to solve mathematics problems through concept mapping and Vee mapping.* Apia, Samoa: National University of Samoa.

Johnson, D. W., Johnson, R. T., & Holubec, E. J. (1988). *Cooperation in the classroom, revised.* Edina, MN: Interaction Book Co.

Kuhn, T. S. (1962). *The structure of scientific revolutions.* Chicago, IL: University of Chicago Press.

Miller, N. L., Cañas, A. J., & Novak, J. D. (2008). Use of the CmapTools recorder to explore acquisition of skill in concept mapping. In A. J. Cañas, P. Reiska, M. Åhlberg & J. D. Novak (Eds.), *Concept mapping: Connecting educators. Proceedings of the third international conference on concept mapping* (Vol. 2, pp. 674–681). Tallinn, Estonia: Tallinn University.

Minemier, L. (1983). *Concept Mapping and educational tool and its use in a college level mathematics skills course*. Unpublished M.S. thesis, Cornell University.

Novak, J. D. (1977). *A theory of education*. Ithaca, NY: Cornell University Press.

Novak, J. D. (1986). The importance of emerging constructivist epistemology for mathematics education. *Journal of Mathematical Behavior, 5*, 181–184.

Novak, J. D. (1990). Concept maps and Vee diagrams: Two metacognitive tools for science and mathematics education. *Instructional Science, 19*, 29–52.

Novak, J. D. (1998). *Learning, creating, and using knowledge: Concept maps as facilitative tools in schools and corporations*. Mahwah, NJ: Lawrence Erlbaum Associates.

Novak, J. D. (2002). Meaningful learning: The essential factor for conceptual change in limited or appropriate propositional hierarchies (liphs) leading to empowerment of learners. *Science Education, 86*(4), 548–571.

Novak, J. D., & Gowin, D. B. (1984). *Learning how to learn*. New York, NY: Cambridge University Press.

Novak, J. D., & Canãs, A. J. (2004). Building on new constructivist ideas and the CmapTools to create a new model for education. In A. J. Canãs, J. D. Novak, & F. M. Gonázales (Eds.), *Concept maps: Theory, methodology, technology. Proceedings of the first international conference on concept mapping*. Pamplona, Spain: Universidad Publica de Navarra.

Novak, J. D., & Musonda, D. (1991). A twelve-year longitudinal study of science concept learning. *American Educational Research Journal, 28*(1), 117–153.

Sánchez, E., Bennett, C., Vergara, C., Garrido, R., & Cañas, A. J. (2008). Who am I? Building a sense of pride and belonging in a collaborative network. In A. J. Cañas, P. Reiska, M. Åhlberg, & J. D. Novak (Eds.), *Concept mapping: Connecting educators. Proceedings of the third international conference on concept mapping*. Tallinn, Estonia & Helsinki, Finland: University of Tallinn.

Tarté, G. (2006). Conéctate al Conocimiento: Una Estrategia Nacional de Panamá basada en Mapas Conceptuales. In A. J. Canãs & J. D. Novak (Eds.), *Concept maps: Theory, methodology, technology. Proceedings of the second international conference on concept mapping* (Vol. 1, pp. 144–152). San José, Costa Rica: Universidad de Costa Rica.

Vygotsky, L. S. (1934/1986). *Thought and language*. In Alex Kozulin (Ed. & Trans.). Cambridge, MA: The MIT Press.

Part II
Primary Mathematics Teaching and Learning

Chapter 2
Analysing the "Measurement" Strand Using Concept Maps and Vee Diagrams

Karoline Afamasaga-Fuata'i

The chapter presents data from a case study, which investigated a primary student teacher's developing proficiency with concept maps and vee diagrams as tools to guide the analyses of syllabus outcomes of the "Measurement" strand of a primary mathematics syllabus and subsequently using the results to design learning activities that promote working and communicating mathematically. The student teacher's individually constructed concept maps of the sub-topics length, volume and capacity are presented here including some vee diagrams of related problems. Through concept mapping and vee diagramming, the student teacher's understanding of the mapped topics evolved and deepened, empowering her to confidently provide mathematical justifications for strategies and procedures used in solving problems which are appropriate to the primary level, effectively communicate her understanding publicly, and developmentally sequence learning activities to ensure future students' conceptual understanding of the sub-topics.

Introduction

Various *Professional Teaching Standards* point to the need for teachers of mathematics to have deep understanding of students' learning, pedagogical content knowledge of the relevant syllabus and the ability to plan learning activities that develop students' understanding, as essential to achieve excellence in teaching mathematics (AAMT, 2006). These Standards therefore imply that student teachers should develop deep knowledge and understanding of principles, concepts and methods they are expected to teach their future students. For example, the underlying theoretical principles of the New South Wales Board of Studies' *K-6 Mathematics Syllabus* (NSWBOS, 2002) encourage the development of students' conceptual understanding through an appropriate sequencing of learning activities and implementation of working and communicating mathematically strategies. To this end, this

K. Afamasaga-Fuata'i (✉)
School of Education, University of New England, Armidale, Australia
e-mail: kafamasa@une.edu.au

K. Afamasaga-Fuata'i (ed.), *Concept Mapping in Mathematics*,
DOI 10.1007/978-0-387-89194-1_2, © Springer Science+Business Media, LLC 2009

chapter proposes that the application of the metacognitive tools of hierarchical concept maps (maps) and vee diagrams (diagrams), and the innovative strategies of concept mapping and vee diagramming can influence (a) the development of students' meaningful learning and conceptual understanding and (b) the dynamics of working and communicating mathematically within a social setting. Therefore, the focus question for this chapter is: *"In what ways do hierarchical concept maps and vee diagrams facilitate the preparation of primary student teachers for teaching mathematics, in particular, the development of a deep understanding of the content of the relevant syllabus?"* This chapter presents the case study of a Bachelor of Education (Primary) student teacher (i.e., Susan) who concept mapped and vee diagrammed over a semester, in her third year mathematics education course in a regional Australian university.

Literature Review of Concept Mapping and Vee Diagrams

Ausubel's theory of meaningful learning, which defines meaningful learning as learning in which students actively make connections between what they already know and new knowledge, underpins concept mapping particularly its principle that learners' cognitive structures are hierarchically organized with more general, superordinate concepts subsuming less general and more specific concepts. Linking new concepts to existing cognitive structures may occur via *progressive differentiation* (reorganization of existing knowledge under more general ideas) and/or *integrative reconciliation* (synthesising many ideas into one or two when apparent contradictory ideas are reconciled) (Ausubel, 2000; Novak & Canãs, 2006). By constructing maps/diagrams, students illustrate publicly their interpretation and understanding of topics/problems. Hierarchical concept maps were first introduced by Novak as a research tool to illustrate the hierarchical interconnections between main concepts (nodes) in a knowledge domain with descriptions of the interrelationships (linking words) on the connecting lines. The basic semantic unit (*proposition*) describes a meaningful relationship as shown by the triad "valid node – valid linking words-> valid nodes" (Novak & Canãs, 2006; Novak & Gowin, 1984). Vee diagrams, in contrast, were introduced by Gowin as an epistemological tool, in the shape of a vee that is contextualised in the phenomenon to be analysed. The vee's left side depicts the philosophy and theoretical framework, which drive the analysis to answer the focus question. On the vee's right side are the records, methods of transforming the records to answer the focus question and value claims. The epistemological vee was later modified (Afamasaga-Fuata'i, 1998, 2005) to one that is focused on guiding the thinking and reasoning involved in solving a mathematics problem (examples are presented later).

Numerous studies examined the use of maps and/or diagrams as assessment tools of students' conceptual understanding over time in the sciences (Novak & Canãs, 2004; Brown, 2000; Mintzes, Wandersee, & Novak, 2000) and mathematics (Afamasaga-Fuata'i, 2004; Hannson, 2005; Liyanage & Thomas, 2002;

Williams, 1998). Investigations of the usefulness of maps/diagrams to illustrate university students' evolving understanding of mathematics topics found students' mapped knowledge structure became increasingly complex and integrated as a consequence of multiple iterations of the processes of presentation → critique → revisions → presentation over the semester (Afamasaga-Fuata'i, 2007a, 2004). Others also demonstrated the value of maps as pedagogical planning tools to provide an overview of a topic (Brahier, 2005; Afamasaga-Fuata'i, 2006; Afamasaga-Fuata'i & Reading, 2007) or to analyse mathematics lessons (Liyanage & Thomas, 2002). Research also demonstrated the usefulness of diagrams to scaffold students' thinking and reasoning and to illustrate their understanding of the interconnections between theory and application in mathematics problem solving (Afamasaga-Fuata'i, 2007b, 2005), scientific inquiry (Mintzes et al., 2000) and epistemological analysis (Novak & Gowin, 1984; Chang, 1994). In summary, the review of the literature shows three uses of maps/diagrams that are particularly relevant to teacher education. Firstly, maps/diagrams as *learning tools* to illustrate students' evolving knowledge and understanding of the conceptual structure of a domain; secondly, as *analytical tools* to scaffold the conceptual analysis of topics or problems; and thirdly, as *pedagogical tools* to organize and sequence teaching and learning activities using the results of the conceptual analyses of syllabus outcomes.

Methodology

The case study reported here, started with a familiarisation phase in which Susan was introduced to the metacognitive strategies of concept mapping and vee diagramming using simple topics such as fractions and operations with fractions. The main project for the course required Susan to construct a comprehensive, hierarchical concept map of a mathematics topic to be selected from the primary mathematics syllabus, and diagrams of related problems, which demonstrate the applications of the mapped concepts. There were three phases to the project. The first phase (Assignment 1) required that Susan compiled an initial list of concepts, based on a conceptual analysis of the relevant syllabus outcomes, and then to construct an initial topic concept map and diagrams of problems. These were presented and critiqued in class before returning for further revision and expansion. The second phase (Assignment 2) involved the presentation of a more structurally complex, expanded concept map and diagrams of more problems. These were socially critiqued and returned for further revision and expansion. The third phase (Assignment 3) was the final submission of a more comprehensive, hierarchical topic concept map and more diagrams of related problems and activities, which extended previous work and incorporating comments from previous critiques, and including a journal of reflections of concept mapping and vee diagramming experiences.

Data collected included maps and diagrams from the familiarization phase, weekly workshops, and three phases of the main project including a journal of reflections. This chapter presents samples of Susan's submitted work to illustrate the application of maps/diagrams as learning, analytical and pedagogical tools for the

Length and *Volume* sub-strands of the *NSW BOS K-6 Mathematics Syllabus* while her work with the *Area* sub-strand is reported in Afamasaga-Fuata'i (2007b). The *Time, Mass, Volume, Capacity, Area,* and *Length* sub-strands make up the *"Measurement"* content strand of the *NSW K-6 Mathematics Syllabus* (NSWBOS, 2002).

Data Collected and Analysis

Susan's map/diagrams presented here illustrate her interpretations of the *Length* and *Volume* Knowledge & Skills (K&S) and Working Mathematically (WM) syllabus outcomes from *Early Stage One to Stage 3* and critical analyses and conceptual understanding of related mathematics problems. Also presented are excerpts from Susan's reflection journal to support her concept map and vee diagram data.

"Length" Concept Maps

Provided in Figs. 2.1 and 2.2 are the results of Susan's conceptual analyses of the *Length* syllabus outcomes for Early Stage 1 to Stage 3. The concept maps are analysed by considering the networks of propositions each displays to determine its meaningfulness and interconnectedness.

For Early Stage 1 and Stage 1 (Fig. 2.1), the corresponding super inclusive (most general) propositions are: (P1) "Measurement" has a substrand "Length" when looked at has very specific ideas and principles "Early Stage 1" (P1a) when something is measured from end to end it is measuring "Length", and (P1b) when something is measured between two points it is known as "Distance"; and (P2) "Measurement" has a substrand "Length" looks at, in "Stage 1" (P2a) the principle of "Length", and (P2b) the idea of "Distance".

The particular propositions for Early Stage 1 are: (P3) "Length" is measured using "Informal" language and ideas to "Predict" whether an object will be long or short and are asked to "Explain" using everyday language (such as) "Long, Short, High, Tall, Low, The Same", for example, "the door is tall"; and (P4) "Distance" is measured using "informal" calculations using appropriate "Unit" to conduct "Direct Comparison" which is also called "One to one correspondence".

Progressive differentiation from the latter node resulted in a number of propositions such as: (P5) "One to one correspondence" when comparing two objects in the same position is called "Equal Length"; (P6) "One to one comparison" can be "Recorded" informally by "Drawing", "Cutting Pasting" and "Tracing"; (P7) "Drawing", "Cutting Pasting" and "Tracing" these visual representations of the attribute of length can be used to make repeating "Patterns", for example, (as shown pictorially); (P8) "One to one correspondence" when describing this, students use everyday language such as "Long, Short, High, Tall, Low, The Same", for example, "the door is tall"; (P9) "One to one correspondence" students can also describe the comparison of two objects using "Comparative Language" such as "Longer,

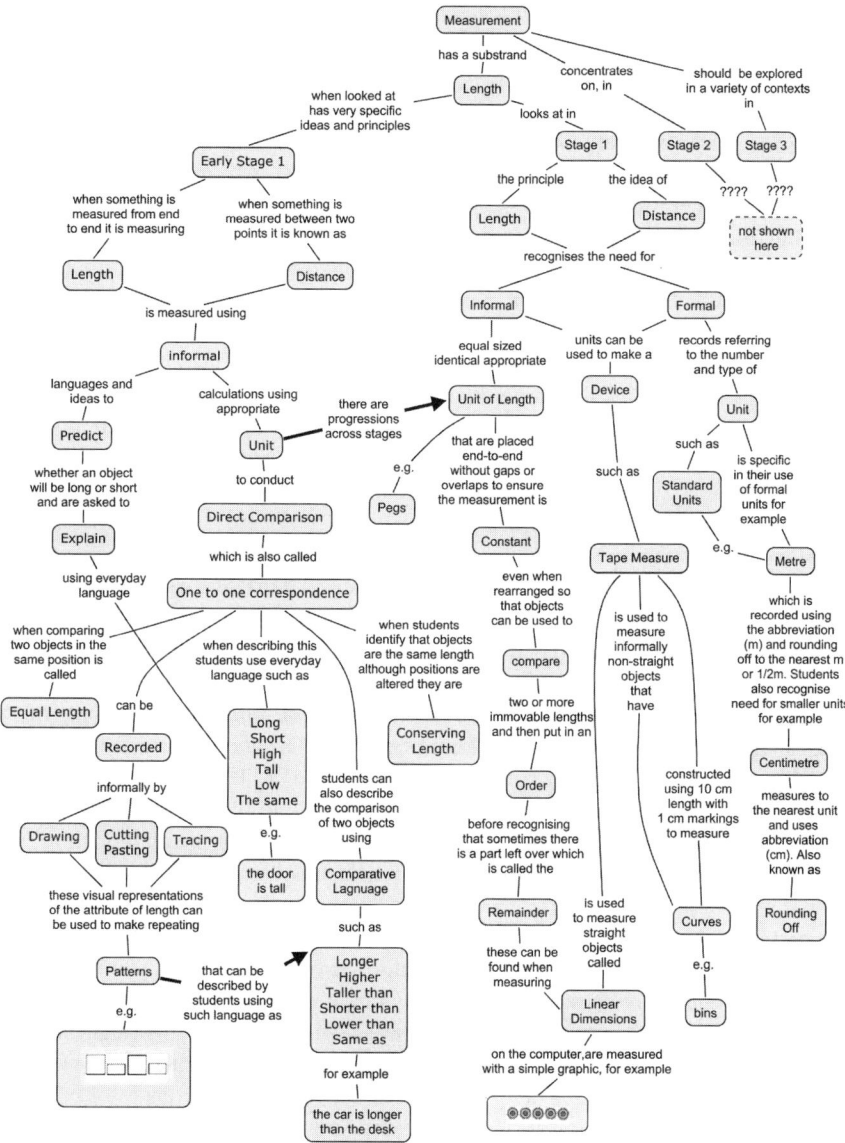

Fig. 2.1 Early Stage 1 and Stage 1 *Length* substrand

Higher, Taller than, Lower than, Same as", for example, "the car is longer than the desk"; and (P10) "One to one correspondence" when students identify that objects are the same length although positions are altered they are "Conserving Length". A cross link (indicated by a thicker, directional link) between two systems of concepts within this branch resulted in the integrative proposition (P11) "Patterns" that can be described by students using such language as "Longer, Higher, Taller than, Lower than, Same as". These propositions encompass Susan's interpretations of the length sub-strand for Early Stage 1. A number of progressive differentiation nodes (indicated by multiple outgoing links usually between a higher more general concept to lower less general concepts such as the "One to one correspondence" node) and integrative reconciliation links (indicated by at least two links from more general concepts merging with a less general concept such as the "Informal" node) including illustrative examples demonstrated Susan's understanding of interconnections between relevant concepts. A cross-link (indicated by a thicker, directional link) between the "Unit" node of Early Stage 1 and "Unit of Length" node of the Stage 1 branch demonstrated integration between the two branches resulting in the proposition (P12) "Unit" there are progressions across stages "Unit of Length".

For Stage 1 (Fig. 2.1), the particular propositions inclusive under the super inclusive P2 are: (P13) "Length" recognises the need for "Informal" equal sized identical appropriate "Unit of Length", for example, "Pegs"; (P14) "Unit of Length" that are placed end-to-end without gaps or overlaps to ensure the measurement is "Constant" even when rearranged so that objects can be used to "compare" two or more immovable lengths and then put in an "order" before recognising that sometimes there is a part left over which is called the "Remainder"; (P15) "Remainder" these can be found when measuring "Linear Dimensions"; and (P16) "Linear Dimensions" on the computer, are measured with a simple graphic, for example, (as pictorially shown). At the top right of this sub-branch inclusive under proposition P2b are the propositions (P17) "Distance" recognises the need for "Formal" units can be used to make a "Device" such as the "Tape Measure"; (P18) "Distance" recognises the need for "Formal" records referring to the number and type of "Unit" such as "Standard Units", for example, "Metre"; (P19) "Unit" is specific in their use of formal units for example "Metre"; (P20) "Metre" which is recorded using the abbreviation (m) and rounding off to the nearest m or $\frac{1}{2}$ m. Students also recognise need for smaller units, for example "Centimetre"; and (P21) "Centimetre" measures to the nearest unit and uses abbreviation (cm). Also known as "Rounding Off".

Progressively differentiating from the "Tape Measure" node are the propositions (P22) "Tape Measure" is used to measure straight objects called "Linear Dimensions"; (P23) "Tape Measure" is used to measure informally non-straight objects that have "Curves"; and (P24) "Tape Measure" constructed using 10 cm length with 1 cm markings to measure "Curves", for example, "bins". Occurrences of progressive differentiation (e.g., at the "Tape Measure" node) and integrative reconciliation (e.g., at the "Device" node) with some illustrative examples demonstrated the interconnectedness of Susan's interpretation and understanding of Stage 1 syllabus outcomes.

For Stage 2, the super inclusive proposition is (Fig. 2.2) (P25) "Measurement" has a substrand "Length" concentrates on, in "Stage 2", (P25a) measures "Length" using units that are "Formal", and (P25b) measures "Distance" using units that are "Formal". The rest of the propositions inclusive under the left sub-branch of the "Formal" node included (P26) "Formal" measurement requires "Units" which are used to estimate, measure and compare lengths and distances using specifically

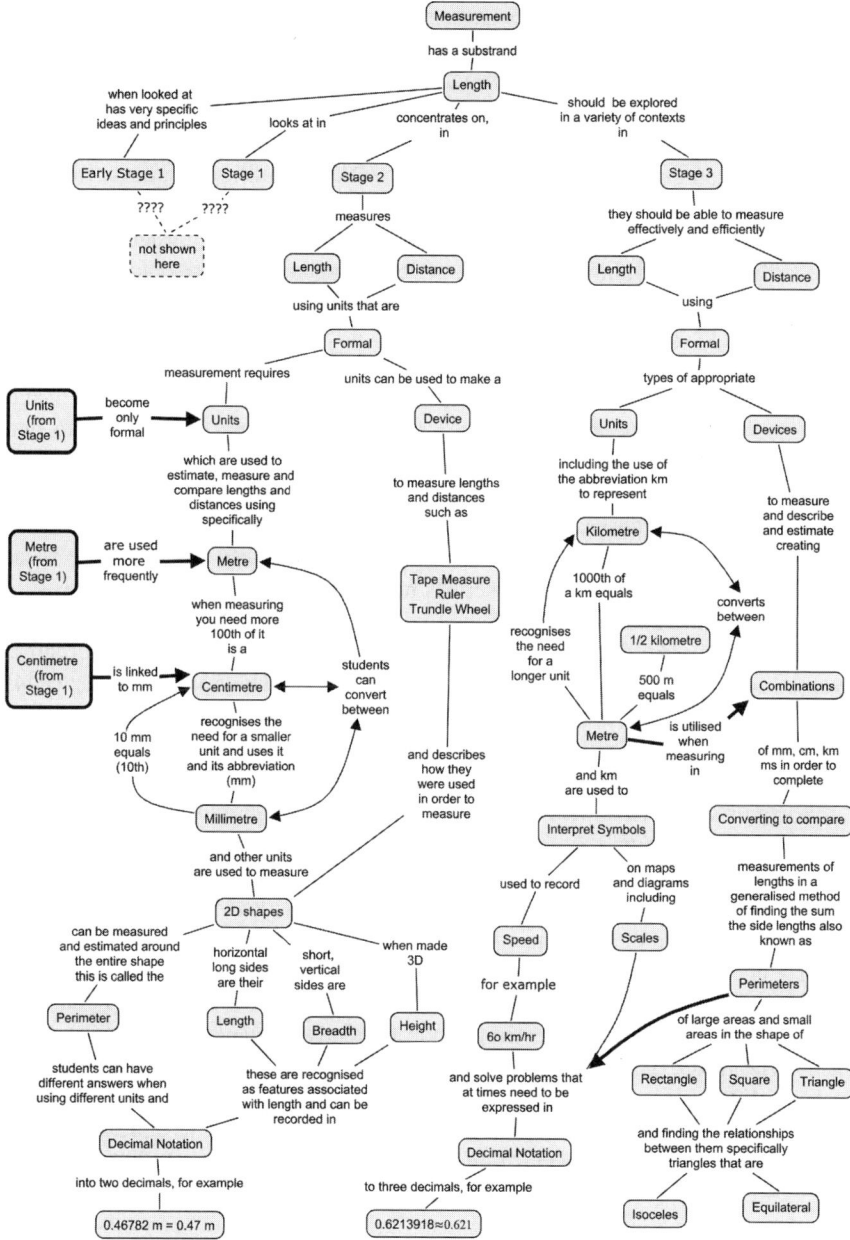

Fig. 2.2 Stages 2 and 3 of the *Length* substrand

"Metre"; (P27) "Metre" when measuring you need more 100th of it is a "Centimetre"; (P28) "Centimetre" recognises the need for a smaller unit and uses it and its abbreviation (mm) "Millimetre"; (P29) "Millimetre" (for example) 10 mm equals (10th) "Centimetre"; (P30) "Metre" students can convert between "Centimetre"; (P31) "Centimetre" students can convert between "Metre"; (P32) "Centimetre" students can convert between "Millilitre"; (P33) "Millilitre" students can convert between "Centimetre"; (P34) "Millimetre" and other units are used to measure "2D shapes"; (P35) "2D shapes can be measured and estimated around the entire shape – this is called the "Perimeter"; (P36) "2D shapes", horizontal long sides are their "Length"; (P37) "2D shapes", short, vertical sides are "Breadth"; (P38) "2D shapes", when made 3D "Height"; (P39) "Perimeter" students can have different answers when using different units and "Decimal Notation"; (P40) "Length", "Breadth", and "Height" these are recognised as features associated with length and can be recorded in "Decimal Notation"; and (P41) "Decimal Notation" into two decimals, for example "0.46782 m ≈ 0.47 m".

In comparison to propositions P26 to P41 of the left "Formal" sub-branch, inclusive under the right sub-branch is the single proposition (P42) "Formal" units can be used to make a "Device" to measure lengths and distances such as "Tape Measure, Ruler, Trundle Wheel" and describes how they were used in order to measure "2D shapes". The incoming cross-links (indicated by thicker, directional links) on the extreme left from Stage 1 to Stage 2 nodes resulted in the integrative propositions (P43) "Units" (from Stage 1) become only formal "Units" (in Stage 2); (P44) "Metre" (from Stage 1) are used more frequently "Metre" (in Stage 2); and (P45) "Centimetre" (from Stage 1) is linked to mm (by) "Centimetre" (in Stage 2). Occurrences of progressive differentiation and integrative reconciliation with some illustrative examples demonstrated the interconnectedness of Susan's interpretation and understanding of Stage 2 syllabus outcomes. Cross-links between Stages such as propositions P43 to P45 demonstrated her comprehension and awareness of the progressive development of informal units initially encountered in the lower stage to more formal units in the subsequent stage such as metre, centimetre and millimetre.

For Stage 3, the super inclusive proposition is (P46) "Measurement" has a substrand "Length" should be explored in a variety of contexts in "Stage 3", (P46a) they should be able to measure effectively and efficiently "Length", and (P46b) "Distance". The next extended proposition inclusive under the "Formal" node is, from the left sub-branch, namely, (P47) "Length" and "Distance" using "Formal" types of appropriate "Units" including the use of the abbreviation km to represent "Kilometre", followed by other propositions such as (P48) "Kilometre" 1000th of a km equals "Metre"; (P49) "Metre" 500 m equals "1/2 kilometre"; (P50) "Metre" recognises the need for a longer unit "Kilometre"; (P51) "Kilometre" converts between "Metre"; (P52) "Metre" converts between "Kilometre"; (P53) "Metre" and km are used to "Interpret Symbols" used to record "Speed" for example "60 km/hr" and solve problems that at times need to be expressed in "Decimal Notation" to three decimal places, for example "0.6213918 ≈ 0.621"; (P54) "Metre" and km are used to "Interpret Symbols" on maps and diagrams including "Scales" and solve problems that at times need to be expressed in "Decimal Notation".

In comparison to propositions P47 to P54 of the left "Units" sub-branch, inclusive under the "Devices" sub-branch to the right are the extended propositions (P55) "Length" and "Distance" using "Formal" types of appropriate "Devices" to measure and describe and estimate creating "Combinations" of mm, cm, km, ms in order to complete "Converting to compare" measurements of lengths in a generalised method of finding the sum of the side lengths also known as "Perimeters"; (P56) "Perimeters" of large areas and small areas in the shape of "Rectangle", "Square" and "Triangle"; and (P57) "Rectangle", "Square" and "Triangle" and finding the relationships between them specifically triangles that are (P57a) "Isosceles" and (P57b) "Equilateral". Cross-links and integrative propositions between the two sub-branches (indicated by thicker, directional links) and inclusive under the "Formal" node are (P58) "Converting to compare" measurements of lengths in a generalised method of finding the sum of the side lengths also known as "Perimeters" and solve problems that at times need to be expressed in "Decimal Notation", and (P59) "Metre" (is) utilised when measuring in "Combinations". Clearly, occurrences of progressive differentiation and integrative reconciliation with some illustrative examples and bi-directional linkages that form cyclic connections demonstrated the interconnectedness of Susan's interpretation and understanding of Stage 3 syllabus outcomes. For example, cross-links between the two sub-branches of the "Formal" branch indicated her comprehension and pedagogical understanding of the use of units of length to measure distances and to record speed, scales on maps and total distance around shapes (perimeters).

Overall, the overview concept map (Figs. 2.1 and 2.2) delineated the incremental development of units of length and distance, initially from informal units and subsequently to increasingly more formal units such as metre, centimetre and kilometre using strategies that varied from direct comparison of two objects in the early stages to increasingly more sophisticated strategies such as using devices to measure linear dimensions and curves. Linkages to other content strands (e.g., decimal notation and geometry) were also displayed.

Volume "Concept Maps"

Provided in Figs. 2.3 to 7 are the results of Susan's conceptual analyses of the Volume syllabus outcomes for Early Stage 1 to Stage 4. For the Early Stage 1 sub-strand (Fig. 2.3), the super inclusive propositions are: (P1) "Early Stage 1" explores the outcome "MES1.3"; (P2) "MES1.3" determines that the amount of space an object or substance occupies is "Volume"; and (P3) "MES1.3" determines that the amount a container can hold is called "Capacity". Inclusive under the "Volume" node, from left to right, are the propositions (P4) "Volume" found by "comparing two piles of materials" (P4a) by "filling two identical containers", and (P4b) by directly "observing the space each object occupies" for example "car vs truck" this is recorded using "drawings, numerals and words"; (P5) "drawings, numerals and words", (P5a) for example "the trucks takes more space than the car. (words)", (P5b) for example

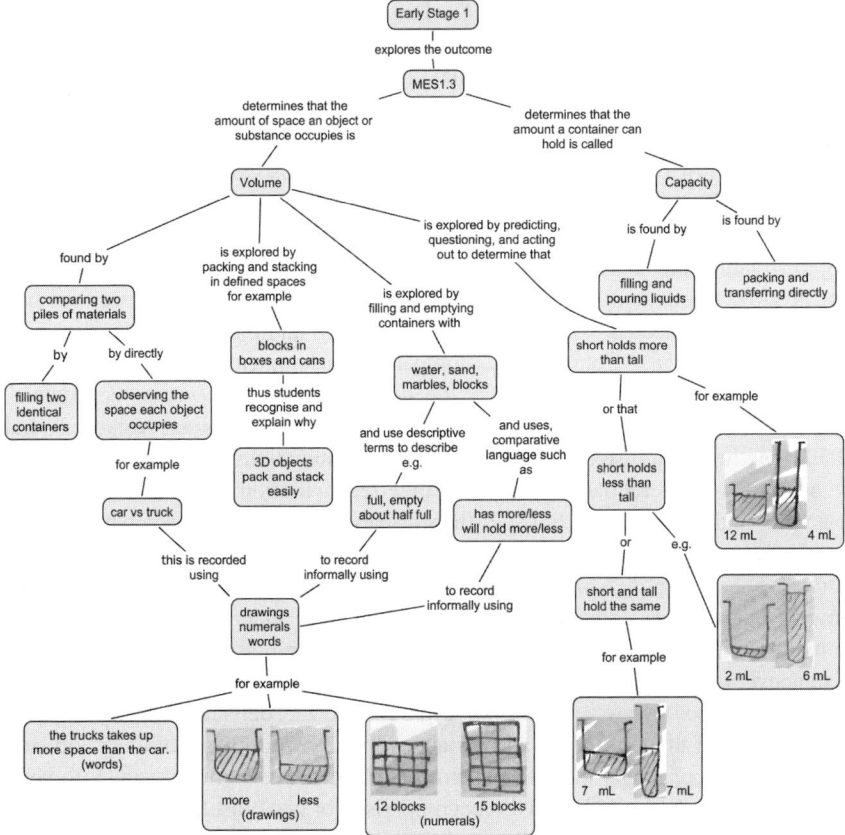

Fig. 2.3 Early Stage 1 of the *Volume* substrand

(shown pictorially with two cylinders to demonstrate "more" and "less"), and (P5c) for example, (shown pictorially with two sets of blocks to demonstrate "12 blocks" compared to "15 blocks").

For proposition P6, "Volume" is explored by packing and stacking in defined spaces, for example "blocks in boxes and cans" thus students recognise and explain why "3D objects pack and stack easily"; (P7) "Volume" is explored by filling and emptying containers with "water, sand, marbles, blocks", (P7a) and use descriptive terms to describe, e.g., "full, empty, about half full" to record informally using "drawings, numerals, words", (P7b) and uses comparative language such as "has more/less, will hold more/less" to record informally using "drawings, numerals, words"; and (P8) "Volume" is explored by predicting, questioning and acting out to determine that "short holds more than tall", (P8a) for example, (as shown pictorially with two cylinders holding "12 mL" and "4 mL"), (P8b) or that "short holds less than tall" for example, (shown pictorially for a short cylinder with "2 mL" and tall thin cylinder with "6 mL"), and (P8c) or "short and tall hold the same", for example, (as shown pictorially with a short cylinder with "7 mL" compared to long thin cylinder with "7 mL").

To the right at the top inclusive under the "Capacity" node, from left to right, are the propositions (P9) "Capacity" is found by "filling and pouring liquids" and (P10) "Capacity" is found by "packing and transferring directly". Evidently, occurrences of progressive differentiation and integrative reconciliation with illustrative examples using drawings, numerals or words, demonstrated the interconnectedness of Susan's interpretation and understanding of the Early Stage 1 syllabus outcomes. Overall, the concept map displayed the focus of this stage on the development of a conceptual understanding of volume as the amount of space an object occupies and capacity as the amount a container can hold through the use of appropriate objects, sand or liquid to stack, pack, or fill respectively, depending on whether it is measuring volume or capacity. Through comparison activities, the importance of appropriate descriptive language is introduced concretely and informally developed.

For subsequent concept maps, ideas that are revisited again (i.e., prior knowledge) for further development and consolidation in the current Stage are distinguished from the new knowledge that is being developed with concept ovals that have bold outlines while connections to other content strands are indicated by ovals with broken outlines. In the original maps, Susan used different colours to distinguish between prior and new knowledge and connections to other topics.

For the Stage 1 sub-strand (Fig. 2.4), the corresponding extended proposition, from left to right, is: (P1) "Stage 1" has the outcome "MS1.3" estimates with "Informal Units" such as the number and type of "discrete objects used" for example "Blocks" leads to use of formal units such as "cubed metres and centimetres"; propositions (P2) "Informal Units" which are used to measure "Volume"; (P3) "Volume" through trial and error strategies can be "compared and ordered"; (P4) "compared and ordered" two or more containers by counting the "number of blocks in each container"; (P5) "compared and ordered" two or more containers by noting the "change in water level when submerged" also known as "Displacement" which can be linked to "Fractions"; (P6) "Volume" through trial and error strategies can be "estimated and measured" including "piles of materials"; and (P7) "Informal Units" when they are small it means that "more units are needed". A reconciliative proposition integrating the "Volume" and "Capacity" systems of concepts inclusive under the respective nodes is (P8) "Volume", "Capacity" different shapes can have the same "Displacement" which can be linked to "Fractions". With the "Capacity" branch, its more inclusive proposition is (P9) "Informal Units" which are used to measure "Capacity" while the inclusive propositions are (P10) "Capacity" can be "compared and ordered", (P10a) by "filling and pouring into a third container and marking level", and (P10b) by "filling and pouring into each other"; (P11) "Capacity" can be found by "filling with informal units and counting", (P11a) for example "cubic units into rectangular containers with no gaps" and (P11b) for example, "such as using a calibrated large container using informal units", for example, "mark water level"; and (P12) "Capacity" can be found by "counting times the smaller container can fill the bigger". To the right of the "Capacity" branch is the extended proposition (P13) "Informal Units" such as the number and type of "continuous materials",

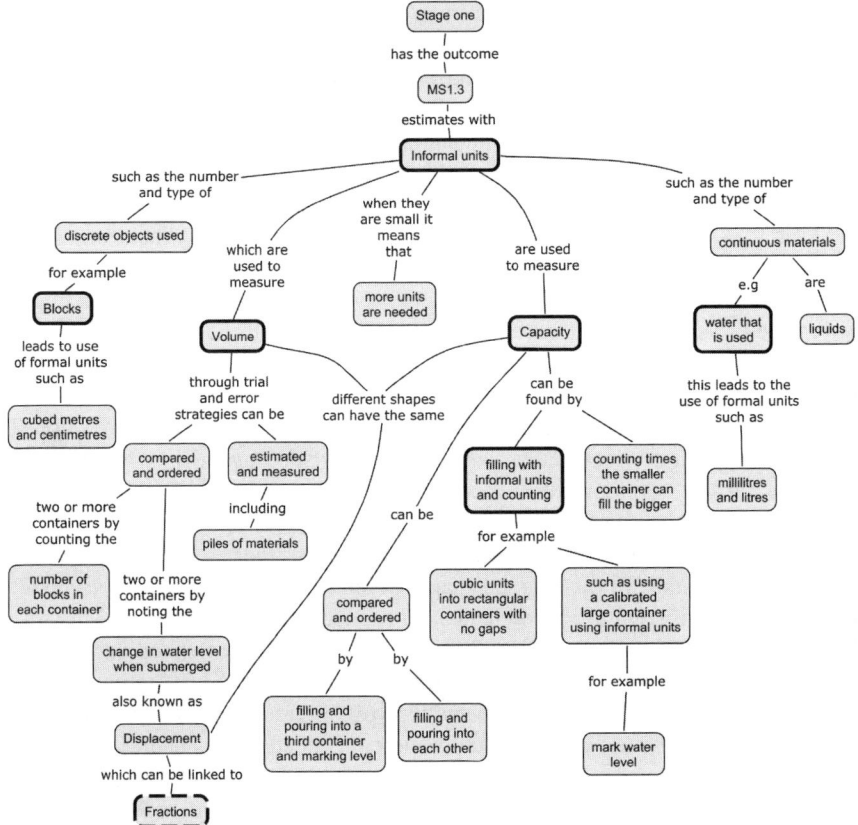

Fig. 2.4 Stage 1 of the *Volume* substrand

(P13a) for example, "water that is used" this leads to the use of formal units such as "millilitres and litres", and (P13b) are liquids.

Occurrences of progressive differentiation and integrative reconciliation with descriptions of illustrative examples demonstrated the interconnectedness of Susan's interpretation and understanding of Stage 1 syllabus outcomes. Overall, the concept map displayed the focus of this stage on the development of a conceptual understanding of informal units of measure in terms of number and type through the use of discrete objects (e.g., blocks) and continuous materials (e.g., water). Displacement is introduced as well as formal units of millilitre and litre. Through comparison activities, the importance of appropriate units for volume and capacity is further developed such as cubic metres and cubic centimetres. Prior knowledge from the previous Stage is indicated by ovals with bold outlines (e.g., "blocks", "informal units", "capacity" etc) while a reference to the "Fractions" strand is indicated by a broken-outlined oval.

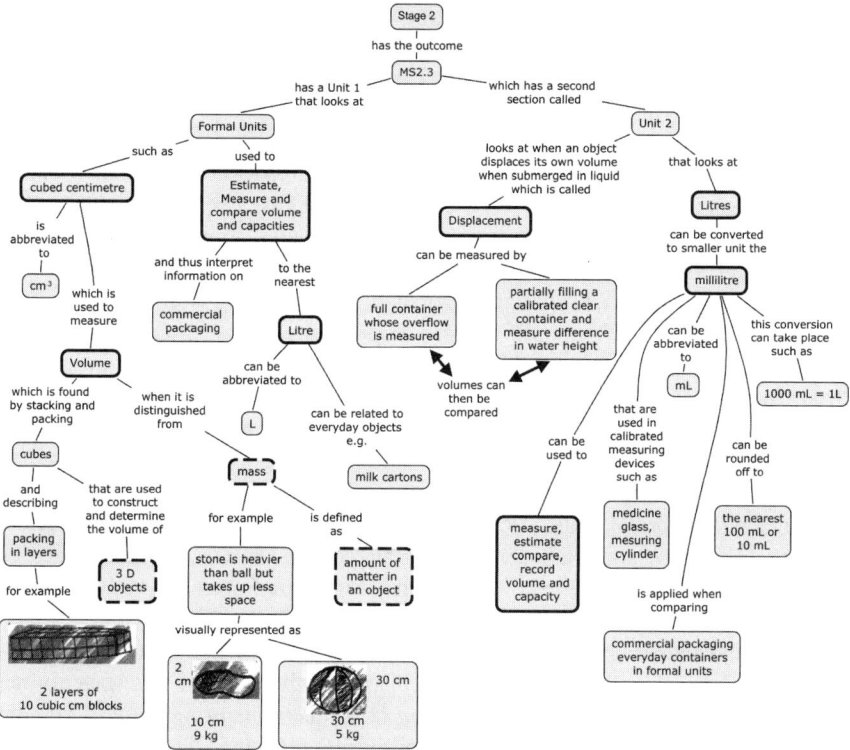

Fig. 2.5 Stage 2 of the *Volume* substrand

For the Stage 2 sub-strand (Fig. 2.5), the super inclusive propositions are (P1) "Stage 2" has the outcome "MS2.3"; (P2) "MS2.3" has a Unit 1 that looks at "Formal Units"; and (P3) "MS2.3" which has a second section called "Unit 2". Again, prior knowledge is indicated with bold-outlined ovals (e.g., "Displacement") while references or connections to other content strands are indicated by broken-outlined ovals such as with "3D objects".

Inclusive within the "Formal Units" branch are the propositions, from left to right, (P4) "Formal Units" such as "cubed centimetre" is abbreviated to "cm^3"; (P5) "Formal Units" such as "cubed centimetre" which is used to measure "Volume"; (P6) "Volume" which is found by stacking and packing "cubes", and describing "packing in layers", for example, as shown pictorially for "2 layers of 10 cubic cm blocks"; (P7) "cubes" that are used to construct and determine the volume of "3D objects"; (P8) "Volume" when it is distinguished from "mass", for example, "stone is heavier than ball but takes up less space" visually represented as (a picture of a stone with side lengths "2 cm by 10 cm" with mass "9 kg") and (a picture of a ball with diameter "30 cm" with mass "5 kg"); (P9) "mass" is defined as "amount of matter in an object"; (P10) "Formal Units" used to "Estimate, Measure and compare volume and capacities" and thus interpret information on "commercial packaging";

and (P11) "Formal Units" used to "Estimate, Measure and compare volume and capacities" to the nearest "Litre", (P11a) can be abbreviated to "L", and (P11b) can be related to everyday objects, e.g., "milk cartons".

Moving to the right and and inclusive under the "Unit 2" node are the propositions, from left to right, (P12) "Unit 2" looks at when an object displaces its own volume when submerged in liquid which is called "Displacement"; (P13) "Displacement" can be measured by "full container whose overflow is measured"; and (P14) "Displacement" can be measured by "partially filling a calibrated clear container and measure difference in water height". The last two propositions are inter-connected through a bi-directional cross-link with linking words "volumes can then be compared". To the right are the propositions (P15) "Unit 2" that looks at "Litres"; (P16) "Litres" can be converted to (a) smaller unit the "millilitre"; (P17) "millilitre" can be used to "measure, estimate, compare, record volume and capacity"; (P18) "millilitre" that are used in calibrated measuring devices such as "medicine glass, measuring cylinder"; (P19) "millilitre" can be abbreviated to "mL"; (P20) "millilitre" is applied when comparing "commercial packaging everyday containers in formal units"; (P21) "millilitre" can be rounded off to "the nearest 100 mL or 10 mL"; and (P22) "millilitre" this conversion can take place such as "1,000 mL = 1 L". Occurrences of progressive differentiation and integrative reconciliation with descriptions and pictures of some illustrative examples, demonstrated the interconnectedness of Susan's interpretation and understanding of Stage 2 syllabus outcomes. Overall, the concept map displayed the focus of this stage on the consolidation of a conceptual understanding of formal units of measure of volume and capacity (e.g, cubic centimetre, litre, millilitre) and consolidating the displacement strategy by extending on the informal approach of the last two Stages – prior knowledge is indicated with bold-outlined ovals. Introduced in this Stage is the concept of mass as an example of making connections to other content strands (indicated by broken-outlined ovals).

For Stage 3 (Fig. 2.6), the super inclusive propositions are (P1) "Stage 3" has a measurement outcome that looks at "MS3.3"; (P2) "MS3.3" involves "Volume"; and (P3) "MS3.3" looks at the selection of appropriate "units", and (P4) "MS3.3" involves "Capacity".

The propositions inclusive under the "Volume" branch, from left to right, are (P5) "Volume" is demonstrated using cubes of side lengths 100 cm which "will displace 1 L" thus "$1,000 \text{ cm}^3 = 1L$"; (P6) "Volume" can be found using displacement when measuring "irregular solids", for example, (as pictorially shown); (P7) "Volume" of rectangular prisms is found through "repeated addition" which can lead to the use of "layers", for example, (as pictorially shown by a block with 3 layers of 15 blocks); (P8) "Volume" can be the same if "the shape is different" this can be seen through "construction", for example, (as pictorially shown for a block of 2 x 4 x 5 and a block of 2 x 10 x 2 each with a volume of "40 units^3"); (P9) "Volume" is found by counting cubic centimetre blocks used to construct "rectangular prisms"; (P10) "Volume" is demonstrated using medicine cups and a cubic centimetre "which displaces 1 mL of water" thus "$1 \text{ cm}^3 = 1 \text{ mL}$"; and (P11) "Volume" demonstrates a relationship between "length, breadth, height of rectangular prisms"

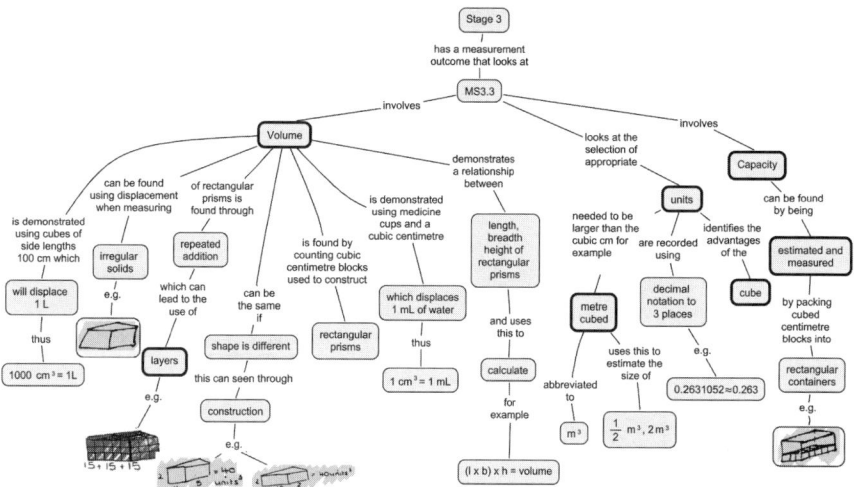

Fig. 2.6 Stage 3 of the *Volume* substrand

and uses this to "calculate", for example " $(l \times b) \times h =$ volume". Those propositions inclusive within the "units" branch are (P12) "units" needed to be larger than the cubic cm for example "metre cubed", (P12a) abbreviated to "m^3", and (P12b) uses this to estimate the size of "$\frac{1}{2}m^3$, 2 m^3"; (P13) "units" are recorded using "decimal notation to 3 decimal places", for example, "$0.2631052 \approx 0.263$"; and (P14) "units" identifies the advantages of the "cube". The rightmost branch shows the single proposition (P15) "Capacity" can be found by being "estimated and measured" by packing cubed centimetre blocks into "rectangular containers", for example, (as shown pictorially).

Overall, occurrences of progressive differentiation and integrative reconciliation with descriptions and pictures of some illustrative examples, demonstrated the interconnectedness of Susan's interpretation and understanding of Stage 3 syllabus outcomes. In particular, the concept map displayed the focus of this stage on a deeper understanding of volume and capacity, formal units, strategies and introduction of a volume formula. Prior knowledge (as indicated by bold-outlined ovals) is further consolidated in this Stage.

For Stage 4 (Fig. 2.7), the super inclusive proposition is (P1) "Stage 4" has the outcome "MS4.2"; (P2) "MS4.2" looks at "Surface Area", and (P3) "MS4.2" looks at "Volume".

The leftmost propositions inclusive within the "Surface Area" branch are (P4) "Surface Area" looks at "Prisms' areas"; (P5) "Prisms' areas" are calculated by identifying Surface Area and edge length of "Rectangular Prisms" by applying practical means such as "nets"; and (P6) "Prisms' areas" are calculated by identifying Surface Area and edge length of "Triangular Prisms" by applying practical means such as "nets" for example, (as pictorially shown); and (P7) "nets" which are "the blueprints of a 3D object". The adjacent "Prisms" branch to the right included

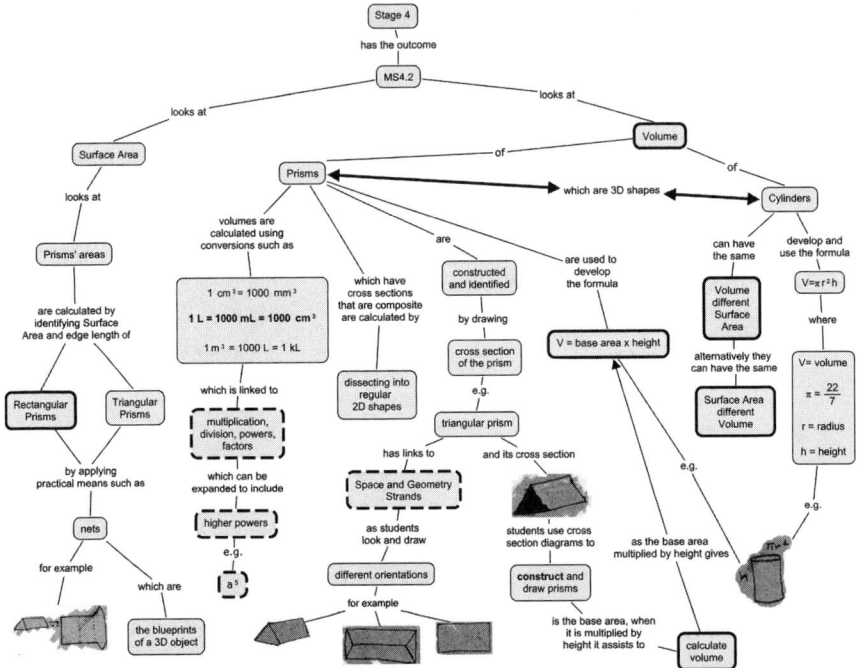

Fig. 2.7 Stage 4 of the *Volume* substrand

the proposition (P8) "Prisms" volumes are calculated using conversions such as "1 cm^3 = 1,000 mm^3, 1 L = 1,000 mL = 1,000 cm^3, 1 m^3 = 1,000 L = 1 kL" which is linked to "multiplication, division, powers, factors" which can be expanded to include "higher powers", for example, a^5; (P9) "Prisms" which have cross sections that are composite are calculated by "dissecting into regular 2D shapes"; (P10) "Prisms" are "constructed and identified" by drawing "cross section of the prism", for example, "triangular prism"; (P11) "triangular prism" has links to "Space and Geometry Strands" as students look and draw "different orientations", for example, (as pictorially shown, a side view, top view and bottom view of a triangular prism); (P12) "triangular prism" and its cross section (as shown pictorially) students use cross section diagrams to "construct and draw prisms"; and (P13) "construct and draw prisms" is the base area, when it is multiplied by height assists to "calculate volume".

An uplink from the latter node results in the proposition (P14) "calculate volume" as the base area multiplied by its height gives "V = base area × height", for example, (as pictorially shown with a cylinder with base area "πr^2" and height "h"). The rightmost sub-branch inclusive under the "Prisms" node is (P15) "Prisms" are used to develop the formula "V = base area × height". A more inclusive proposition (P16) integrates the two branches, namely, "MS4.2 looks at "Volume", (P16a) of "Prisms", and (P16b) of "Cylinders" with a bi-directional cross link between the last two nodes and linking words "which are 3D shapes" describing the nature of their inter- relationship. Inclusive under the "Cylinders" node are the propositions (P17) "Cylinders" can have the same "Volume different Surface Area" alternatively they can have the same "Surface Area different Volume" and (P18) "Cylinders" develop and use the formula "$V = \pi r^2 h$" where "$V = $ volume, $\pi = \frac{22}{7}$, $r = $ radius, $h = $ height", for example, (as pictorially shown with a cylinder with base area "πr^2" and height "h"). Overall, occurrences of progressive differentiation and integrative reconciliation with descriptions and pictures of some illustrative examples, demonstrated the interconnectedness of Susan's interpretation and understanding of Stage 4 syllabus outcomes. In particular, the concept map displayed the focus of this stage on the extension of students' deeper understanding of volume and capacity, formal units, strategies, volume formula and the introduction of surface area of prisms. A number of ideas encountered in previous Stages (bold-outlined ovals) and connections to other content strands (broken-outlined ovals) continued to be consolidated and synthesized with the formal concepts of "volume" and "surface area".

The concept map data presented above suggested that the student teacher became competent and confident in her critical abilities to conceptually analyse the *Length* and *Volume* sub-strands of the "Measurement" Strand for its key and subsidiary ideas and strategies. By concept mapping the conceptual analysis results for each Stage, Susan generated visual maps which highlighted developmental trends and increasingly sophisticated means of estimating, measuring, comparing, and recording length, volume and capacity using a variety of informal strategies and units before introducing more formal strategies, units and formulas. Susan tracked the progressive development and consolidation of previously encountered ideas by using colours in her handwritten maps (but by using bold-outlined ovals here) with connections to other content strands (as broken-outlined ovals here).

Vee Diagrams of Mathematics Problems

Examples of Susan's' vee diagrams are presented to illustrate how the vee diagrams were applied in problem solving as a means to make explicit the interconnections between mathematical principles (i.e. general statements of relationships between concepts) and methods of solutions. A vee diagram (see Fig. 2.8a) illustrates the conceptual and methodological information of a mathematics problem. The conceptual (thinking) aspects (i.e., principles and concepts) of the problem are on the left with the methodological (doing) aspects (i.e., given information, methods and answers) on the right while the tip of the vee is anchored in the problem to be solved and the focus question to be answered at the top. When completing the vee, the conceptual analysis information is displayed on the thinking side as the mapper's responses to the guiding questions *What do I know already?* (e.g., mathematical principles) and *What are the main ideas?* (main concepts) with a statement of mathematical beliefs as a response to *Why I like mathematics?* On the doing side, is the given information (*What is the information given?*) and the methods of transforming

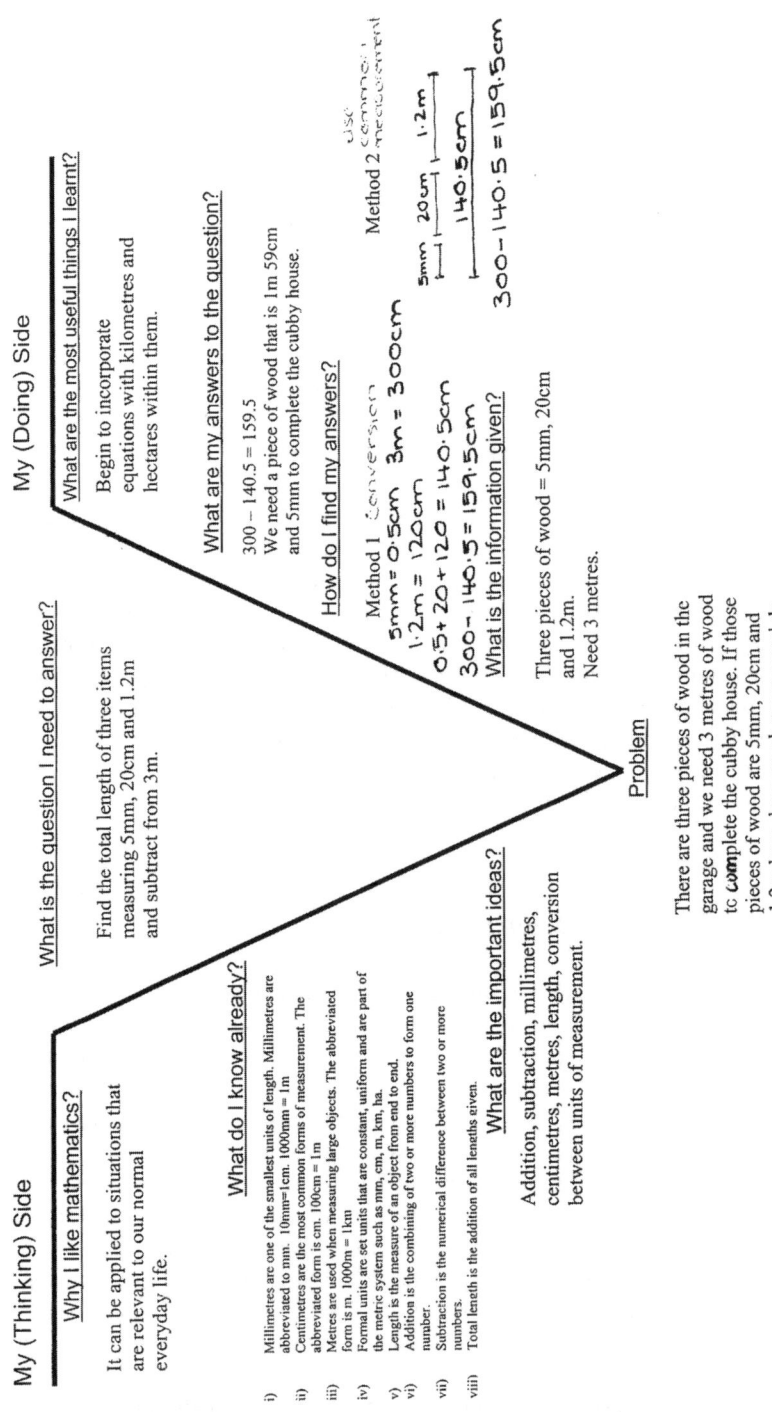

Fig. 2.8a A *Length* mathematics problem

the given information (*How do I find my answers?*) by applying the listed principles. One's reflections on the educational value of solving the problem, ideas for subsequent learning or connections to other topics are stated as responses to *What are the most useful things I learnt?* (Or alternatively *Where do I go from here?* or *What is an interesting extension to the current learning activity?*). The following analysis of vee diagrams is done by considering the cohesiveness and relevance of the displayed information to solving the given problem.

Figures 2.8a and 2.9 show Susan's two examples of vee diagrams from her second assignment while Figs. 2.8b and 2.10 are examples from the third assignment. One of the main differences between the first two vees and the last two is in the phrasing of the "focus question"; specifically, the entries were not phrased as questions to be answered but as statements in the first two vee diagrams. For example, Fig. 2.8a included a statement and the methodological phrase "*... subtract from 3 m*"; the latter would be more appropriate on the right side under "*How do I find my answers?*" while the corresponding entry for Fig. 2.9 needs to be phrased as a question to be answered not a statement and without the methodological pointers. The latter would be more suitable under methods. As a result of feedback to Susan, this concern was appropriately addressed by the third assignment as evident by the entries "*What is the length of the pencil, desk and book? Which is shortest?*" and "*What is the volume of a rectangular prism with dimensions 12 m × 6 m × 4 m?*" respectively, in Figs. 2.8b and 2.10.

A second main difference between the 4 vee diagrams is the absence of cross-referencing of principles under *How do I find my answers?* in the first two vee diagrams (Figs. 2.8a and 2.9). The inclusion of cross references to the listed principles on the right side in Figs. 2.8b and 2.10 made it easier to explicitly identify the principle that represented the conceptual basis of a main step. For example, in Fig. 2.8b, mathematical justifications for Method 1 are the principles 1 (P1), 2 (P2), and 4 (P4) while principle 5 (P5) is the underlying conceptual basis for Method 2. Similarly for Fig. 2.10, Susan referenced principle (v) (i.e., pv) as the mathematical justification for Line 1 of Method 1 while principles (ii) and (iv) (i.e., pii and piv) underpinned Line 2 of Method 1. The most significant similarity between all four vee diagrams is the depth of conceptual details evident in the entries for *What do I know already?*. For example, Susan's responses for all four vee diagrams reflected the depth and richness of her interpretations and conceptual analyses of the relevant syllabus outcomes and as concept mapped in Figs. 2.1 to 2.7. The listed principles represented Susan's mathematical justifications for the multiple methods on the right hand side. These are phrased in terms of the appropriate mathematical language, recommended strategies and multiplicity of approaches most suitable for primary level.

For the "*What are my answers to the question?*" sections of the 4 vee diagrams, entries also exhibited improvement by the third assignment. For example, the entry for Fig. 2.8a showed part of the method (i.e., $300 - 140.5 = 159.5$) whilst those in Figs. 2.8b, 2.9 and 2.10 consisted mainly of the answers, not the methods, to the respective focus question.

The entries for *Why I like mathematics?* revealed Susan's perceptions of why mathematics is important given the contexts of the problems while on the opposite

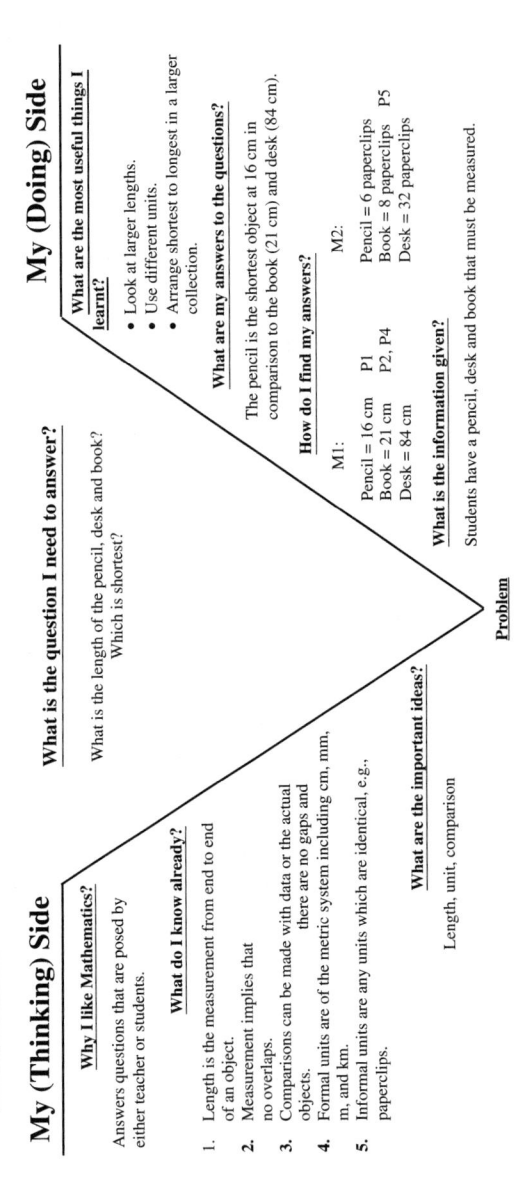

Fig. 2.8b A *Length* mathematics problem

My (Thinking) Side

Why I like Mathematics?

Answers questions that are posed by
either teacher or students.

What do I know already?

1. Length is the measurement from end to end
   of an object.
2. Measurement implies that
   no overlaps.
3. Comparisons can be made with data or the actual
                        there are no gaps and
   objects.
4. Formal units are of the metric system including cm, mm,
   m. and km.
5. Informal units are any units which are identical, e.g.,
   paperclips.

What are the important ideas?

Length, unit, comparison

What is the question I need to answer?

What is the length of the pencil, desk and book?
                    Which is shortest?

Problem

What is the shortest object between a pencil, a
desk and a book?

My (Doing) Side

**What are the most useful things I
learnt?**

- Look at larger lengths.
- Use different units.
- Arrange shortest to longest in a larger
  collection.

What are my answers to the questions?

The pencil is the shortest object at 16 cm in
comparison to the book (21 cm) and desk (84 cm).

How do I find my answers?

M1:   M2:

Pencil = 16 cm   P1   Pencil = 6 paperclips
Book = 21 cm   P2, P4   Book = 8 paperclips   P5
Desk = 84 cm   Desk = 32 paperclips

What is the information given?

Students have a pencil, desk and book that must be measured.

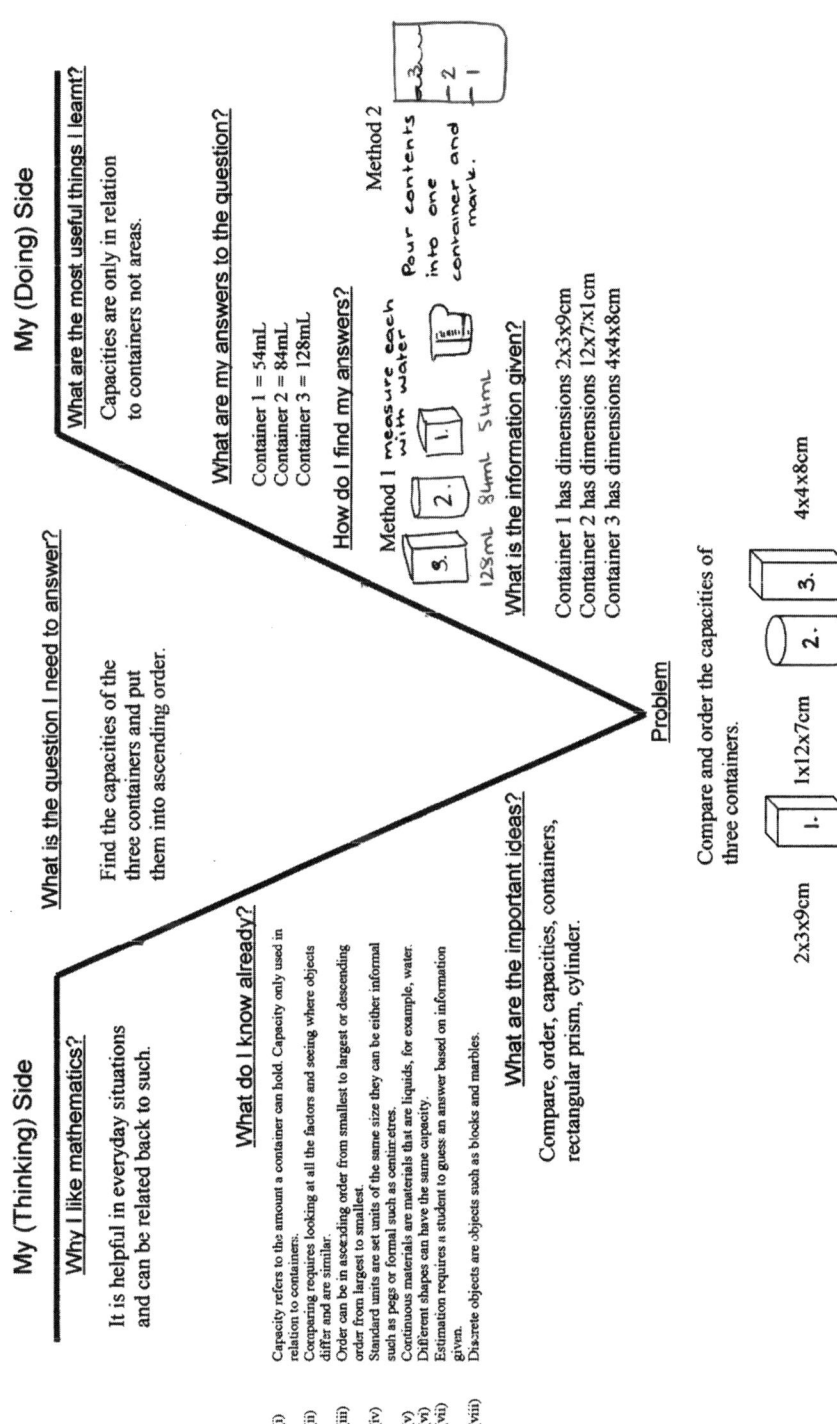

My (Thinking) Side

Why I like mathematics?

It is helpful in everyday situations and can be related back to such.

What do I know already?

(i) Capacity refers to the amount a container can hold. Capacity only used in relation to containers.
(ii) Comparing requires looking at all the factors and seeing where objects differ and are similar.
(iii) Order can be in ascending order from smallest to largest or descending order from largest to smallest.
(iv) Standard units are set units of the same size they can be either informal such as pegs or formal such as centimetres.
(v) Continuous materials are materials that are liquids, for example, water.
(vi) Different shapes can have the same capacity.
(vii) Estimation requires a student to guess an answer based on information given.
(viii) Discrete objects are objects such as blocks and marbles.

What are the important ideas?

Compare, order, capacities, containers, rectangular prism, cylinder.

What is the question I need to answer?

Find the capacities of the three containers and put them into ascending order.

My (Doing) Side

What are the most useful things I learnt?

Capacities are only in relation to containers not areas.

What are my answers to the question?

Container 1 = 54mL
Container 2 = 84mL
Container 3 = 128mL

How do I find my answers?

Method 1   I measure each with water

Method 2   Pour contents into one container and mark.

128mL 84mL 54mL

What is the information given?

Container 1 has dimensions 2x3x9cm
Container 2 has dimensions 12x7x1cm
Container 3 has dimensions 4x4x8cm

Problem

Compare and order the capacities of three containers.

2x3x9cm   1x12x7cm   4x4x8cm

Fig. 2.9 A *Capacity* mathematics problem

My (Thinking) Side

Why I like mathematics?

It can be demonstrated visually but can be worked through mentally depending on the difficulty of the problem.

What do I know already?

(i) Cross section is found by visually cutting a prism in half and observing the shape of the cut section.
(ii) Area of a rectangle is found by multiplying the length by the breadth.
(iii) Units in volume are cubed millimetres, centimetres, metres etc.
(iv) Perpendicular height is the height of a shape when the height is measured at a ninety degree angle.
(v) Students learn the formula and use Volume = base x perpendicular height
(vi) Displacement relies on the fact that an object displaces its own volume when totally submerged in liquid.
(vii) Volume is the amount of space occupied by an object or substance.
(viii) Length is the longest side of the rectangle. Breadth is the width of the rectangle.

What are the important ideas?

Volume, rectangular prism, length, breadth, cross section, area, metres, squared and cubed metres.

What is the question I need to answer?

What is the volume of a rectangular prism with dimensions 12m x 6m x 4m?

Problem

Find the volume of this rectangular prism.

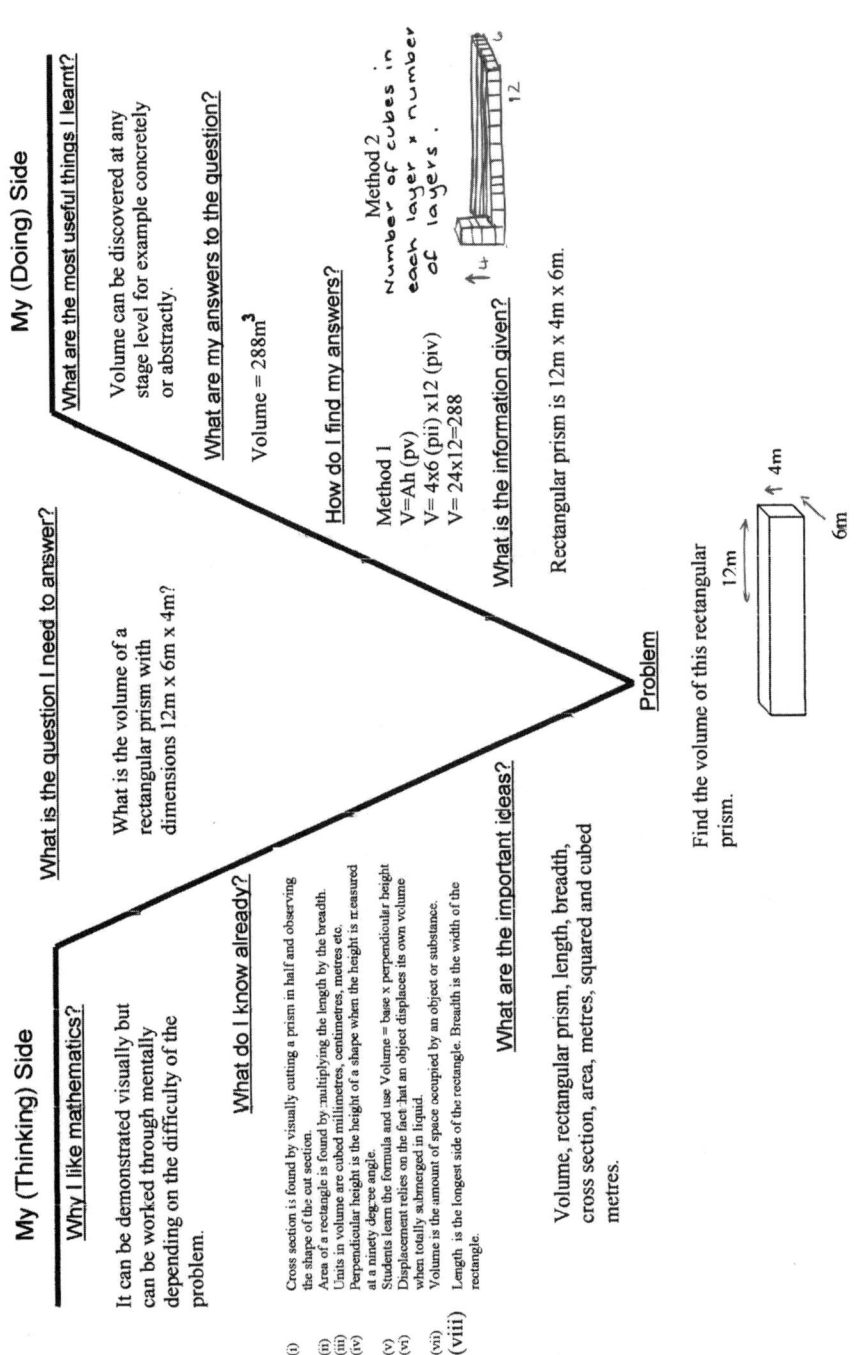

My (Doing) Side

What are the most useful things I learnt?

Volume can be discovered at any stage level for example concretely or abstractly.

What are my answers to the question?

Volume = $288m^3$

How do I find my answers?

Method 1
V=Ah (pv)
V= 4x6 (pii) x12 (piv)
V= 24x12=288

Method 2
Number of cubes in each layer x number of layers.

What is the information given?

Rectangular prism is 12m x 4m x 6m.

Fig. 2.10 A *Volume* mathematics problem

side on the right under *What are the most useful things I learnt?* are Susan's suggestions for subsequent learning activities (Figs. 2.8a and 2.8b), affirmation of capacities as related to containers only (Fig. 2.9) and general pedagogical belief for discovering volume at any Stage (Fig. 2.10).

As was encouraged in the course, all 4 vee diagrams displayed more than one method of solving each problem. The listed concepts on the other hand, indicated Susan's perception of the key and subsidiary ideas most pertinent for comprehending and solving each problem. On the right side under *What is the given information?* are lists of the given quantities as extracted from problem statements and diagrams listed under *Problem*. Overall, each vee diagram provided a comprehensive record of the most relevant conceptual and methodological information required in the generation of an answer to the focus question. Of significance is the depth and richness of prior knowledge (or principles) used as mathematical justifications for the multiple methods.

Journal of Reflections

Susan's journal of reflections documented her developing proficiency with concept mapping and vee diagramming during the course (excerpts are in italics). She recorded that her main reason for taking the unit was to help her develop a better understanding of the mathematics syllabus and of the teaching of its concepts to primary students. Rather than the year twelve (end of secondary level) style that she *"was used to, of formulas are everything"*, she perceived the need to approach primary mathematics from a primary perspective. In so doing, she felt she *"was taking a major risk in that (she) was approaching something that would take apart all (her) previously attained ideas and approaches."* Organised around four themes, namely, (1) critical ability to analyse syllabus topics and problems, (2) solve mathematics problems, (3) communicate effectively, and (4) develop a deep conceptual understanding of the sub-strands, are Susan's reflections presented below.

Critical Ability to Analyse Topics and Problem

At the beginning of the unit, Susan viewed problems as simply questions to be answered and topics as containing a lot of information around one idea that needed to be taught to students. However, upon completing workshop activities and the first assignment, it became clear that *"there was more to a problem than a formula and an answer."* Instead, *"(p)roblems consisted of a wide variety of factors that contributed to the understanding and subsequent answer"* such as the kinds of prior knowledge one possessed, which influenced the methods, and through reflection, the value of the learning experience, subsequent learning or extensions to the current activity. The presented maps/diagrams provided evidence of Susan's critical and conceptual analyses of Staged outcomes, which facilitated the mapping of interconnecting concepts, identification of multiple strategies and illustration of the theoretical bases of multiple methods.

Solve Mathematics Problems

As a result of constructing and completing vee diagrams, Susan realised that solving a problem became more than just *"an answer finder"*. Initially, she found it difficult to complete the thinking side because *"(she) did not know how (she) constructed the answer on the right hand side … and thus, did not know what principles (she) had to list nor the important ideas"*. However, she wrote, *"I struggled with this as, as a student I had only been taught the formulas never what was behind them."* With this self-realisation, Susan chose to challenge herself, namely, *"before finding the answer in future diagrams, first, (she) would look only at the question and think what (she needed) to know about it before (she actually solved) the problem."*

Communicate Effectively

Susan admitted it was always difficult for her to explain problems to others, *"I always had difficulty in explaining what I wanted them to do and it frustrated me that they did not understand when I explained it the first time."* Through her reflections though, she said that her *"communication skills verbally (had) been assisted greatly by (her) written communication in both concept maps and vee diagrams."* She claimed, *"I now have the basic skills written before me and because it was me that had to construct the written version I was able to explain what I did verbally better than I had done before."*

Develop a Deep Conceptual Understanding of a Topic

Through concept mapping over time, Susan eventually realised that a topic has a number of key and relevant concepts and recommended strategies that should be introduced, developed and extended for students through a suitable selection of learning activities to ensure a developmental conceptual understanding of the topic. For example, she wrote, *"(the length and volume sub-strands) have many connections that linked across a broad range of topics and through the construction of maps and diagrams a deep understanding of the topics was achieved."*

Overall, over the semester, Susan developed critical (i) proficiency completing the *Thinking Side* of diagrams *"as quickly and as effectively as the (Doing Side)"* and (ii) analytical skills, *"… to see where a problem is going before the actual completion … there are problems I can work backwards (from solution to principles) to see where I am going"*. She also found concept maps useful guides for completing vee diagrams, *"I could see the links and the next step (more clearly) in the solving of problems in relation to the sub-strand."*

Discussion

Findings suggested Susan became competent and confident in her critical abilities to analyse syllabus outcomes and problems and displaying the results appropriately on concept maps and vee diagrams. She analysed the syllabus outcomes for key concepts, strategies and illustrative examples before placing the results in a conceptual, developmental order within each Stage and from left to right on concept maps. Making connections between and within Stages was achieved by colour coding nodes (in the original maps but with bold- and broken-outlined ovals here) and cross-linking to differentiate between prior and new knowledge. Using vee diagrams, she systematically analysed problems to make explicit both the conceptual and methodological information involved in generating plausible solutions.

Through her maps and diagrams, Susan was able to communicate effectively with her audience. For example, because she had individually constructed the maps/diagrams, she was in a better and stronger position to explain and justify her ideas publicly due to the critical process she undertook in constructing and completing a concept map or a vee diagram. At the conclusion of the concept mapping and vee diagramming activities, Susan realised that she was able to *see the connections that infiltrated the topic*", consequently gaining a better understanding of sequencing learning activities. For example, *"(she) now understands what needs to be taught first and where (she) needs to go from there*" through the connections that she made visible on maps and diagrams. Furthermore, over time, completing the thinking side of vee diagrams eventually became much easier and done as efficiently as she did the right hand side. At times, she challenged herself by using the thinking side first to guide the development of her methods, which was something she did not use to do before. This was a significant development in her critical approach to problem solving. In her journal, she recorded that the principles guided her development of appropriate solutions, and sometimes, if the method is done first, she could flexibly use the solutions to infer what the principles should be. Over time, Susan became increasingly confident in critically using her concept map to identify subsequent learning and the next developmentally appropriate strategy or method guided by the propositions on the concept maps.

The presented data is only a sample of Susan's work over the semester. However, they explicitly illustrated the richness of information that can be captured by the combined usage of maps and diagrams in analysing syllabus outcomes and mathematical problem solving. Through her statements of mathematical principles using the appropriate mathematical language (on vee diagrams), Susan captured the conceptual and developmental essence of length, volume and capacity as recommended in the syllabus outcomes. The concept maps on the other hand, visually illustrated the recommended strategies to be promoted in primary mathematics, as well as making explicit visual connections between length, volume, capacity and surface area. The rich linking words describing the nature of the interrelationships between nodes resulted in valid propositions. The latter transformed the hierarchical concept maps into networks of meaningful, interconnecting propositions that coherently described the focus and scope of each Stage.

The presented maps/diagrams demonstrated that a deep understanding of the "length" and "volume" sub-strands of the "Measurement" Strand was developed and reinforced through the construction of maps/diagrams. The vee diagram structure provided not only the space to express one's mathematical beliefs and critical reflections, but also projections for future learning as evident from Susan's projection for subsequent learning. Overall, constructing maps/diagrams evidently encouraged Susan to move beyond a procedural view to a more conceptually based justification of methods and a purposeful and clearer understanding of sequencing prior, new and future learning to promote students' developmental and conceptual understanding.

Finally, Susan concluded that constructing maps/diagram had begun "*a new chapter in (her) understanding and teaching of mathematics.*" She felt confident and her understanding of the sub-strands had deepened particularly in "*how each and every one of (the concepts and strategies) builds upon the prior knowledge of the last*". Findings from this case study contribute empirical data to support the development of primary teachers' deep understanding of mathematics of the relevant syllabus through concept mapping and vee diagramming and the pedagogical use of maps/diagrams to make explicit the developmental trends of key ideas across multiple Stages and to highlight the critical synthesis of conceptual and methodological knowledge in problem solving.

Implications

The visual displays of networks of propositions on concept maps and theoretical and procedural information of problems on vee diagrams effectively encapsulated the interconnection between the Knowledge & Skills and Working Mathematically Syllabus Outcomes. This suggests the potential educational value of regularly exposing primary students to the strategies of concept mapping and vee diagramming to enhance and develop both their conceptual and methodological understanding of mathematics. Doing so would necessarily enable working and communicating mathematically amongst students in the classroom. Having students regularly construct their own maps and diagrams before and after a topic, as part of their normal mathematics classroom practices can lead to a more integrated and interconnected understanding of mathematics concepts and strategies as well as the development of critical thinking and reasoning.

As the data demonstrates, constructing concept maps engenders a deep understanding of how concepts and recommended strategies are developmentally progressed and consolidated across the Stages while constructing vee diagrams enhances the critical synthesis of the relevant mathematical principles and methods of generating solutions to problems. Collectively the impact engenders a better appreciation of how the main concepts and recommended strategies are developmentally progressed within each sub-strand, and how its mathematical principles are applied in methods as demonstrated by the sample data presented. These findings imply that concept maps and vee diagrams are potentially useful tools for learning mathematics and solving problems more meaningfully and more conceptually. The classroom applications of these tools are worthy of further investigation.

References

Afamasaga-Fuata'i, K. (1998). *Learning to solve mathematics problems through concept mapping & Vee mapping.* National University of Samoa.

Afamasaga-Fuata'i, K. (2004, September 14–17). Concept maps and vee diagrams as tools for learning new mathematics topics. In A. J. Canãs, J. D. Novak, & F. Gonázales (Eds.), *Concept maps: Theory, methodology, technology. Proceedings of the first international conference on concept mapping* (pp. 13–20). Spain: Dirección de Publicaciones de la Universidad Pública de Navarra.

Afamasaga-Fuata'i, K. (2005, January 17–21). Students' conceptual understanding and critical thinking? A case for concept maps and vee diagrams in mathematics problem solving. In M. Coupland, J. Anderson, & T. Spencer, (Eds.), *Making mathematics vital. Proceedings of the twentieth biennial conference of the Australian Association of Mathematics Teachers (AAMT)* (pp. 43–52). Sydney, Australia: University of Technology.

Afamasaga-Fuata'i, K. (2006). Innovatively developing a teaching sequence using concept maps. In A. Canãs & J. Novak (Eds.), *Concept maps: Theory, methodology, technology. Proceedings of the second international conference on concept mapping* (Vol. 1, pp. 272–279). San Jose, Costa Rica: Universidad de Costa Rica.

Afamasaga-Fuata'i, K. (2007a). Communicating students' understanding of undergraduate mathematics using concept maps. In J. Watson & K. Beswick (Eds.), *Mathematics: Essential research, essential practice. Proceedings of the 30th annual conference of the Mathematics Education Research Group of Australasia* (Vol. 1, pp. 73–82). University of Tasmania. Australia, MERGA. Also available from http://www.merga.net.au/documents/RP12007.pdf

Afamasaga-Fuata'i, K. (2007b). Using concept maps and vee diagrams to interpret "area" syllabus outcomes and problems. In K. Milton, H. Reeves, & T. Spencer (Eds.), *Mathematics essential for learning, essential for life. Proceedings of the 21st biennial conference of the Australian Association of Mathematics Teachers, Inc.* (pp. 102–111). University of Tasmania, Australia, AAMT.

Afamasaga-Fuata'i, K., & Reading, C. (2007). Using concept maps to assess preservice teachers' understanding of connections between statistical concepts. Published on IASE site, http://www.swinburne.edu.au/lss/statistics/IASE/CD˙Assessment/papers/IASE˙SAT˙07˙Afamasaga˙Reading.pdf

Australian Association of Mathematics Teachers (AAMT). (2006). *Standard for excellence in teaching mathematics in Australian schools.* Retrieved February 12, 2006, from http://www.aamt.edu.au/standards/standxtm.pdf

Ausubel, D. P. (2000). *The acquisition and retention of knowledge: A cognitive view.* Dordrecht; Boston: Kluwer Academic Publishers.

Brahier, D. J. (2005). *Teaching secondary and middle school mathematics* (2nd ed.). New York: Pearson Education, Inc.

Brown, D. S. (2000). The effect of individual and group concept mapping on students' conceptual understanding of photosynthesis and cellular respiration in three different levels of biology classes. *Dissertation Abstracts International AADAA-I9970734*, University of Missouri, Kansas City.

Chang, T. (1994). Taiwanese students' epistemological commitments and research knowledge construction (Chinese). *Dissertation Abstracts International AAD94-16683*, Cornell University, New York.

Hannson, O. (2005). Preservice teachers' views on $y = x + 5$ and $y = \pi x^2$ expressed through the utilization of concept maps: A study of the concept of function. Retrieved March 3, 2008, from http://www.emis.de/proceedings/PME29/PME29RRPapers/PME29Vol3Hansson.pdf

Liyanage, S., & Thomas, M. (2002, July 7–10). *Characterising secondary school mathematics lessons using teachers' pedagogical concept maps. Proceedings of the 25th annual conference of the Mathematics Education Research Group of Australasia (MERGA-25)* (pp. 425–432). New Zealand: University of Auckland.

Mintzes, J. J., Wandersee, J. H., & Novak, J. D. (Eds.) (2000). *Assessing science understanding: a human constructivist view.* San Diego, California, London: Academic.

New South Wales Board of Studies (NSWBOS). (2002). *K-6 mathematics syllabus.* Sydney, Australia: NSWBOS.

Novak, J. D., & Cañas, A. J. (2004). Building on new constructivist ideas and the CmapTools to create a new model for education. In A. J. Cañas, J. D. Novak, & F.M. Gonázales (Eds.), *Concept maps: Theory, methodology, technology. Proceedings of the first international conference on concept mapping.* Pamplona, Spain: Universidad Publica de Navarra.

Novak, J. D., & Canãs, A. J. (2006). *The theory underlying concept maps and how to construct them* (Technical Report IHMC CmapTools 2006-01). Florida Institute for Human and Machine Cognition, 2006. Available at http://cmap.ihmc.us/publications/ ResearchPapers/TheoryUnderlyingConceptMaps.pdf

Novak, J. D., & Gowin, D. B. (1984). *Learning how to learn.* Cambridge, UK: Cambridge University Press.

Williams, C. G. (1998). Using concept maps to assess conceptual knowledge of function. *Journal for Research in Mathematics Education*, 29 (4), 414–421.

Chapter 3
Concept Mapping as a Means to Develop and Assess Conceptual Understanding in Primary Mathematics Teacher Education

Jean Schmittau and James J. Vagliardo

Psychologists such as Vygotsky and Skemp indicate that as a superordinate concept the understanding of positional system requires knowledge of several bases for its adequate development. However, current elementary mathematics curricula fail to adequately develop the concept of positional system, attempting instead to teach operations in base ten in isolation. This paper exhibits the power of concept mapping to reveal to teachers the centrality of this concept in elementary mathematics. The map presented here, constructed by Maryanne, a pre-service teacher, also features the pedagogical content knowledge required to successfully teach the concept of positional system and the other mathematics concepts to which it is related. Maryanne's pedagogical treatment is neither simplistic nor reductionist, but reveals the conceptual essence of the concepts in question and the complexity of their relationships within elementary mathematics when taught as a conceptual system.

Introduction

The teaching of multiple bases to develop the concept of positional system that figured so prominently in the US mathematics education reform during the 1960s and 1970s was swept out of favor with the advent of the back to basics movement that succeeded the "new math" era. Unfortunately, its former prominence in the elementary mathematics curriculum has yet to be restored. It is not included in the current reform effort of the National Council of Teachers of Mathematics (NCTM, 1989, 2000), and is absent from most elementary school mathematics and elementary mathematics methods textbooks as well (cf. Burris, 2005, for a notable exception). The multi-base blocks invented by Zoltan Dienes some fifty years ago have virtually disappeared not only from US classrooms but from the catalogs of suppliers of mathematics manipulatives as well (the Prairie Rainbow Blocks developed by George Gagnon constitute a rare exception). Only base ten blocks are in common

J. Schmittau (✉)
State University of New York at Binghamton, Binghamton, New York, USA
e-mail: jschmitt@binghamton.edu

K. Afamasaga-Funata'i (ed.), *Concept Mapping in Mathematics*,
DOI 10.1007/978-0-387-89194-1_3, © Springer Science+Business Media, LLC 2009

use in US classrooms. And yet not only Dienes, but Skemp (1987) and Vygotsky (1962) stressed the importance of teaching multiple basic level concepts for the formation of a superordinate concept such as positional system.

> As long as the child operates with the decimal system without having become conscious of it as such, he has not mastered the system but is, on the contrary, bound by it. When he becomes able to view it as a particular instance of the wider concept of a scale of notation, he can operate deliberately with this or any other numerical system. The ability to shift at will from one system to another. . .is the criterion of this new level of consciousness, because it indicates the existence of a general concept of a system of numeration (Vygotsky, 1962, p. 115).

Skemp (1987) stresses that a superordinate concept can not be developed in the mind on the basis of a single basic level concept. At least several basic level concepts are required. A concept of positional system can not be developed through the teaching of base ten alone. And Vygotsky (1962) asserts that a child can not master the decimal system without attaining a mastery of the more general concept of positional system. Thus, as these leading psychologists in the theory of learning attest, one cannot understand base ten without learning other bases as well.

Developing the Concept of a Positional System in Teacher Education

Not only is the knowledge of multiple bases essential for understanding the concept of a positional system, but the concept itself provides a conceptual foundation for the four fundamental operations on whole numbers and for the development of the concept of variable, exponent, polynomial and polynomial operations, decimals, fractions, and area and volume experienced through the geometric modeling of positional systems. Consequently, the omission of this important means to concept development has multiple consequences for the learning of mathematics beyond the elementary level.

While the case for teaching multiple bases has been well documented in the psychological literature (Vygotsky, 1962; Skemp, 1987), its omission from elementary school mathematics textbooks presents difficulties for prospective and practicing teachers who may acknowledge its importance but be inclined to protest that "it's not in the curriculum". Assigning pre-service and in-service teachers the task of constructing a concept map exploring the connections of the concept of positional system to other important mathematical concepts, reveals the centrality of this concept both in elementary mathematics and as a foundation for concepts encountered at the middle school level and beyond.

The concept map shown in Fig. 3.1 was made by a pre-service teacher (pseudonym Maryanne) in response to such an assignment. When the concept of positionality was addressed in her elementary mathematics methods course, concept mapping was also introduced, and students were asked to begin to draw a map centered on the role of positionality in school mathematics.

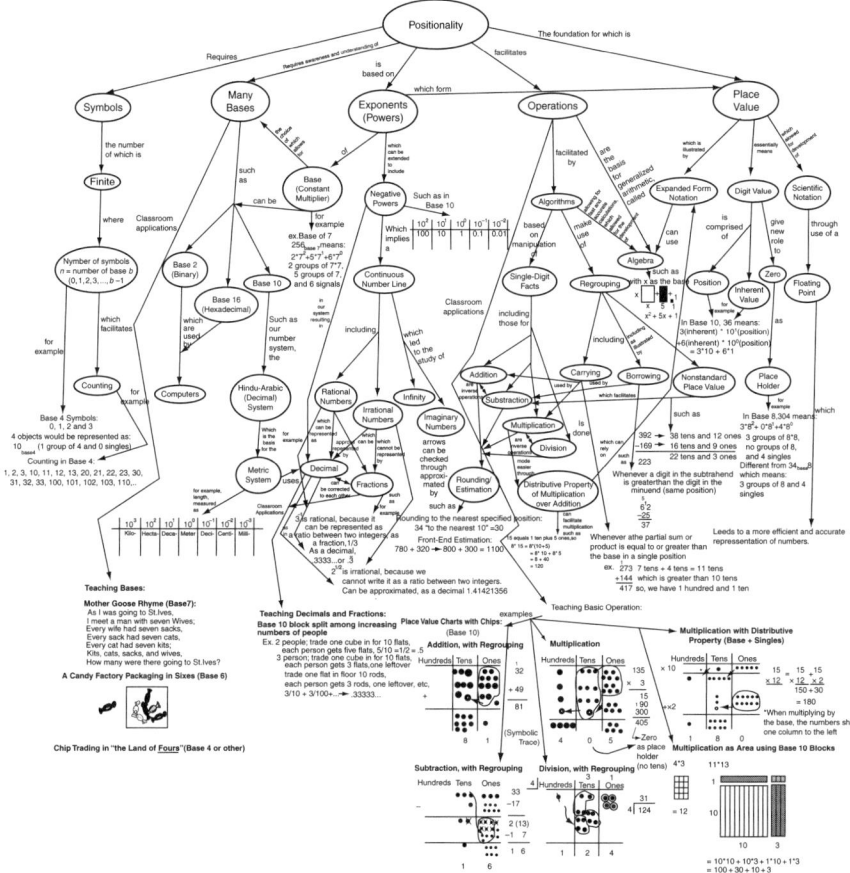

Fig. 3.1 Maryanne's concept map

As additional mathematics concepts were explored, students were encouraged to continue adding to their maps, so that the construction of the concept map became a project that continued to develop over the semester. The assignment revealed to students the multiple connections the concept of positional system has with mathematics concepts that children will study in the later elementary and middle school years and even beyond. These include such concepts as decimals, exponentials, area, volume, and polynomials and their operations.

Maryanne's Map

Figure 3.1 presents a view of Maryanne's concept map in its entirety. Figure 3.2 focuses on the extreme left section of the map, which reveals her understanding of the need to establish the concept of positional system on the foundation of multiple

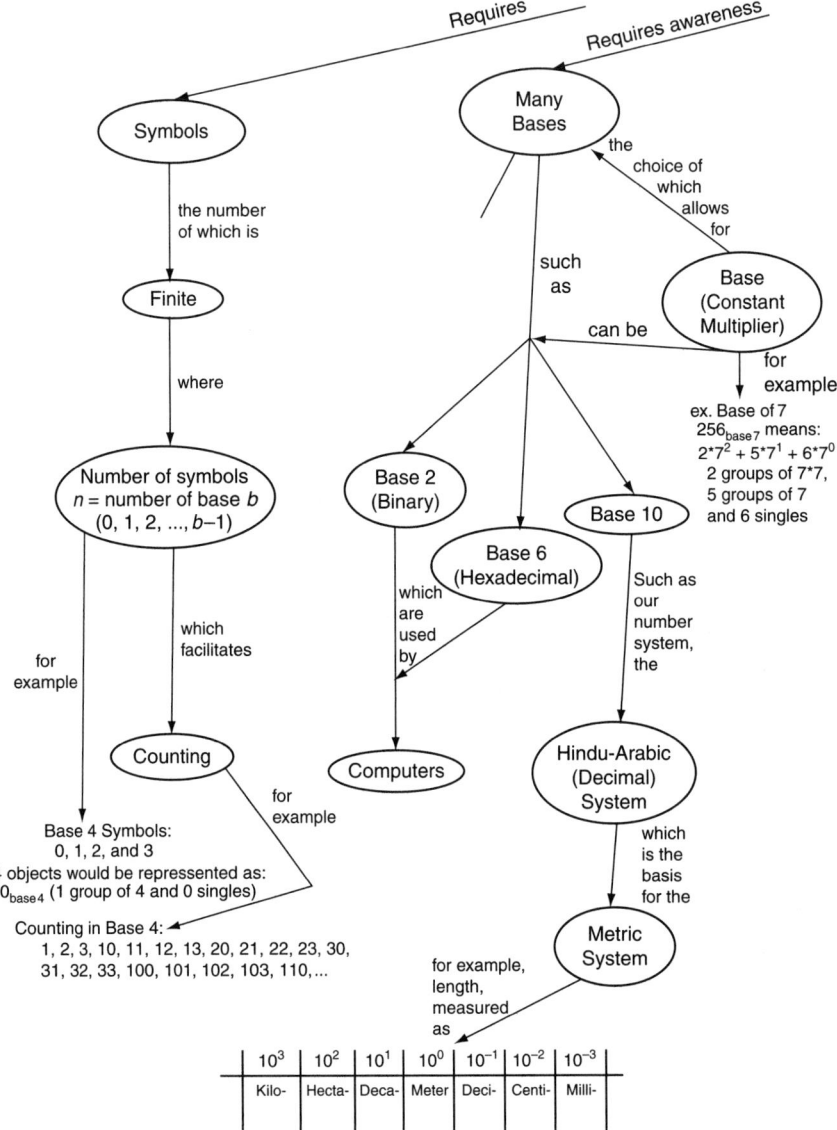

Fig. 3.2 Left side of Maryanne's concept map showing "Symbols" and "Many Bases" as subsumed concepts

bases. Under the designation "Symbols" we see that she understands that for any base b, numerals from 0 through 1, 2, 3, ..., b-1 are necessary to designate the numbers in the system. Base 4 is provided as an example, with 0, 1, 2, and 3 as numerical symbols, and counting in base 4 is illustrated.

Moving to the right across the map (Fig. 3.2), Maryanne further develops these ideas, beginning with the initial proposition that the concept of "positionality requires awareness and understanding of many bases", including base 2 and 16 used in computers, and base ten employed in both the metric and Hindu Arabic numeration systems. Her illustrations of the powers of ten designated in the metric system and the powers of 7 in the expanded form of the number 256 $_{base\ 7}$ show the link to exponentiation that derives from a consideration of meaning of numbers across systems employing diverse bases.

Figure 3.3 identifies the section of Maryanne's concept map that includes three classroom applications involving different bases. A Mother Goose rhyme in base 7, a packaging example in base 6, and chip trading in base 4, are examples of methods that can be used with elementary students to develop the concept of positional system from the study of multiple bases. Pre-service teachers were encouraged to include pedagogical methods (and these are found throughout Maryanne's map), as exemplary pedagogical practices that embody important conceptual content.

Chip trading in base 4 is treated in Burris' (2005) text for elementary teachers Understanding the Math You Teach, a feature that renders this text unique. The use of chip trading to teach the fundamental operations of addition, subtraction, multiplication, and division appears in the bottom right section of Maryanne's concept map, denoted as "Teaching Basic Operations" (see Fig. 3.1). This section is specifically highlighted in Fig. 3.6 and represents a pedagogical approach that connects the algorithmic operations with their conceptual genesis.

Teaching Bases:

Mother Goose Rhyme (Base 7):
As I was going to St. Ives,
I met a man with seven wives;
Every wife had seven sacks,
Every sack had seven cats,
Every cat had seven kits;
Kits, cats, sacks, and wives,
How many were there going to St. Ives?

A Candy Factory Packaging in Sixes (Base 6)

Chip Trading in "the Land of Fours" (Base 4 or other)

Fig. 3.3 Classroom applications involving bases

Considerations of exponentiation are shown in Fig. 3.4, including the extension to negative integer and fractional powers, which designate fractional and irrational numbers, respectively, and to their repeating (rational) and infinite non-repeating (irrational) decimal representations. Maryanne goes further to relate to the real number line the combined infinite set of all numbers capable of being expressed as a base raised to an integer or fractional power, and illustrates the generating of terminating and repeating decimals from partitive division using base ten blocks, an effective teaching strategy.

Figure 3.5 shows the interrelationships among the four fundamental operations of addition, subtraction, multiplication, and division of whole numbers, together with attendant processes of regrouping as required. Embedded within Fig. 3.5 is an important link between the standard algorithms for these processes and modeling of similar operations with polynomials.

The standard algorithms are unfortunately downplayed in the current reform movement (NCTM, 1998) but are, as Maryanne's map reveals, fully conceptual. She illustrates the ease of transition to algebra from a consideration of multiple bases, as it is an easy step from the notion of a variable base b, where b can be any integer greater than 1, to the variable x which can be any real number. One simply removes the positive integer restriction to produce a real valued variable. Further, the familiar expanded form for numbers in various bases (such as $304_{\text{base 8}}$ shown here and $256_{\text{base 7}}$ in Fig. 3.3) is isomorphic to the form in which a polynomial (such as $x^2 + 5x + 1$) must of necessity be written, since the value of x is unknown and hence, its terms cannot be added together. Maryanne uses an area model for the polynomial. In addition, she integrates both estimation by rounding and scientific notation into this section of the map.

In Fig. 3.6, Maryanne reveals the pedagogical content knowledge necessary for proper instruction in positionality employing chip trading (Davidson, Galton, & Fair, 1975) with the requisite trades for regrouping as required in performing the four fundamental multi-digit operations on integers. She also employs an area model for multiplication.

It is noteworthy that the conceptual and the algorithmic are connected without a separation of "concepts" and "procedures". Under "Teaching Basic Operations" the use of chip trading as a pedagogical tool is displayed. In the "Addition with Regrouping" model, Maryanne illustrates the removal of ten "ones" chips and their trade for a "ten" chip in adding $32 + 49 = 81$. In the "Subtraction with Regrouping" model, she shows the manner in which a "ten" chip is traded for ten "ones" chips, and then 7 "ones" chips are removed from the resulting 13 "ones" chips. Then one "ten" chip is removed from the remaining two "tens" chips in the subtraction problem $33 - 17 = 16$.

In the depiction of multiplication using chip trading, Maryanne solves the problem $135 \times 3 = 405$ by tripling the 5 "ones" chips and trading ten of the resulting 15 chips for a "ten" chip, then tripling the 3 "tens" chips and trading the resulting 9 +1 "tens" chips for a single "hundred" chip. Tripling the original "hundred" chip and adding this additional "hundred" chip results in the solution of 4 "hundreds", no "tens" and 5 "ones" chips. She then shows how in multiplying 15×12, the 12

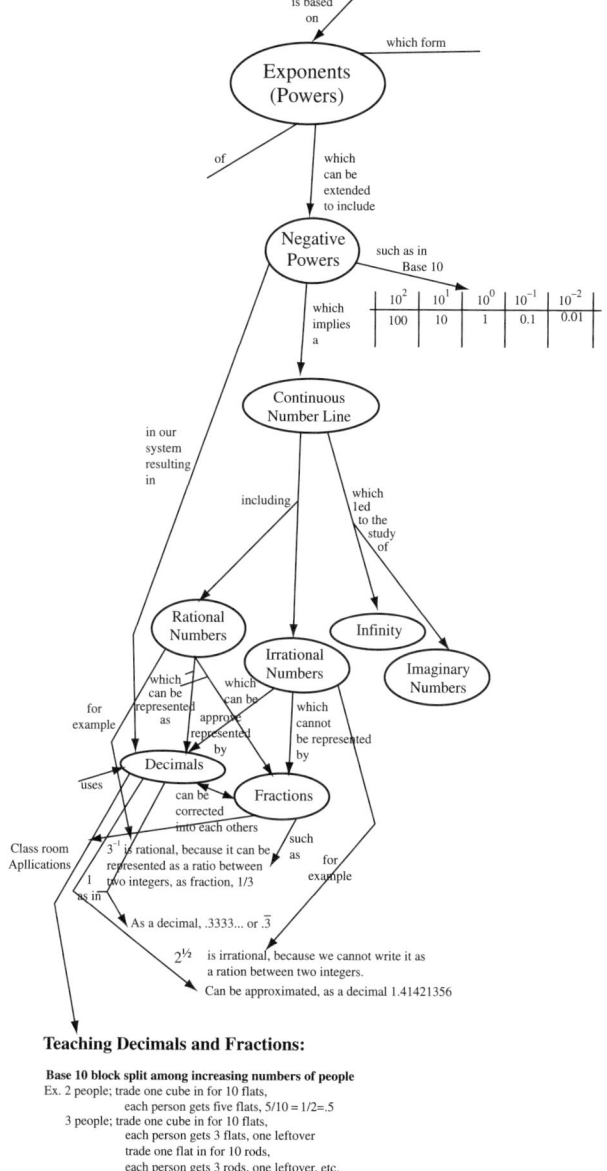

Fig. 3.4 Center of Maryanne's concept map showing "Exponents" as a related concept

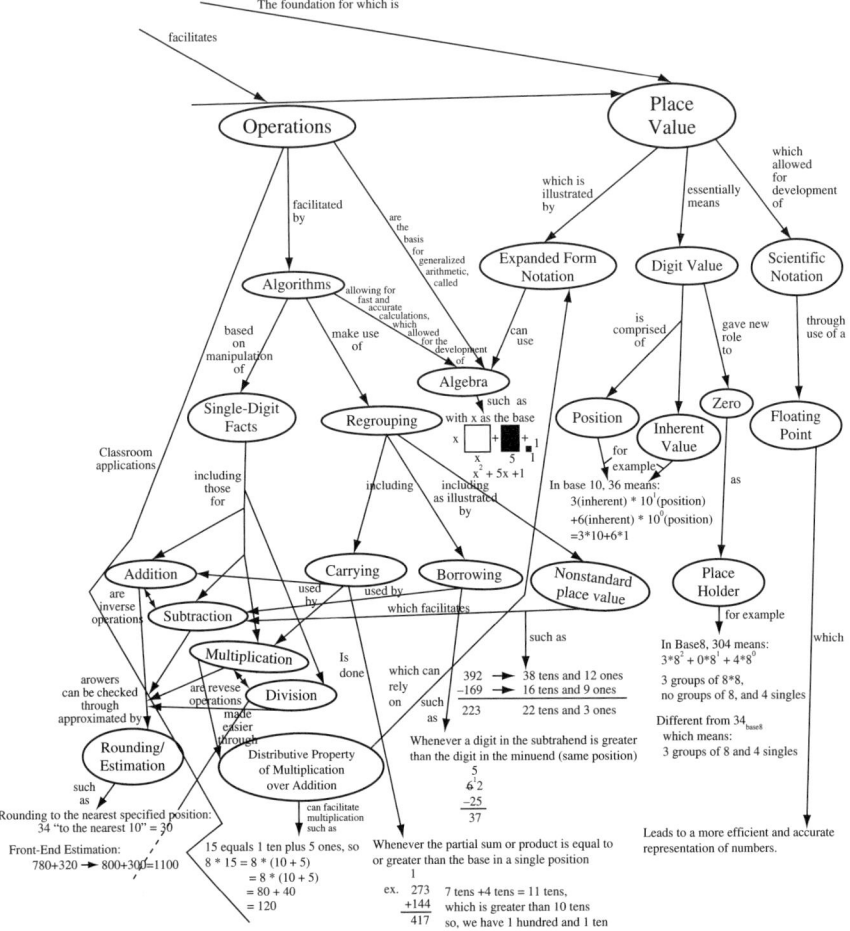

Fig. 3.5 Right side of Maryanne's concept map showing "Operations" and "Place Value" as subsumed concepts

is first split into 10 + 2, then each addend separately multiplies 15, and finally the resulting partial products are added to obtain 180. The distributive property is the underlying conceptual mechanism here, and this is emphasized in the area model for multiplication of 13 × 11.

Finally, Maryanne illustrates the use of chip trading in the division of 124 by 4. First the "hundred" chip is traded for 10 "tens" chips to provide 12 "tens" chips. Twelve tens divided by 4 is 3 tens, and finally four "ones" divided by four is 1 "one". Therefore the quotient is 31 which is 3 tens and 1 unit.

The power of chip trading is that it requires the child to employ the same cognitive processes that are required for computing efficiently with multi-digit numbers in our base ten positional system. As these processes are employed, the conceptual

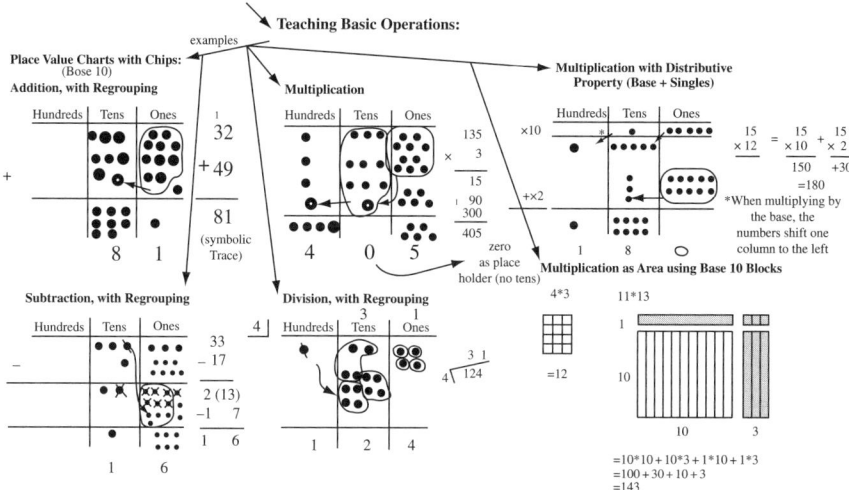

Fig. 3.6 Maryanne's pedagogical approach connecting the algorithmic with the conceptual

content of the algorithms is continually reinforced; they are not arbitrary "rules" to be memorized and executed by rote, but meaningful processes that rely for their power on the concept of positional system and the properties of actions such as the distributive property of multiplication over addition. Eventually, children have no need for the chips and simply invoke the representation of the trades. Finally, they are able to perform the requisite mathematical actions on the numbers alone, and to see that the algorithms are merely the symbolic trace of the meaningful mathematical actions they formerly performed on objects. When they attain this level of competence and understanding they will not find actions on large numbers (which would be too cumbersome to be easily performed with objects) to be daunting, nor will they need a calculator to perform them.

Maryanne's map thus shows connections of the concept of positional system with its genesis from a consideration of multiple bases (perhaps begun with simple nursery rhymes, and pictures for young children as shown in Fig. 3.3), through connections with exponentiation, decimals, fractions, the real number line, estimation and rounding, the metric system and scientific notation so important for measurement, area models, the concept of variable, and the operations of addition, subtraction, multiplication, and division of both integers and polynomials.

Death by Decimal

Concept mapping reveals the centrality of the concept of positional system in the conceptually dense system of concepts that comprise elementary school mathematics. Not only does it connect to many important concepts that students will study concurrently or for which it will prepare them for study in the future, it is also a prerequisite for any real understanding of the base ten system. It is significant that the mathematics study group of the Mathematical Association of America recently affirmed the concept of positional system to be foundational in school mathematics. Indeed the consequences of failure to adequately grasp this concept in real world applications range from measurement inaccuracies in trades such as construction to those in professions such as medicine. The first can be costly; the second deadly.

In her study of the mathematical errors of student nurses, Pirie (1987) documents the extent to which student nurses fail to correctly use mathematics to make mindful decisions in such tasks as unit conversions, dosage calculations, and fluid monitoring. Fragile and/or incorrect conceptual development in mathematics often invites the use of procedural shortcuts that increase the potential for error and the possibility of disastrous results for a patient whose life may depend on the correctness of the calculation. In the absence of fundamental conceptual grounding, the same mathematical procedures that could be used to promote the health and well-being of a patient become unreliable.

The severity of miscalculation is evident in the simulation study published in the American Journal of the Diseases of Children. Perlstein et al., whose study involved the staff of a neonatal intensive care unit working with simulated physicians' orders, reported that "56% of the errors tabulated would have resulted in administered doses ten times greater or less than the ordered dose" (cited in Pirie, 1987, p. 145.) Lesar (2003) studied and classified the 200 tenfold errors in medication dosing that occurred in an 18 month period at a 631-bed teaching hospital citing such errors as a misplaced decimal point, adding an extra zero, or omitting a necessary zero. Przybycien (2005, p. 32) reports that a physician ordered morphine 0.5 mg IV for a 9-month-old baby but because of a missed decimal an inexperienced nurse gave the baby 5 mg of morphine IV, and the baby died. It is an example of "death by decimal" and the lack of meaningful understanding of positionality continues to lead to such tragedy.

Summary

Using concept mapping in the development of the concept of positional system substantively reveals the need to teach multiple basic level concepts for the formation of a superordinate concept. Maryanne's carefully considered concept map of positionality is in sharp contrast to the superficial treatment often found in popular mathematics texts which display a decimal number with digits to the right and left of the decimal point labeled to indicate the name and relative value of each position. Maryanne's concept map clearly indicates what must be taught to students for a meaningful understanding of positionality to develop. It reflects her in-depth exploration of the meanings associated with and underlying the concept of positional system, its antecedent concepts, and the complexity of their interrelationships within a conceptual system. The map addresses a serious deficiency in current elementary mathematics programs and provides a reliable direction for future mathematics curriculum development. The inadequate development of positional system inhibits future learning in mathematics and has important consequences for societal applications that require knowledge of the decimal system. Research in the field of medicine corroborates the number of deadly errors attributable to misplaced decimals in fluid monitoring and drug dosage calculations. It would be difficult to imagine a nursing student who understands the concepts and relationships depicted in Maryanne's map ever making the devastating errors chronicled in Pirie's (1987) and Przybycien's (2005) research.

Acknowledgement   Our special thanks to the pre-service student who graciously provided the concept map discussed in this chapter.

References

Burris, A. (2005). *Understanding the math you teach*. Upper Saddle River, NJ: Prentice Hall.

Davidson, P. S., Galton, G. K., & Fair, A. W. (1975). *Chip trading activities*. Fort Collins, CO: Scott Resources.

Lesar, T. S. (2003). Tenfold medication dose prescribing errors. *The American Journal for Nurse Practictioners, 7*(2), 31–32, 34–38, 43.

National Council of Teachers of Mathematics. (1989). *Curriculum and evaluation standards for school mathematics*. Reston, VA: Author.

National Council of Teachers of Mathematics. (1998). *The teaching and learning of algorithms in school mathematics*. Reston, VA: Author.

National Council of Teachers of Mathematics. (2000). *Principles and standards for school mathematics*. Reston, VA: Author.

Pirie, S. (1987). *Nurses and mathematics: Deficiences in basic mathematical skills among nurses*. London: Royal College of Nursing.

Prairie Rainbow Blocks. Oakland, CA: Prairie Rainbow Company.

Przybycien, P. (2005). S*afe meds: An interactive guide to safe medication practice*. St. Louis: Elsevier Mosby.

Skemp, R. (1987). *The psychology of learning mathematics*. Mahwah, NJ: Lawrence Erlbaum.

Vygotsky, L. S. (1962). *Thought and language*. Cambridge, MA: MIT Press.

Chapter 4
Using Concept Maps and Vee Diagrams to Analyse the "Fractions" Strand in Primary Mathematics

Karoline Afamasaga-Fuata'i

The chapter presents data from Ken, a post-graduate student who participated in a case study to examine the value of concept maps and vee diagrams as means of communicating his conceptual analyses and developing understanding of the "Fractions" content strand of a primary mathematics syllabus. Ken's work required that he analysed syllabus outcomes and related mathematics problems and to display the results on concept maps and vee diagrams (maps/diagrams) to illustrate the interconnectedness of key and subsidiary concepts and their applications in solving problems. Ken's progressive maps/diagrams illustrated how his pedagogical understanding of fractions evolved over the semester as a consequence of social critiques and further revision. Progressive vee diagrams also illustrated his growing confidence to justify methods of solutions in terms of mathematical principles underlying the main steps.

Introduction

As mathematics teachers, it is incumbent upon us to ensure that we have a deep understanding of the content of the syllabus we are going to teach, that we can pedagogically and effectively mediate the development of school students' understanding and meaningful learning of mathematical concepts and processes by providing students with support as they engage with appropriately designed learning activities that challenge their mathematical thinking and reasoning. Further, teachers should have the capacity to diagnose instances of significant learning and, when it is not occurring, to provide appropriate support to assist students along their developmental learning trajectories (e.g., AAMT, 2007; NCTM, 2007). This requires that we are, not only familiar with the psychology and epistemology of learning to empower us to appropriately assist individuals coming to know, understand and learn new ideas meaningfully, but that we are also familiar with the range of socio-cultural factors that impact on learning in a social milieu so that we can support students'

K. Afamasaga-Fuata'i (✉)
School of Education, University of New England, Armidale, Australia
e-mail: kafamasa@une.edu.au

interactions and exchange of ideas to further the development of their conceptual understanding of mathematical situations, concepts and processes.

Shulman's taxonomy of knowledges for effective quality teaching includes knowledge of the content of the discipline, i.e., subject matter knowledge (SMK), and knowledge of teaching mathematics, i.e., pedagogical content knowledge (PCK). SMK is further defined as consisting of substantive knowledge (i.e., knowledge of principles and concepts of the discipline) and syntactic knowledge (i.e., knowledge of the discipline's methods of generating and validating knowledge) (Shulman, 1986). While knowledge has the properties of a commodity – that is, it is categorical, codifiable, and can be traded or exchanged (Lyotard, 1979, as cited in Feldman (1996)), understanding is the result of meaning-making in situations (Bruner, 1990) and requires that students actively organize knowledge hierarchically to show interconnections between relevant concepts, with the most general and most inclusive concepts superordinate to less general and most specific concepts (Ausubel, 2000; Novak & Gowin, 1984; Novak, 2002). That the conceptual interconnections may be described, in accordance with the discipline knowledge, to generate propositions that are mathematically correct, indicate the occurrence of learning that is meaningful and conceptually based.

Through social interactions in a classroom setting, students and teacher collaboratively negotiate meaning and shared understanding as they deliberate about, and engage with, a learning activity. According to Ausubel's theory of meaningful learning, students' understanding is developed through the construction of their own patterns of meanings and through participation in social interactions and critiques. When new knowledge is meaningfully learnt, the student decides which established ideas in his/her cognitive structure of meanings are most relevant to it. If there are discrepancies and conflicts, the student reorganizes and reconstructs existing patterns of meanings, reformulates propositions, or forms new patterns to allow for the effective assimilation of new meaning. For example, if the student could not reconcile the apparent contradictory ideas, then a degree of synthesis (integrative reconciliation) or reorganization of existing knowledge under more inclusive and broadly explanatory principles would be attempted (progressive differentiation). In contrast, rote learning is learning where students tend to accumulate isolated propositions rather than developing integrated, interconnected hierarchical frameworks of concepts (Ausubel, 2000; Novak & Cañas, 2006). Students' conceptual understanding of a domain may be displayed on hierarchical concept maps and vee diagrams.

Definitions of Concept Maps and Vee Diagrams

Concept maps are hierarchical graphs of interconnecting concepts (i.e., nodes) with linking words on connecting lines to form meaningful propositions. Concepts are arranged with the most general and most inclusive concepts at the top with less general and less inclusive concepts towards the bottom. Vee diagrams, in contrast, are vee structures with its vee tip situated in the problem or event to be analysed with its left side displaying the conceptual aspects (i.e., theory, principles, and concepts) of the problem/activity while the methodological aspects (i.e., records (given information), transformations and knowledge claims) are on the left side. A completed vee diagram represents a record of the conceptual and methodological information of solving a problem to answer some focus questions. The theoretical basis of these meta-cognitive tools is Ausubel's theory of meaningful learning (Ausubel, 2000; Novak & Gowin, 1984). Examples of maps/diagram are provided later.

Case Study

The case study of a post-graduate student, Ken, presented here, is about his concept maps and vee diagrams (maps/diagrams) constructed to illustrate and communicate his analysis for key and subsidiary concepts (knowledge), comprehension and critical interpretation (meaning) of the "Fractions" syllabus outcomes holistically at the macro level, from early primary to early secondary (i.e., Early Stage 1 to Stage 4 of the *NSWBOS K-10 Mathematics Syllabus*) (NSWBOS, 2002), and then at the micro level in the context of solving fraction problems. Over the semester, Ken transformed his evolving understanding of the relevant conceptual and methodological interconnections of key and subsidiary concepts of fractions and its progressive development across the primary years of schooling (NSWBOS, 2002) into visual displays of hierarchical conceptual interconnections on concept maps on one hand while on the other, of the synthesis of conceptual and methodological information in solving problems on vee diagrams.

Draft maps/diagrams were presented to the lecturer-researcher on a weekly basis in 1-hour workshops for 12 weeks. Through interactive discussions and negotiations of meaning in a social setting, Ken explained, justified and elaborated his constructed maps/diagrams while the lecturer-researcher challenged his explanations and critiqued his presented maps/diagrams. In between meetings, Ken revised his maps/diagrams to accommodate critical comments of the previous workshop in preparation for the next presentation.

Context

Ken was enrolled in a post-graduate mathematics education one-semester course which introduced the meta-cognitive strategies of concept mapping and vee diagramming to analyse and explicate the conceptual structure of a domain, in terms of the hierarchical interconnections of its key and subsidiary concepts as an abstract overview displayed on concept maps and to make visible the connections between methods of solving mathematics problems and the relevant concepts and principles underpinning the methods on vee diagrams. Ken practiced constructing maps/diagrams of different syllabus outcomes and mathematics problems. The set of maps/diagrams presented here constituted part of his assignments for the course. Only data from two tasks are presented.

Task 1 was for Ken to analyse the treatment of fractions in the NSW K-10 Mathematics Syllabus, specifically, the developmental learning trend of "fractions of the form $\frac{a}{b}$, equivalence and operations with fractions" before conducting a small pilot study to examine some students' conceptions of fractions/equivalence/operations (i.e., pilot study content). The students were selected from Years 4, 6 and 8 of a local school to correspond to the end years of Stages 2, 3 and 4 of the *K-10 NSW Mathematics Syllabus* (NSWBOS, 2002). Whilst the results of this pilot study is reported elsewhere (Jiygel & Afamasaga-Fuata'i, 2007), Task 1 focused on Ken's conceptual analyses of the "Fractions" content most relevant to his pilot study.

Task 2 required Ken to construct (a) an overview concept map of the "Fractions" content strand and (b) vee diagrams of fraction problems to demonstrate the application of some of the mapped conceptual interconnections in (a).

Ken was an international student enrolled in the Master of Education program in a regional Australian university. Although, he was an experienced primary teacher from his own country, it was important that he fully understood the development of the "Fractions" content strand in the primary mathematics and early secondary mathematics (PESM) syllabus implemented at the local school. Ultimately, his concept mapping and vee diagram tasks were intended to make explicit, for discussion and evaluation, his conceptual analyses and pedagogical understanding of (a) the content specific to his pilot study and (b) the overall "Fractions" content strand of the PESM syllabus. Therefore, the focus question of this chapter is: *In what ways do concept maps and vee diagrams facilitate the conceptual analyses and pedagogical understanding of syllabus outcomes?*

Data Collected and Analysis

The following sections present Ken's concept map and vee diagram data as required for Tasks 1 and 2 followed by a discussion of the results. Concept maps are analysed by considering the *propositions* formed by strings of connected nodes and linking words, the presence of *cross-links* between concept hierarchies which indicate integrative reconciliation between groups or systems of concepts and *multiple branching* nodes which indicate progressive differentiation between more general and less general concepts. Further, Ken's pedagogical content knowledge and understanding of fractions in accordance with the requirements of Tasks 1 and 2 (and as displayed on maps/diagrams) are compared to the relevant syllabus outcomes of the *K-10 NSW Mathematics Syllabus* (NSWBOS, 2002).

Task 1 Data and Analysis

Early Stage 1 and Stage 1 Concept Maps

Ken's analysis of Early Stage 1 (Kindergarten) syllabus outcomes in Fig. 4.1 shows that "half" or "halves" are introduced concretely from everyday context by the

Fig. 4.1 Early Stage 1
Fractions concept map

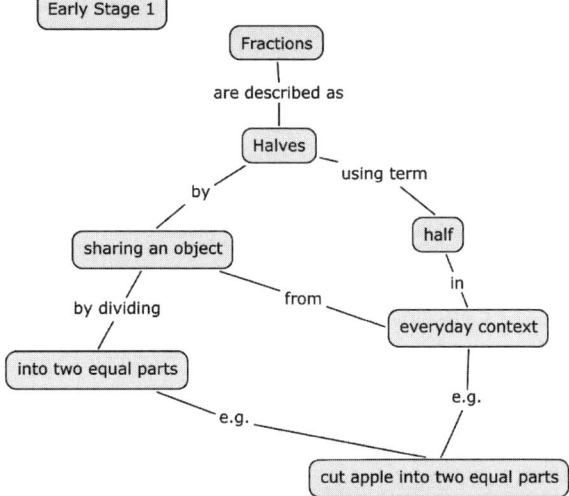

Fig. 4.2 Stage 1 *Fractions*
concept map

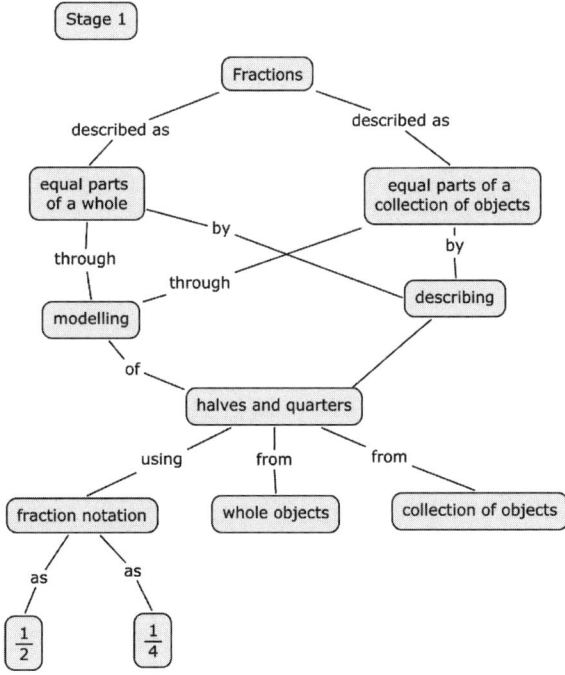

sharing of an object and by dividing it into "two equal parts". The emphasis at this stage is that the two parts are equal to ensure fairness.

Figure 4.2 indicates that, at Stage 1 (Years 1–2), fractions are used in two different ways, to describe "equal parts of a whole" and to describe "equal parts of a collection of objects". Fractions also expand to include "quarters". Modelling and describing halves and quarters using whole objects and collection of objects continues and the notations "$\frac{1}{2}$" and "$\frac{1}{4}$" are introduced to represent "half" and "quarter" respectively. According to the syllabus (NSWBOS, 2002, p. 61), it is not necessary for students at this stage to distinguish between the roles of the numerator and denominator. Subsequently, students may use the symbol "$\frac{1}{2}$" as an entity to mean "one-half" or "a half" and similarly for "$\frac{1}{4}$". These last two points (Stage 1) and the "fairness" basis of Early Stage 1 were not included in Ken's maps.

Stage 2 Concept Map

Figure 4.3 shows that, at Stage 2 (Years 3–4), fractions are described in different ways such as "equal parts of a whole" and "equal parts of collection of objects" and students' repertoire of fractions increase to include those with denominators 8, 5, 10

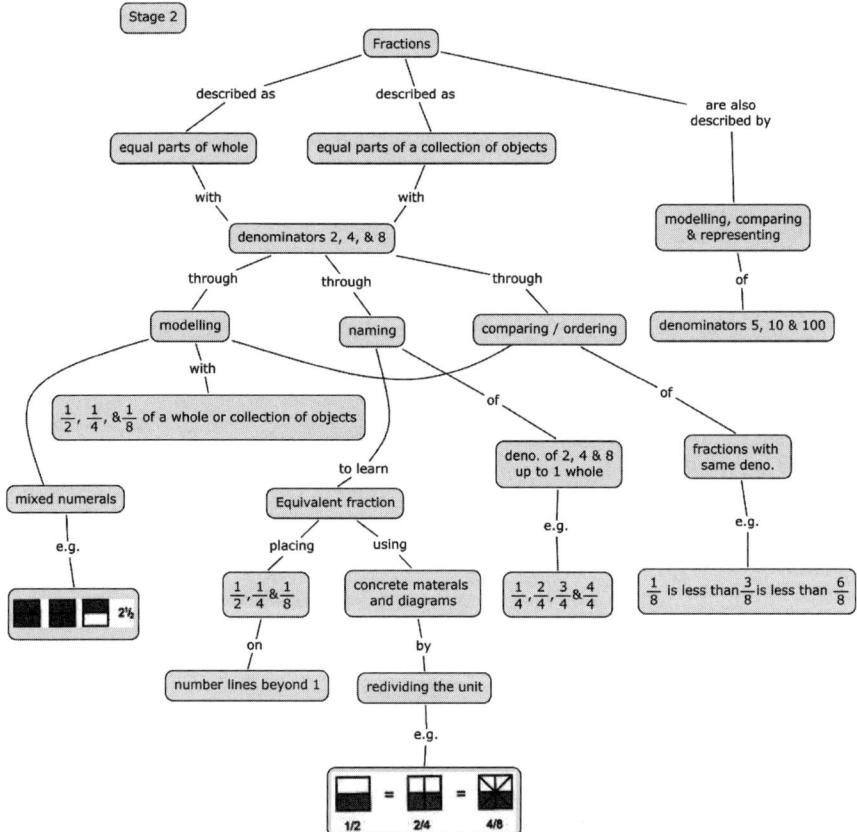

Fig. 4.3 Stage 2 *Fractions* concept map

and 100. Students learn about "modelling", "naming", and "comparing/ordering" fractions with "denominators 2, 4 & 8".

These fractions are modelled with "$\frac{1}{2}$, $\frac{1}{4}$ & $\frac{1}{8}$" of a whole or collection of objects and "mixed numerals" are modelled using diagrams as shown for $2\frac{1}{2}$. Naming fractions with "(denominators) 2, 4, & 8 up to 1 whole" is also developed, for example, $\frac{1}{4}$, $\frac{2}{4}$, $\frac{3}{4}$ and $\frac{4}{4}$ for quarters. Naming is also used to learn about "equivalent fraction" (e.g., between half, quarters and eighths) by placing "$\frac{1}{2}$, $\frac{2}{4}$ & $\frac{4}{8}$" on "number lines beyond 1" and using "concrete materials and diagrams" by "redividing the unit" as shown diagrammatically for $\frac{1}{2}$, $\frac{2}{4}$ and $\frac{4}{8}$. Furthermore, fractions with the same denominators are compared and ordered such as "$\frac{1}{8}$ is less than $\frac{3}{8}$ is less than $\frac{6}{8}$".

The rightmost branch illustrated that the "modelling, comparing & representing" of fractions with "denominators 5, 10 and 100" (i.e., fifths, tenths and hundredths) is also developed by extending the knowledge and skills illustrated by the branches to the left with halves, quarters and eighths. Whilst this last point is explicitly mentioned in the syllabus outcomes, Ken did not explicate this on his map, either by cross-linking to the left branches or extending by adding more nodes. Alternatively, it is a pedagogical concern that could be specifically addressed when planning lessons and designing classroom activities (not covered here).

Omitted also from the concept map is the notion of "numerator" although "denominator" is explicitly mentioned. However, syllabus notes caution that "(a)t this Stage, it is not intended that students necessarily use the terms 'numerator' and 'denominator'" (NSWBOS, 2002, p. 62). Other missing ideas are the use of fractions as operators related to division and the term "commonly used fractions" to refer to those with denominators 2, 4, 8, 5, 10 and 100 as recommended in the syllabus documentation. Although the introduction of decimals (to two decimal places), place value, money (as an application of decimals to two decimal places), and simple percentage are to occur at this stage, Ken did not include them in his concept map in Fig. 4.3. However, given the specific content of the pilot study and focus of Task 1, the omission was to be expected. It was nonetheless, a point that needed consideration for Task 2.

Stage 3 Concept Map

Ken's analysis of Stage 3 (Years 5–6) syllabus outcomes as concept mapped in Fig. 4.4 shows that the new fractions introduced are the thirds and sixths to supplement the halves, quarters, eighths, fifths, tenths and hundredths from previous stages. The "mixed numerals" branch illustrated fractions may be expressed with "mixed numerals" as "improper fractions" through the use of diagrams and number lines leading to a mental strategy. Adjacent to the right of this branch, are interconnections indicating that the "modelling" of thirds, sixths and eighths are to be done with "whole/collection of objects" and by "placing" them on a "number line between 0 and 1" to "develop equivalence" as illustrated by the 3 number lines provided in the middle of the map. The rightmost branch of the map (Fig. 4.4), which is

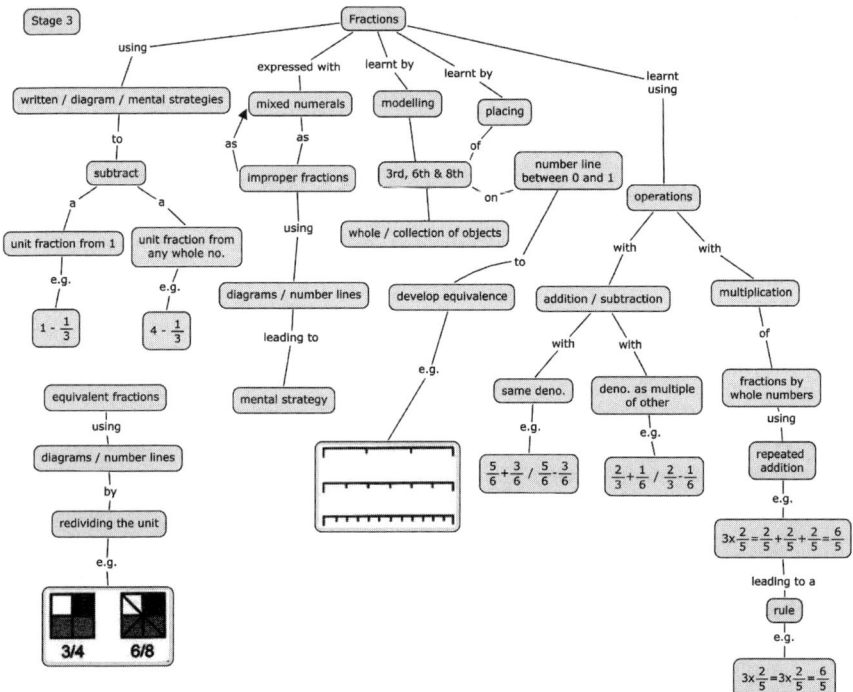

Fig. 4.4 Stage 3 *Fractions* concept map

inclusive under the node "operations" are two concept hierarchies (or sub-branches). Whereas the left hierarchy displayed "addition/subtraction" of fractions with the "same denominator" (e.g., $\frac{5}{6} + \frac{3}{6}/\frac{5}{6} - \frac{3}{6}$) and with "(denominators) as a multiple of the other" (e.g., $\frac{2}{3} + \frac{1}{6}/\frac{2}{3} - \frac{1}{6}$), the right hierarchy illustrated "multiplication" of "fractions by whole numbers" using "repeated addition" (as shown by the illustrative example) which led to a "rule" as shown by the last node.

In comparison, the leftmost branch inclusive under the "written/diagram/mental strategies" node are illustrative examples for subtraction of a "unit fraction from 1" (e.g., $1 - \frac{1}{3}$) and "unit fraction from any whole number" (e.g., $4 - \frac{1}{3}$). A concept sequence subsumed by the "equivalent fractions" node is isolated from the rest of the map. It depicted the "redividing the unit" idea that was viewed in Fig. 4.3 (for the equivalence of half, two-quarters and four-eighths) but is now illustrating the case of three-quarters and six-eighths.

Twelfths were not explicitly mentioned in the upper hierarchical levels of Fig. 4.4 but it is diagrammed on the number line shown in the middle. In this Stage, the label "simple fractions" referred to those with denominators 2, 3, 4, 5, 6, 10, 12, and 100 but was missing from the map. Also missing were "mental strategies" for finding equivalent fractions and for reducing them to lowest term and calculating unit fractions of a collection (e.g., $\frac{1}{3}$ of 30). Decimal and percentage coverage for this stage is omitted from the concept map. Instead it focused specifically on the case of fractions of the form $\frac{a}{b}$ and equivalence given the emphasis of the pilot study (e.g., Task 1).

Stage 4 Concept Map

Ken's analysis of the Stage 4 (Years 6–7) fraction outcomes in Fig. 4.5 shows a focus on "operations" involving "addition" and "multiplication/division" of "fractions & mixed numerals" and "substraction" of "fractions from a whole number". Illustrative examples are shown for the four operations. The rightmost branch illustrated that fractions may be expressed as "improper fraction" and as mixed numerals shown by a cross-link to the more general "fractions & mixed numerals" node. The leftmost branch, in comparison, depicted the idea that "equivalent fractions" may be reduced to its "lowest term".

The Stage 4 outcomes about decimals, percentages, ratios and rates are omitted from the concept map at this early stage of his mapping experiences. Instead, Ken focussed on the development of fractions of the form $\frac{a}{b}$, equivalent fractions and operations with fractions $\frac{a}{b}$ where the denominators are 2, 3, 4, 5, 6, 8, 10, 12 and 100 (defined in the syllabus as "simple fractions") most relevant to the focus of his pilot study.

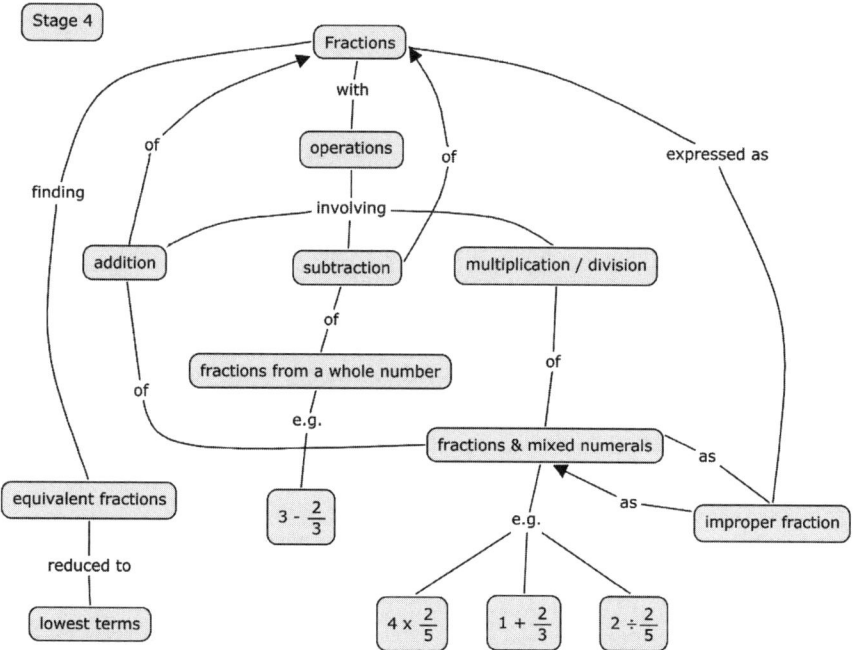

Fig. 4.5 Stage 4 *Fractions* concept map

Overall, Ken's five concept maps traced the introduction, development, and consolidation of (a) the views of simple fractions as "part of a whole", "part of a collection of objects" and a "number" on the number line; (b) equivalence between, and ordering of, simple fractions; and (c) operations with fractions and mixed numerals using concrete materials, illustratively with diagrams and the $\frac{a}{b}$ notation.

Whilst his objective for Task 1 is met with the presentation of the 5 concept maps (Figs. 4.1–4.5), it was incumbent upon him as a teacher to also develop a big picture overview of the coverage of fractions across the four stages of the PESM syllabus (NSWBOS, 2002), as described for Task 2.

Task 2 Data and Analysis

Overview "Fractions" Concept Map

Ken's overview concept map of the "Fractions" content strand, provided in Fig. 4.6, had over 100 nodes. Close-up views of the top, middle and bottom sections of the full map are shown in Figs. 4.7, 8 and 9. Structurally, all three sections are interconnected. For example, integratively reconciled links from the "Fractions" (Level 1) node and Level two nodes ("part of a whole", "part of collection of objects", "ratio", "decimals", "quotients", "percents", "probability", and "rates") all merge at the "$\frac{a}{b}$" node (Figs. 4.6 and 4.7), collectively illustrating the interconnections of the top section to the "$\frac{a}{b}$" node of the middle section (Fig. 4.8).

Interconnections between the middle (Fig. 4.8) and bottom sections (Fig. 4.9) are through the two progressive differentiation links from the node "computation", at the bottom of the middle section (Fig. 4.8), to link to "single operations" at the top of the bottom section and "mixed operations" towards the bottom of the bottom section (Fig. 4.9). Hence, the three sections (although split up for legibility and ease of discussion) are appropriately linked to provide a single overview concept map as requested for Task 2. In contrast to the early maps in Task 1, this overview concept map was completed towards the end of the study period, and as such, it was expected that Ken would accommodate some of the syllabus omissions raised during the presentations of Figs. 4.1 to 4.5. Each section is further examined below.

Top Section – Inclusive under the "Fractions" node (Fig. 4.7) are 8 branches subsumed under 8 Level 2 nodes, which, collectively, represented different ways of describing or using fractions. While the 2-leftmost branches was the focus of Task 1, Fig. 4.7 displays the full range of forms and applications of fractions based on Ken's critical analysis of the relevant syllabus outcomes. An inspection of the interconnections within each branch revealed Ken's attempts at elaborating, representing and communicating his understanding of the meaning and/or application of each of the Level 2 nodes. For example, the "part of a whole" branch included the extended proposition (P1): "Fractions" represent "part of a whole" which can be represented by "area models" such as "squares", "rectangles" and "circles". Further linear linking from each of the last 3 three nodes (i.e., "squares", "rectangles" and "circles"), showed illustrative examples descriptively and diagrammatically.

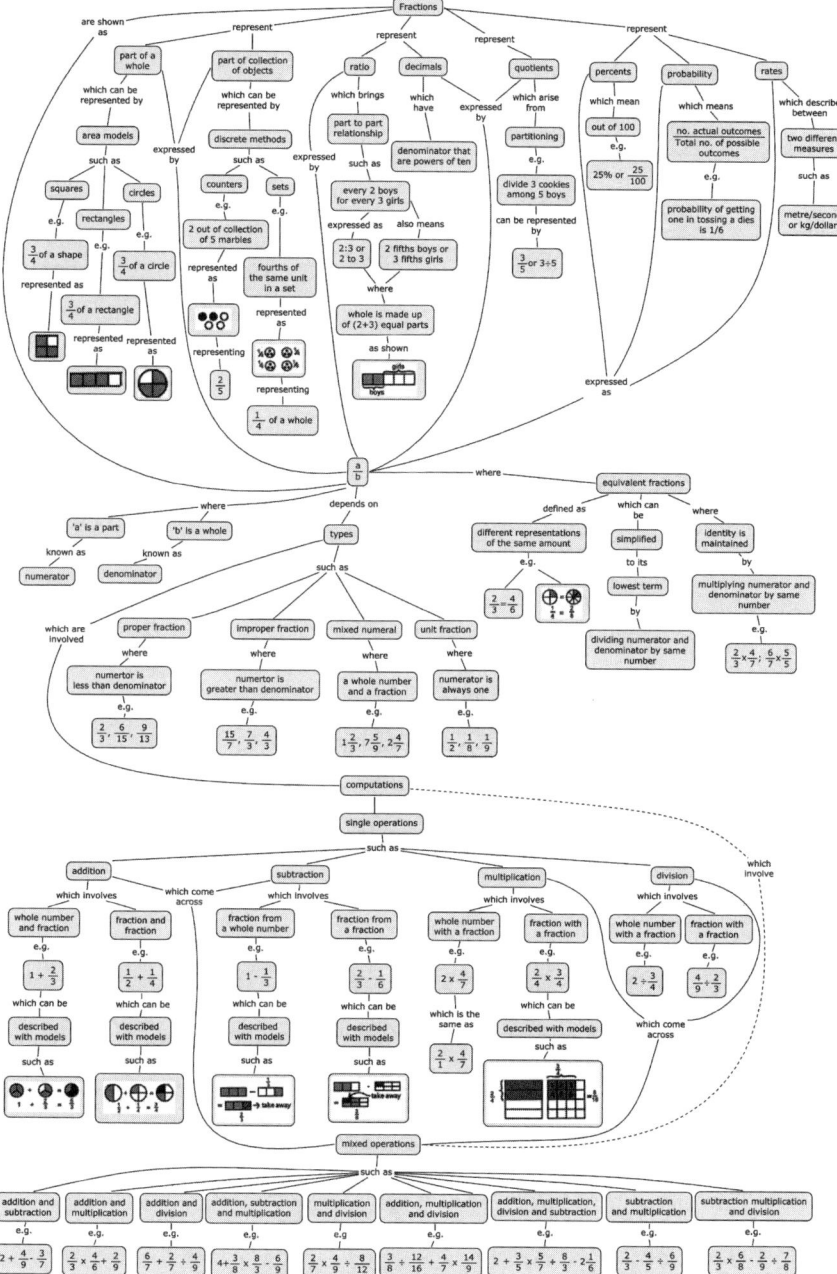

Fig. 4.6 Ken's overview *Fractions* concept map

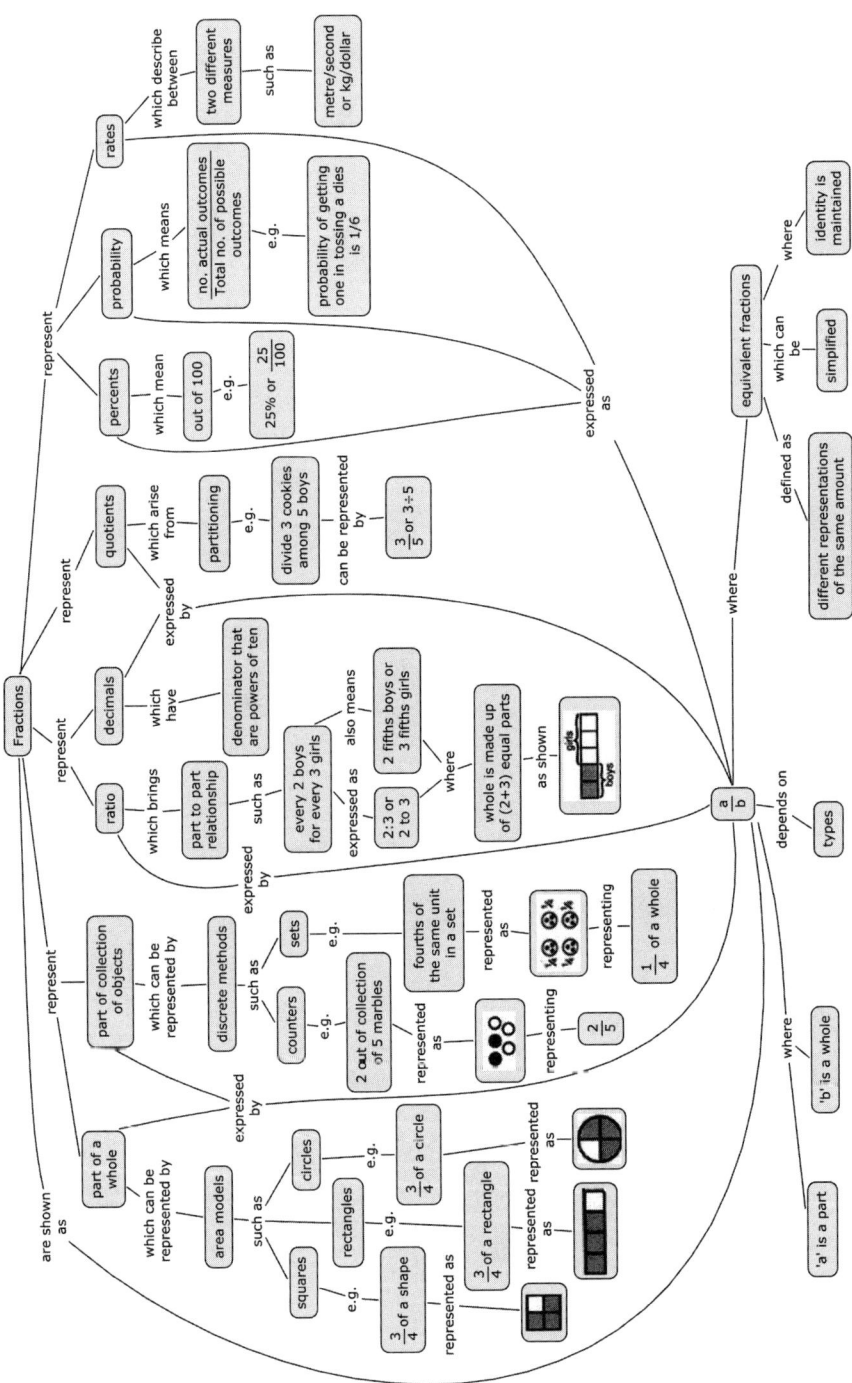

Fig. 4.7 Top section – Ken's overview *Fractions* concept map

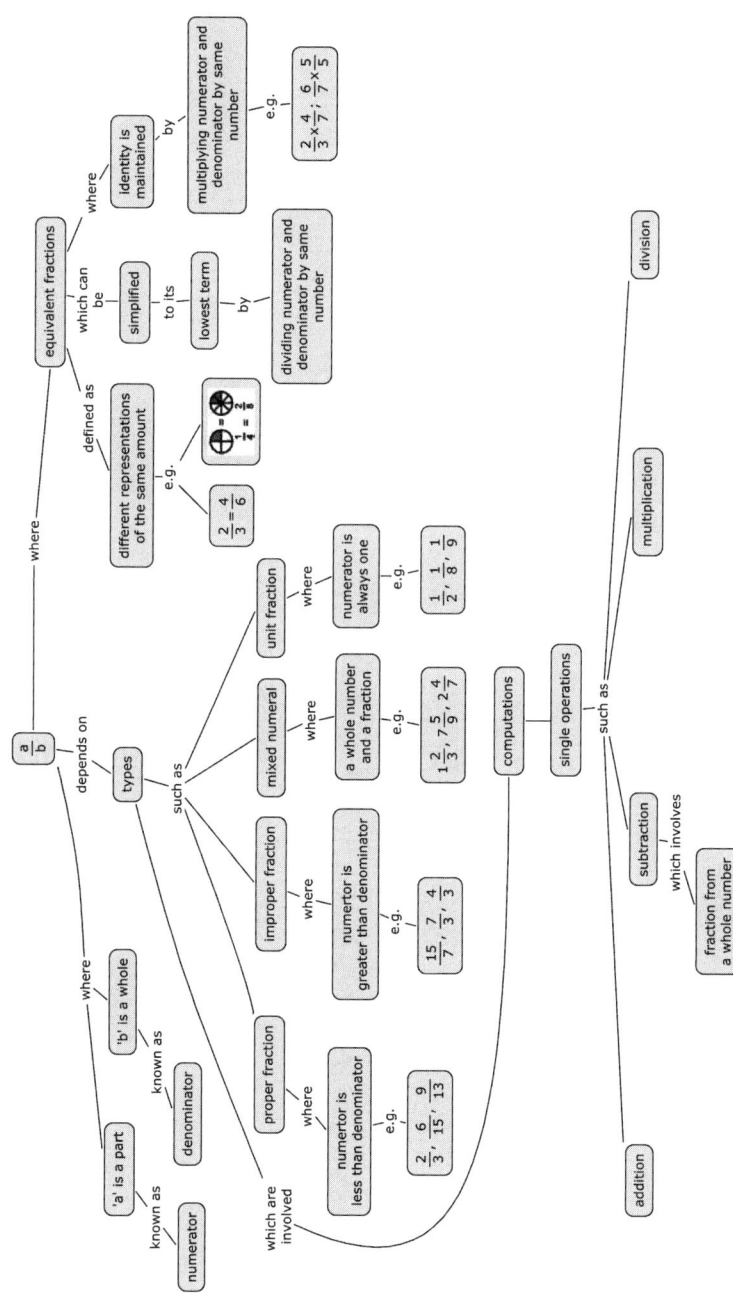

Fig. 4.8 Middle section – Ken's overview *Fractions* concept map

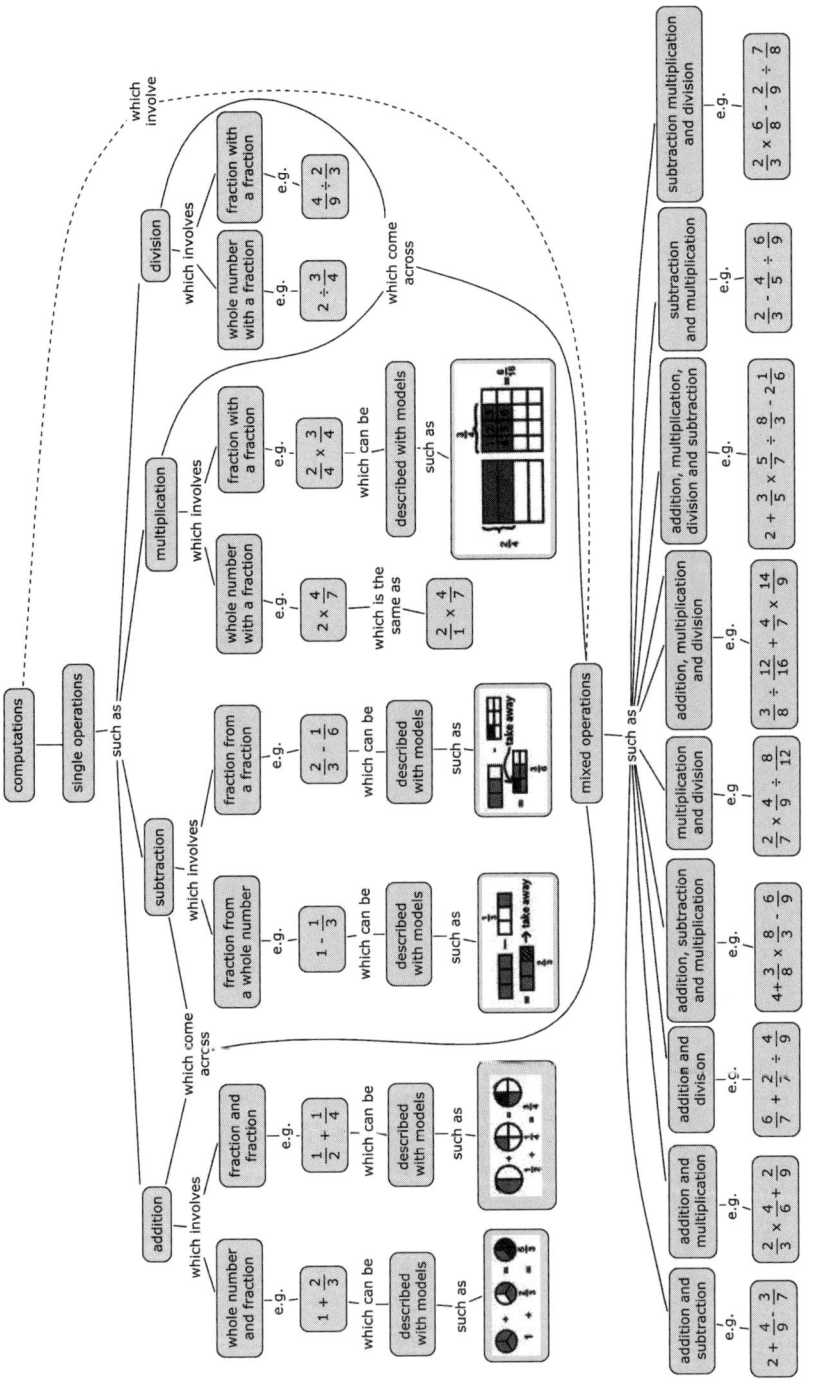

Fig. 4.9 Bottom section – Ken's overview *Fractions* concept map

For the next branch to the right, the extended proposition (P2) is: "Fractions" represent "part of collection of objects" which can be represented as "discrete methods" such as "counters" and "sets". Using counters, an illustrative example is (P2a): "2 out of (a) collection of 5 marbles" represented as (diagrammatically shown) representing "$\frac{2}{5}$". The illustrative example for "sets" is: (P2b): "fourths of the same unit in a set" represented as (diagrammatically shown) representing "$\frac{1}{4}$ of a whole".

The extended proposition (P3a) for the "ratio" branch is: "Fractions" represent "ratio" which brings "part to part relationship" such as "every 2 boys for 3 girls" expressed as "2:3 or 2 to 3" where "whole is made up of (2+3) equal parts" as shown (diagrammatically). A second extended proposition (P3b) of the "ratio" branch is the result of the progressive differentiating links at the node "every 2 boys for 3 girls" node and integrative reconciliation links merging at the node "whole is made up of (2+3) equal parts". Whereas the proposition (P4) for the "decimals" branch is: "Fractions" represent "decimals" which have "denominator(s) that are powers of ten" such as "$\frac{1}{10} = 0.1$", that for the "quotients" branch is: (P5) "Fractions" represent "quotients" which arise from "partitioning", for example, "divide 3 cookies among 5 boys" can be represented as "$\frac{3}{5}$ or 3 ÷ 5".

Towards the right, the extended proposition (P6) for the "percents" branch is: "Fractions" represent "percents" which means "out of 100", for example, "25% or $\frac{25}{100}$" while that for the "probability" branch is: (P7) "Fractions" represent "probability", which means "$\frac{\text{no actual outcomes}}{\text{Total no. of possible outcomes}}$", for example, "probability of getting one in tossing a die is $\frac{1}{6}$". Lastly, the rightmost branch displayed the extended proposition (P8): "Fractions" represent "rates" which describe relationship between "two different measures" such as "metre/second or kg/dollar".

A single link at the extreme left of the map resulted in the proposition (P9) "Fractions" are shown as "$\frac{a}{b}$". Each of the Level 2 nodes cross-linked to the "$\frac{a}{b}$" node resulting in 8 more propositions. Some examples are (P10): "Fractions" represent "part of a whole" expressed by "$\frac{a}{b}$"; (P11): "Fractions" represent "quotients" expressed by "$\frac{a}{b}$"; and (P12): "Fractions" represent "rates" expressed by "$\frac{a}{b}$". Overall, the top section presented a connected web of knowledge displaying the different meanings, uses, applications and notation $\frac{a}{b}$ of fractions. These are more comprehensive and reflective of the "Fractions" syllabus outcomes in the PESM syllabus than Figs. 4.1 to 4.5.

Middle Section – Inclusive under the node "$\frac{a}{b}$" (Fig. 4.8) are four progressive differentiating links to nodes: "'a' is part", "'b' is a whole", "types" and "equivalent fractions". The first proposition (P13) is: "$\frac{a}{b}$" where "'a' is part" known as "numerator" while the adjacent proposition (P14) is: "$\frac{a}{b}$" where "'b' is a whole" known as "denominator". These two sub-branches addressed the information that was missing from Ken's analysis of Stage 2 in Fig. 4.3.

The extended propositions from the "Fractions" node of Level 1 that are inclusive under "types", from left-to-right, are (P15): "Fractions" are shown as "$\frac{a}{b}$" depends on "types" such as "proper fraction" where "numerator is less than denominator", for example, "$\frac{2}{3}, \frac{6}{15}, \frac{9}{13}$"; (P16): "Fractions" are shown as "$\frac{a}{b}$" depends on "types" such as "improper fraction" where "numerator is greater than denominator", for example, "$\frac{15}{7}, \frac{7}{3}, \frac{4}{3}$"; (P17): "Fractions" are shown as "$\frac{a}{b}$" depends on "types" such as "mixed numeral" where "a whole number and a fraction", for example, "$1\frac{2}{3}, 7\frac{5}{9}, 2\frac{4}{7}$"; and (P18): "Fractions" are shown as "$\frac{a}{b}$" depends on "types" such as "unit fraction" where "numerator is always one", for example, "$\frac{1}{2}, \frac{1}{8}, \frac{1}{9}$".

These propositions provided some concepts and definitions that were missing from Figs. 4.1 to 4.5 such as numerator and unit fraction. Inclusive within the "equivalent fractions" branch are the rightmost sub-branches of the middle section (Fig. 4.8). The relevant propositions are: (P19a): "Equivalent fractions" (are) defined as "different representations of the same amount", for example, "$\frac{2}{3} = \frac{4}{6}$"; (P19b): "Equivalent fractions" (are) defined as "different representations of the same amount", for example, (as shown diagrammatically for $\frac{1}{4} = \frac{2}{8}$); (P20): "Equivalent fractions" which can be "simplified" to its "lowest term" by "dividing the numerator and denominator by same number"; and (P21): "Equivalent fractions" where "identity is maintained" by "multiplying numerator and denominator by same number", for example, "$\frac{2}{3} \times \frac{4}{4}; \frac{6}{7} \times \frac{5}{5}$". Overall, this section defined the notation $\frac{a}{b}$ as well as defined and illustrated the different types of fractions.

Bottom Section – Two extended propositions from the top of the overview concept map (Fig. 4.6) connected all three sections (Figs. 4.7, 4.8 and 4.9). These are (P22): (1) "Fractions" are shown as " $\frac{a}{b}$" depends on "types" which are involved in "computation" using "single operations" such as "addition", "substraction", "multiplication", and "division" and (2) (P23): "Fractions" are shown as "$\frac{a}{b}$" depends on "types" which are involved in "computation" which involves "mixed operations" such as listed at the second to last level of the bottom section (see Fig. 4.9) from left to right including illustrative examples for each type.

Four branches particular to the bottom section (Fig. 4.9) are inclusive under the node "single operations" and subsumed under the nodes: "addition", "subtraction", "multiplication", and "division". Inclusive under the "addition" node of proposition P22 are two concept hierarchies. Reading from left-to-right, the left extended proposition (P22a) is: "single operations" such as "addition" which involves "whole number and fraction", for example, "$1 + \frac{2}{3}$" which can be "described with models" such as (diagrammatically shown); and the right one (P22b) is: "single operations" such as "addition" which involves "fraction and fraction", for example, "$\frac{1}{2} + \frac{1}{4}$" which can be "described with models" such as (shown diagrammatically) for "$\frac{1}{2} + \frac{1}{4} = \frac{3}{4}$". For the "subtraction" branch, the extended propositions are (P22c): "single operations" such as "subtraction" which involves "fraction from a whole number", for example, "$1 - \frac{1}{3}$" which can be "described with models" such as (shown diagrammatically) and (P22d): "single operations" such as "subtraction" which involves "fraction from a fraction", for example, "$\frac{2}{3} - \frac{1}{6}$" which can be "described with models" such as (shown diagrammatically). This pattern of propositional links continued all the way across the map to the rightmost node "division".

The second half of the bottom section is the result of merging the integratively reconciled links from the 4 operation nodes ("addition", "subtraction", "multiplication" and "division") and a progressively differentiating link from the "computation" node (bottom of the middle section) at the "mixed operations" node.

Some propositions are (P24): "single operations" such as "multiplication" which come across "mixed operations" and part of P23, namely, "computation" which involves "mixed operations". Emanating from the "mixed operations" node are multiple progressive differentiating links to describe and illustrate the different types of mixed operations at the bottom of Fig. 4.9. These 10 concept hierarchies (inclusive under "mixed operations") represented the final branch of the overview concept map.

Overall, the overview concept map showed hierarchical networks of concepts with the most general concepts at the top (e.g., Level 2 nodes) and progressively less general ones (e.g., "single operations") towards the middle with the more specific ones towards the bottom (e.g., "mixed operations"). Most (sub-)branches terminate with illustrative examples at the bottom. This generality pattern (i.e., most general -> most specific -> illustrative example) was consistently evident with most concept hierarchies.

Concept Maps and Vee Diagrams of "Fraction" Problems

Two "Fractions" problems are presented to illustrate Ken's use of maps/diagrams to communicate his thinking and reasoning when solving problems. His conceptual and process analysis results are provided in Fig. 4.10 for the first problem: "*How many $3\frac{1}{2}$ m lengths of rope can be cut from a length of 35 m?*" (Problem 1).

Concept Maps – Ken envisaged that a useful proposition (P1) is: "Fractions" (should be) understood as "Part of a whole" and "Whole" is "1" helps in solving "Word Problems", for example, (problem 1) while a second proposition (P2) is: "Fractions" involve "operations" involving "Fractions & numerals" for example, "$1 - \frac{2}{5}$" and "$2 \times \frac{3}{4}$." A multi-branched proposition (P3) is: "Fractions" involve "operations" like "Addition", "Subtraction", "Multiplication", "Division" which are involved in "Problem Solving" of "Quantities" expressed as "Word problems", for example, (problem 1). Proposition P3 demonstrated examples of a progressive differentiating link from the "Operations" node and integrative reconciliation links from the four operations before linking to the "Problem Solving" node. A short proposition (P4) is: "Quantities", for example, "$\frac{3}{4}$ of 40 cm". Displayed at the "Word Problems" node is a proposition (P5): "Word Problems" (are) solved using "Strategies" of "Problem Solving" which also illustrated an example of an uplink from a less general concept to a more inclusive one towards the top of the map. Critiques concerned the possibility of elaborating further on what is meant by strategies.

In subsequent workshops, Ken considered a second fraction problem, namely, "*If $\frac{1}{4}$ of a post is below ground level, and 150 cm remains above the ground then find the total length of the post*" (Problem 2). This time, instead of constructing a new concept map, he revised and expanded his previous draft map (Fig. 4.10) to incorporate his thinking and reasoning about the two problems. The revised version combined the main concepts and strategies for the two problems (see Fig. 4.11).

Reading from left-to-right (Fig. 4.11), the leftmost branch displayed the proposition (P6): "Fractions" should be understood as "part of the whole" visualised as (shown diagrammatically) and where the "whole" implies "1" which is evidently a

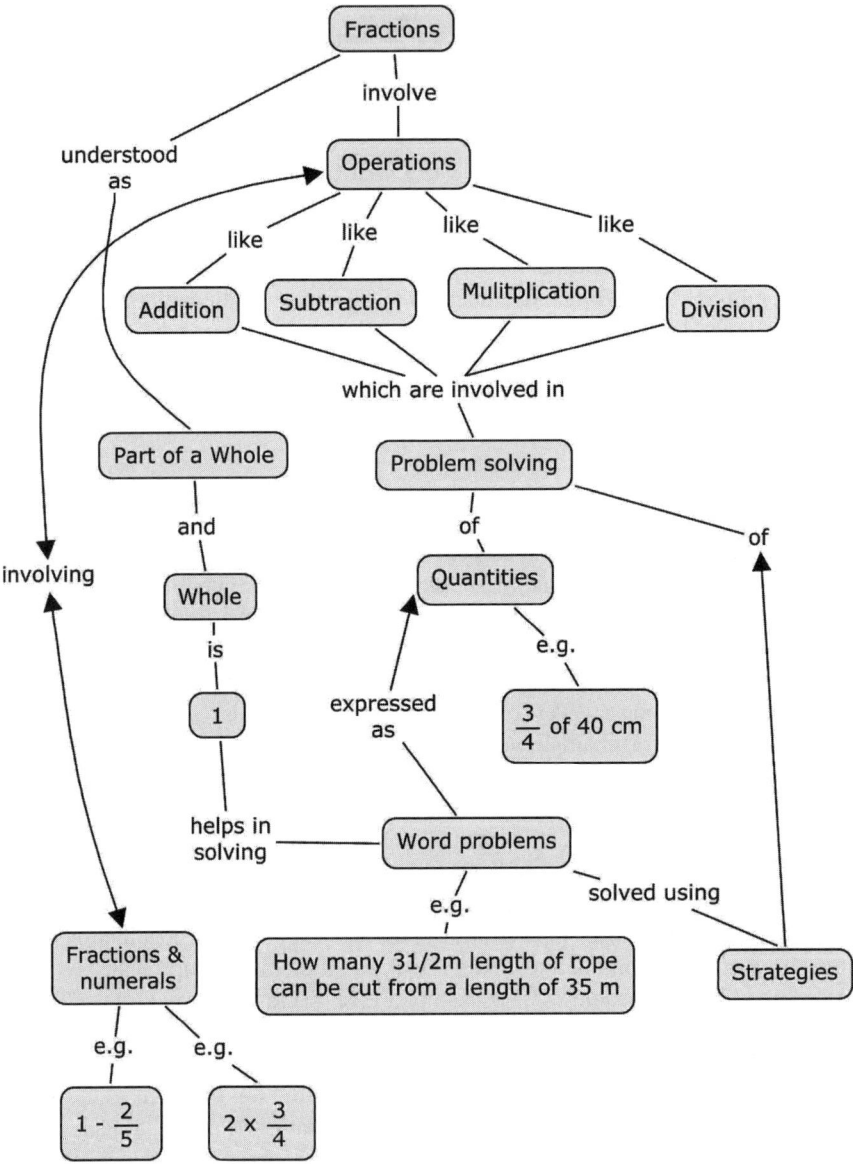

Fig. 4.10 Ken's draft concept map of the rope problem

better and expanded revision of proposition P1 of the previous map (Fig. 4.10) with
the inclusion of a diagram to illustrate the relationship between "part" and "whole".
A cross-link from this branch at node "1" connected to the adjacent "word problems"
branch. The next proposition (P7) is: "Fractions" are used in "word problems"
involving "quantities", for example, "$\frac{3}{4}$ of \$40". Some of the extended propositions,

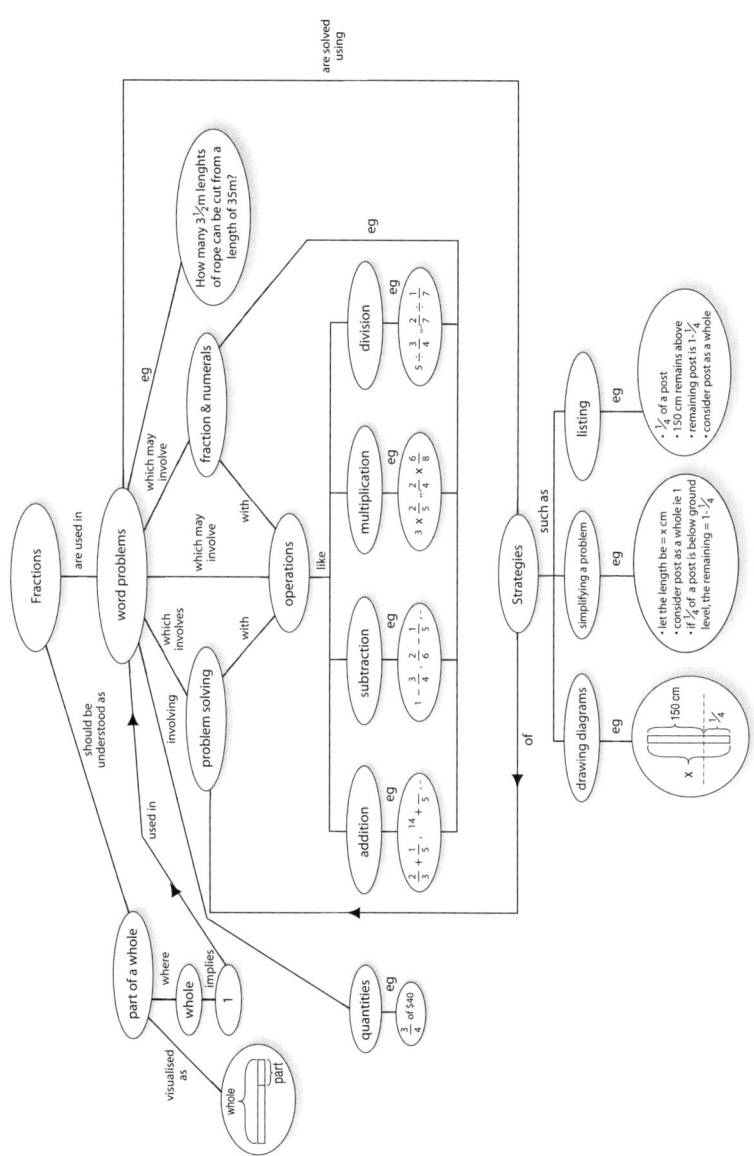

Fig. 4.11 Ken's revised concept map of the two problems

displayed in the middle, emanate from the multi-branching node "word problems". For example, (P8): "Fractions" are used in "word problems" which involves "problem solving" with "operations" like "addition", "subtraction", "multiplication", and "division" with each of the operation node linking to an illustrative example as shown; (P9): "Fractions" are used in "word problems" which may involve "operations"; (P10): "Fractions" are used in "word problems", for example, (problem 1); (P11): "Fractions" are used in "word problems" which may involve "fractions & numerals" with cross-links to the illustrative examples of proposition P8; (P12): "Fractions" are used in "word problems" which may involve "fractions & numerals" with "operations" like "addition", "subtraction", "multiplication", and "division".

Propositions P8, P9 and P12 are integrated at the "operations" node with P11 cross-linking to the propositions' illustrative examples. A number of integrative reconciliation links to, and progressive differentiating links from, nodes: "operations" and "strategies" are displayed.

The rightmost proposition (P13) involving the "strategies" node is: "word problems" are solved using "strategies" of problem solving such as "drawing diagrams" (P13a), "simplifying a problem" (P13b), and "listing" (P13c).

The three sub-branches inclusive under the "strategies" node represented the main additions in this revised map. Proposition P13a is an illustration of the "drawing diagrams" strategy for problem 1 while the second one (P13b) illustrated the "simplifying a problem" strategy. The latter demonstrated a meaningful interpretation and transformation of the given information as situated in the problem's context.

The third concept hierarchy (P13c) illustrated the "listing" strategy using the given information of the problem (i.e., first two bullet points) and application of fraction knowledge to the given information (i.e., last two bullet points). Taken together, the three concept hierarchies (P13a, b & c) depicted the results of his processes of representing, transforming and listing/interpreting the given problem. Overall, the revised and combined concept map for the two problems displayed the key ideas that were applied (e.g., P6, P7 and P8) with the "word problems" node shifting to a more general more inclusive level than the case was in Fig. 4.10, the results of the thinking and reasoning from given information and the key strategies applied as evidenced by propositions P13, and P13a, b, & c respectively. The diversity of propositions delineated above directly resulted from progressive differentiation such as at nodes "word problems", "operations" and "strategies" and integrative reconciliation at the last two nodes.

Vee Diagrams – Only one set of draft and revised vee diagrams are presented here, namely, that for problem 2. Figure 4.12 showed the given problem statement listed in the "Activity (Event)" section at the tip of the vee with the focus question (*What is the total length of the post?*) listed at the top of the vee. On the "Conceptual (Thinking) Side" on the left are the conceptual aspects relevant to the problem. For example, Ken identified three relevant outcomes (as listed under "Outcomes") with the appropriate "Prior Knowledge (Principles)" as shown and 5 relevant concepts listed under "Language (Concepts)". On the "Methodological (Doing) Side" on the right are the given information under "Data (Records)", interpretations and transformations of given information guided by the principles and as displayed under

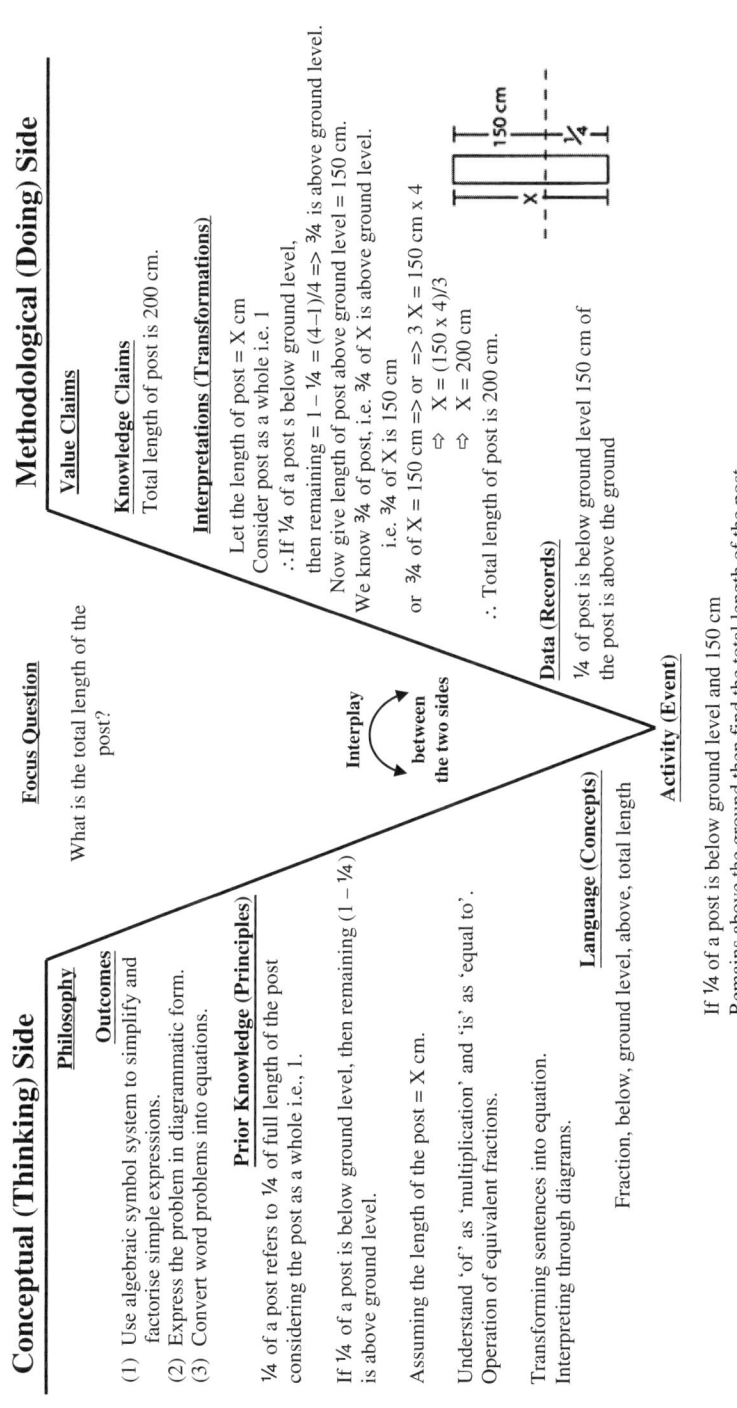

Fig. 4.12 Ken's first vee diagram of the *post* problem

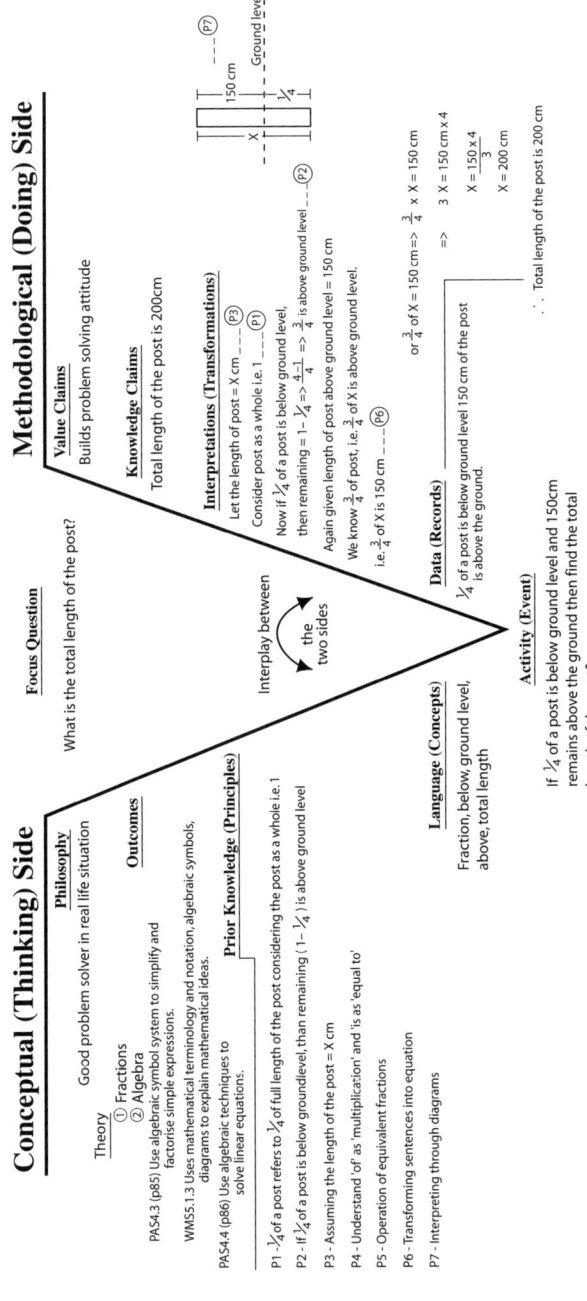

Fig. 4.13 Ken's revised vee diagram of the *post* problem

"Interpretations (Transformations)" with the answer to the focus question under "Knowledge Claims". Ken left the sections "Philosophy" and "Value Claims" blank.

Critiques in class challenged the relevance and appropriateness of the statements of "Prior Knowledge (Principles)" as conceptual statements, and the need to provide philosophical and value statements given the context of the problem. Other comments emphasised the need to ensure that there was a one-to-one correspondence between the listed principles and main steps of the methods. For example, which principles justify which main steps?

The revised vee diagram (Fig. 4.13) showed a number of additions such as the inclusion of references to the relevant syllabus outcomes of the *K-10 NSW Mathematics Syllabus* (NSWBOS, 2002) and reference labels for the listed principles. The "Philosophy" and "Value Claims" sections now included entries (albeit they could be better statements) with the addition of a "Theory" section to display the two main topics most relevant to the problem. For the "Prior Knowledge (Principles)" section, the principle list was now organised and labeled from P1 to P7. While the actual (P1 to P7) statements remained more or less the same as in Fig. 4.12, they would require further elaboration into more suitable theoretical justifications for each main step of the solution (displayed on the right side of the vee). For example, Principles P1 and P2 are not general statements about fractions, which would be more consistent with propositions P6 (Fig. 4.11) and P1 (Fig. 4.6). Instead, they represented Ken's interpretation of the given situation and application of proposition P6 (Fig. 4.11). As stated, these would be more suitable on the right side of the vee in the "Interpretations (Transformations)" section. Each principle from P4 to P7 would need to be rephrased and elaborated more fully so that they are more conceptual (i.e., general statements of relationships between concepts) and less procedural (e.g., P4, P5 and P6) to make them more suitable mathematical justifications as principles or formal statements of conceptual relationships.

Discussion and Implications

The presented concept maps and vee diagrams displayed the results of Ken's conceptual analyses, comprehension and pedagogical understanding of the "Fractions" content strand of the PESM syllabus as required for the two tasks. Whilst Figs. 4.1 to 4.5 represented his early interpretations of a subset of the content strand and early attempts at concept mapping, Fig. 4.6 provided a more macro level and comprehensive, summative concept map which evolved throughout the semester as a result of multiple cycles of presentations, social critiques and revisions. Figure 4.11, in contrast, provided a more situated view of fractions in the context of two problems. Taken together, the 3 sets of concept maps were qualitatively different in terms of their purpose, situation and therefore focus.

For Task 1, the situation was the pilot study and the purpose was for Ken to conceptually analyse the set of syllabus outcomes most relevant to the content of his pilot study. Subsequently, the focus was to make explicit the results of his conceptual analyses and to explicate his comprehension and depth of pedagogical understanding of the interconnections between, and meanings of the identified key and subsidiary ideas visually on concept maps. The quality of his conceptual analyses and pedagogical understanding was assessed by considering the hierarchical levels of organization of the selected concepts, grouping of concepts into coherent hierarchies and their interconnections, and richness of linking words to describe the interrelationships, which collectively formed networks of propositions. In addition, the accuracy of his conceptual analyses and pedagogical understanding was judged by whether or not he had analysed all of the relevant syllabus outcomes most pertinent to the identified situation based on syllabus documentation. The resulting concept maps for Task 1 collectively explicated the progressive development of the fraction concept using the different stages of cognitive development, (i.e., the concrete, iconic, and symbolic ($\frac{a}{b}$)) as part of a whole, part of a collection and a number across the different stages of the PESM syllabus. The demonstrated increasing structural complexity of the five maps not only reflected the incremental depth and breadth of content coverage (according to Ken's interpretations of syllabus documentation) but also his understanding of the interconnectedness between fraction concepts and its multiple models, representations and examples as evidenced by the progressive differentiation of concepts between more inclusive and less general ones and integrative reconciliation between coherent groups of ideas. As expected at primary level, a number of illustrative examples were selected from every day contexts especially for Early Stage 1, and increasingly more use, in subsequent stages, of concrete objects, the $\frac{a}{b}$ notation with selective denominators, diagrams, and number line to model, represent and demonstrate fraction types, equivalence, order and simple operations. Individually, each of the five concept maps illustrated the extent and depth of coverage to be expected for each stage in terms of the meanings to be developed using multiple models and increasingly more sophisticated development of order, equivalence and operations towards the upper stages (i.e., Stages 3 and 4). As Hierbert and Wearne (1986) argue, "conceptual knowledge grows as additional connections are made via assimilation and integration (as) . . . related bits of knowledge are related to earlier ideas" (p. 200). Although a number of key ideas were omitted as irrelevant to the situation for Task 1, there were still some significant relevant concepts missing especially towards the upper stages.

For Task 2, the situation for the first sub-task was the entire "Fractions" content strand in the PESM syllabus and the purpose was for Ken to conceptually analyse the content to be covered up to Stage 4 and the focus was to construct an overview concept map to include key and subsidiary ideas. Subsequently, over the semester, as a result of multiple cycles of presentations, social critiques and revisions, the Task 2 overview concept map evolved into a map that was organised around three main sections, each with its own particular emphasis. Focussing on the different definitions and applications of fractions in the top section, the middle section was on the definition of the notation $\frac{a}{b}$ and fraction types while the bottom section elaborated computations with fractions. The hierarchical levels of generality within each sub-branch appeared clearly defined, each following a basic sequence that constituted a (i) concept label, (ii) brief description of concept meaning, (iii) example description, and (iv) illustrative model/representation using diagrams, pictures, word descriptions and/or $\frac{a}{b}$ notation as demonstrated by propositions P1 to P12 of Fig. 4.7, P13 to P21 of Fig. 4.8 and P22 to 24 of Fig. 4.9. In addition, the overview concept map illustrated integration and interconnectedness between the 3-sections as demonstrated by the extended propositions P22 and P23. Demonstrating Ken's growth of comprehension and pedagogical understanding of the interconnectedness of "Fractions" syllabus outcomes were multiple occurrences of progressive differentiation nodes and integrative reconciliation links which produced a more comprehensive overview of fraction concepts, computation types and illustrative examples with over a hundred nodes, and structurally more complex concept map than the earlier ones.

For the second sub-task of Task 2 (i.e., Figs. 4.10 and 4.11), the situation was mathematical problems while the purpose was for Ken to conceptually analyse the problems and the focus was to construct a concept map of the key and subsidiary ideas pertinent to solving the particular problems. In contrast to the more abstract (i.e., general) concept maps provided in Figs. 4.1 to 4.5, the situation for this sub-task was more contextualised. The most significant difference between the general and contextualised concept map was the inclusion of the "Problem Solving" concept hierarchy in Fig. 4.10, which introduced the concept of "strategies" for solving word problems, while the rest of the Fig. 4.10 nodes were similar to those previously viewed in the earlier abstract maps. Furthermore, although the "computation" branches of Fig. 4.6 displayed various single and mixed operations complete with illustrative examples, the distinction was that the "word problems" branch in the revised map (i.e., Fig. 4.11), in contrast, conveyed both the relevant conceptual propositions and problem solving strategies including the critical synthesis and application of these (hence providing the evidence of the critical thinking and reasoning involved) in the particular situation of problem 2.

Overall, the three sets of concept maps varied in the extent of their selection of concepts, as determined by the purpose, situation and focus of the task, and the hierarchical organisation and structural complexity of the interconnectedness of concepts and richness of its propositions. The various nodes indicating progressive differentiation and integrative reconciliation, hierarchical networks of concepts from most general to most specific, and richness of the resulting propositions evidenced the interconnectedness of Ken's knowledge and growth in conceptual and pedagogical understanding. The conceptual details (labels and meanings) and linking relationships apparent in the final overview concept map were substantively more enriched than the initial attempts of Figs. 4.1 to 4.5. Figure 4.6 is not only relatively more comprehensive conceptually, but it is also organisationally and structurally more differentiated and integrated than Figs. 4.1 to 4.5. In contrast to the general maps (Figs. 4.1 to 4.6), Fig. 4.11 captured the essential synthesis of, and interplay between, concepts, principles, generalisations and strategies most relevant in solving the two problems. Collectively, the concept maps displayed the evidence of Ken's conceptual analyses and his pedagogical content knowledge in terms of the substantive (or conceptual) knowledge of the PESM "Fractions" syllabus. As defined by Hierbert and Lefevre (1986), conceptual knowledge is "knowledge that is rich in relationships ... a connected web of knowledge, in which the linking relationships are as prominent as the discrete pieces of information. Relationships pervade the individual facts and propositions so that all pieces of information are linked to some network" (p. 3–4).

Interestingly, a comparison of Figs. 4.6 and 4.11 suggested a difference in the cognitive loading and processing, in terms of the critical thinking and reasoning involved for "computations" and "problem solving" as encapsulated by the propositions inclusive under "computation" in Fig. 4.6 in contrast to those under "word problems" and in particular under "strategies" in Fig. 4.11. More importantly, while proficiency in computations is desirable and equally important, Fig. 4.11 highlighted that being exposed to problem solving demands a much greater level of cognitive processing and critical thinking. Such higher level of reasoning would be required for constructing a vee diagram especially when completing the "Prior Knowledge (Principles)" and "Interpretations (Transformations)" sections not only to ensure a one-to-one correspondence between the listed principles on the left and the main steps of the solutions on the right but that the listed principles were appropriate general statements of relationships between concepts as mathematical justifications for the steps. According to Blanton and Kaput (2000), justification in any form is a significant part of algebraic (or mathematical) reasoning because it induces a habit of mind whereby one naturally questions and conjectures to establish a generalisation, or in the case of Ken, to establish a principle that underlies a main step in the solution. In addition, they argued that a classroom focus on justification could encourage students to conjecture in order to establish generalisations. The same can also be said for justification using principles to make explicit the conceptual bases of methods on vee diagrams and similarly to creatively structure concepts and linking words to form propositions on concept maps. Thus the data presented in this chapter demonstrated how, through the routine use of concept maps and vee diagrams, a student can develop habits of mind to conceptually and critically analyse mathematical situations, thinking and reasoning from situations and justifying interpretations and transformations in terms of the relevant substantive and syntactic knowledge of the discipline. In so doing, teachers and student teachers can develop a deeper more conceptual understanding of the structure of the relevant mathematics to pedagogically mediate meaning in an educational context.

Overall, the richness of the linking words on the connecting lines and consequently the conceptual richness of the propositions in Figs. 4.6 and 4.11 could be further improved to convey more enriched descriptions of interrelationships than had been shown. As Baroody, Feil, and Johnson (2007) proposed, "depth of understanding entails both the degree to which procedural and conceptual knowledge are interconnected and the extent to which that knowledge is otherwise complete, well-constructed, abstract and accurate" (p. 123). Similarly for further improvement are the statements of principles in Fig. 4.13 to make them more theoretical, less contextualised and less procedural statements, as principled mathematical justifications for the main steps. As Ellis (2007) pointed out for justifications and generalisations, (which is equally viable for justifications and methods of solutions), "learning mathematics in an environment in which providing justifications for one's generalisations (or methods of solutions) is regularly expected can promote the careful development of generalisations (or methods of solutions) that make sense and can therefore be explained" (p. 196). Furthermore, "a focus on justification may help students not only to better establish conviction in their generalisations (or methods of solutions) but also aid in the development of subsequent, more powerful generalisations" (Ellis, 2007, p. 196) (or more powerful methods of solutions).

Findings from this case study contributes empirical data to the literature on the use of concept maps and vee diagrams as viable tools that, through their routine construction, can encourage students to engage in the processes of critical analysis and synthesis, organising, thinking and reasoning, justifying and explaining their knowledge and understanding of a situation publicly for social critiques, discussion and evaluation. Further research is necessary to examine how these ideas could be implemented in a whole class situation in the classroom.

References

Australian Association of Mathematics Teachers (AAMT). (2007). *AAMT standards for excellence in teaching mathematics in Australian schools.* Retrieved October 6, 2007, from http://www.aamt.edu.au/standards

Ausubel, D. P. (2000). *The acquisition and retention of knowledge: A cognitive view.* Dordrecht, Boston: Kluwer Academic Publishers.

Baroody, A. J., Feil, Y., & Johnson, A. R. (2007). Research commentary: An alternative reconceptulization of procedural and conceptual knowledge. *Journal for Research in Mathematics Education, 38*(2), 115–131.

Blanton, M., & Kaput, J. (2000, October). Generalizing and progressively formalizing in a third-grade mathematics classroom: Conversations about even and odd numbers. In M. L. Fernandez (Ed.), *Proceedings of the 22nd annual meeting of the North American chapter of the International Group for Psychology of Mathematics Education* (Vol. 1, pp. 115–119). Columbus, OH: The ERIC Clearinghouse for Science, Mathematics, and Environmental Education.

Bruner, J. (1990). *Acts of meaning.* Cambridge, MA: Harvard University Press.

Ellis, A. B. (2007). Connections between generalizing and justifying: Students' reasoning with linear relationships. *Journal for Research in Mathematics Education, 38*(3), 194–229.

Feldman, A. (1996). Enhancing the practice of physics teachers: Mechanisms for the generation and sharing of knowledge. Retrieved October 2, 2007 from http://www-unix.oit.umass.edu/~afeldman/ActionResearchPapers/Feldman1996.PDF

Hiebert, J., & Lefevre, P. (1986). Conceptual and procedural knowledge in mathematics: An introductory analysis. In J. Hiebert (Ed.), *Conceptual and procedural knowledge: The case of mathematics* (pp. 1–27). Hillsdale, NJ: Lawrence Erlbaum Associates.

Hierbert, J., & Wearne, D. (1986). Procedures over concepts: The acquisition of decimal number knowledge. In J. Hierbert (Ed.), *Conceptual and procedural knowledge: The case of mathematics* (pp. 199–223). Hillsdale, NJ: Erlbaum.

Jiygel, K., & Afamasaga-Fuata'i, K. (2007). Students' conceptions of models of fractions and equivalence. *The Australian Mathematics Teacher, 63*(4), 17–25.

National Council of Teachers of Mathematics (NCTM). (2007). *Executive summary. Principles and standards for school mathematics.* Retrieved October 4, 2007 from http://standards.nctm.org/document/chapter3/index.htm

New South Wales Board of Studies (NSWBOS). (2002). *K-10 Mathematics syllabus.* Sydney, Australia: NSWBOS.

Novak, J. D. (2002). Meaningful learning: The essential factor for conceptual change in limited or appropriate propositional hierarchies (LIPHs) leading to empowerment of learners. *Science Education, 86*(4), 548–571.

Novak, J. D., & Cañas, A. J. (2006). *The theory underlying concept maps and how to construct them* (Technical Report IHMC Cmap Tools 2006-01). Florida Institute for Human and Machine Cognition, 2006, available at http://cmap.ihmc.us/publications/ResearchPapers/ TheoryUnderlyingConceptMaps.pdf

Novak, J. D., & Gowin, D. B. (1984). *Learning how to learn.* Cambridge, UK: Cambridge University Press.

Shulman, L. S. (1986). Those who understand: Knowledge growth in teaching. *Educational Researcher, 15*(2 February), 4–14.

Chapter 5
Concept Maps as Innovative Learning and Assessment Tools in Primary Schools

Karoline Afamasaga-Fuata'i and Greg McPhan

The introduction of concept maps to primary teachers as tools to guide and scaffold their planning of learning activities in mathematics and science, or, alternatively as an assessment tool for student learning, was treated with some trepidation and reservations. That the tools have the potential to scaffold primary students' learning and understanding of mathematics and science concepts was an idea that needed empirical testing in primary classrooms. Over a period of five school terms, through professional development, on-going professional support and collaborations between teachers and university researchers, an incremental introduction of *semi-structured* concept maps was initiated in two primary classrooms. This classroom trial occurred over a period of time until a more receptive and conducive learning environment was established with primary students using concept maps to review their understanding of *Position* in the K-Year 1 classroom and *Fish's Adaptive Features* and *Fractions* in the Year 5/6 classroom. This chapter documents the professional journey of two primary teachers and their students as they struggled, persevered and succeeded in incorporating concept maps as learning and assessment tools, as part of their normal classroom practices during the year. The ultimate highlight of the innovative strategy was the initiative by the two primary teachers and their students to come together for peer tutoring and peer collaborations as the older students mentored and assisted the younger ones in using the software *Inspiration*^{TM} to construct concept maps.

Introduction

Ausubel's theory of meaningful learning proposes that learners' cognitive structures are hierarchically organized with more general, superordinate concepts subsuming less general and more specific concepts where meaningful learning is defined as the active assimilation of new knowledge onto existing knowledge through a process of progressive differentiation between ideas and/or integrative reconciliation across systems of ideas as a student's cognitive structure or patterns of meanings

K. Afamasaga-Fuata'i (✉)
School of Education, University of New England, Armidale, Australia
e-mail: kafamasa@une.edu.au

K. Afamasaga-Fuata'i (ed.), *Concept Mapping in Mathematics*,
DOI 10.1007/978-0-387-89194-1_5, © Springer Science+Business Media, LLC 2009

are reorganized or reformulated to accommodate new knowledge (Ausubel, 2000; Novak & Gowin, 1984). Through the meta-cognitive process of concept mapping, students can illustrate publicly their interpretation and understanding of a knowledge domain by constructing concept maps, which are hierarchical networks of interconnecting concepts (nodes) with linking words describing the nature of interconnections subsequently forming propositions (Novak, 2002). The latter are meaningful statements formed by connecting strings of "*node –linking-words → node*" that are visually displayed on a concept map. Nodes (or contents of nodes) represent key ideas, concepts or strategies most relevant to the focus of the concept map. The richness of the meaning of a particular concept (or node) is dependent on its interconnections to other surrounding concepts or the system of concepts within which it is situated or cross-linked to (examples are provided later).

Using concept maps to assess students' conceptual understanding of a topic requires, according to Ruiz-Primo (2004), making three criteria transparent. First the *mapping task* should invite students to provide evidence of their understanding and knowledge of the domain; second, there should be a clear *format* for students' responses; and third, there must be a *scoring system* to consistently evaluate the maps.

In this chapter, the nature of the professional, collegial and pedagogical processes that ultimately led to, and facilitated, the introduction and incorporation of the innovative strategy of concept mapping in two primary classrooms is examined. Whilst vee diagrams (see Chapter 2 for a definition) was the other innovative strategy, only concept mapping data is presented here based on the final choice made by two primary teachers from a local primary school.

Methodology

The research project, that is reported here, began with a professional development workshop in the fourth term of the 2005 school year, to introduce two innovative strategies to participating teachers with sufficient time soon after for them to reflect and experiment with the strategies in their own classrooms. Reflection sessions, scheduled fortnightly after the workshop, enabled teachers to report back to the group regarding their progress and to discuss any emerging concerns. Over four terms in the 2006 school year, additional on-going professional support was provided in the form of site visits to the schools by the researcher (first author), consultative meetings with the teachers and more reflection sessions as the need arose, while teachers grappled with the introduction and incorporation of the innovative strategies as part of their regular classroom practices. The final concept mapping activity was the development and implementation of a teacher-initiated mapping task (i.e., final mapping project) to demonstrate the teachers' and students' evolving and increasing proficiency in constructing hierarchical concept maps as part of their classroom teaching-learning-assessment practices. Outputs from the research project were showcased in a one-day conference at the end of the 2006 school year.

Two regional Australian central (i.e., has both primary and secondary levels) and primary schools with a total of nine primary and secondary teachers and their students participated in a research project that investigated the impact of innovative teaching and learning strategies on students' engagement in learning mathematics and science. Only data from Matilda Primary School (MPS) (a pseudonym) is presented here. Three of the nine participating teachers were from MPS.

Results

Professional Development Workshops and Reflection Sessions

Teachers' Professional Development Workshop

A two-day professional development (PD) workshop introduced the nine participating teachers to the meta-cognitive tools of concept maps and vee diagrams (maps/diagrams). Presentations from the two authors were interactive allowing teachers to field questions for clarifications of ideas and critical comments of the presented maps/diagrams. The emphasis was on introducing the innovative tools using examples previously constructed by student teachers, teachers and secondary students to illustrate the various applications of maps/diagrams in mathematics and science (a) as *learning* and *assessment* tools for the analysis of problems and activities including the illustration and communication of students' mathematical and/or scientific understanding of topics and activities, and (b) as *planning* tools to design teaching sequences, lesson plans and activities for teaching, learning and assessment (see Afamasaga-Fuata'i, 2005, 2004, 1998 for more details). A prepared schedule of group activities ensured teachers were progressively initiated into a different and innovative way of thinking with sufficient time and practice to experience for themselves the strategies of concept mapping and vee diagramming firstly as a community of learners during and after the PD and before they can be expected, in professional practice, to pedagogically introduce the strategies in their classrooms. For small group activities, the teachers collaboratively and cooperatively co-constructed science and mathematics concept maps, by brainstorming and negotiating ideas about what counts as key concepts and strategies of a topic or activity, compiling the brainstormed ideas into concept lists, ranking them from most general to most specific, and then meaningfully organizing them into coherent hierarchies before linking concepts and describing the nature of inter-relationships using "linking words", which consequently formed hierarchical networks of interconnecting propositions. Group presentations of concept maps and peer critique followed each small-group activity thus providing critical feedback to further improve the clarity and transparency of conceptual connections.

Towards the end of the PD workshop, participants reflectively considered how they might incorporate the innovative strategies in their future planning and classroom activities through their responses to an evaluation questionnaire. Workshop outputs included maps/diagrams co-constructed by teachers during small group activities. Teachers were requested to experiment with the use of concept maps and vee diagrams for up to four weeks before meeting again to reflect on their trials.

In summary, the workshop introduced and familiarised the teachers with the innovative tools whilst group activities enabled the teachers to have hands-on experience in concept mapping and vee diagramming topics, problems and activities. Participating teachers engaged in professional discourse as they collaboratively constructed hierarchical concept maps of topics, syllabus outcomes, activities and problems to illustrate interconnections between concepts, strategies and formulas. Whilst these were mainly for planning learning activities, the teachers were also mindful that maps/diagrams be Stage (or level) appropriate and meet the needs of the targeted students particularly those with numeracy and literacy problems in understanding the languages of mathematics and science. Teachers' evaluative responses demonstrated that they found maps/diagrams useful and efficient means of facilitating interactions, collaboratively negotiating and clarifying meanings, and communicating common understanding of topics/activities/problems.

Reflection Sessions

Following the PD was a four-week period of reflection and classroom trials including fortnightly reflection sessions. Vitally important was the need for the teachers to have sufficient time and space to grapple with the innovative strategies themselves, firstly as learners by reflecting upon their workshop experiences and experimenting making connections between these PD experiences and regular classroom activities and secondly, as experienced teachers pedagogically mediating the use of the innovative strategies to design learning activities and to assess students' conceptual understanding of taught topics.

During the first fortnight after the PD, teachers reflected and experimented with how best to introduce the innovative strategies to students. Thus the first reflection session focused on the reporting of teachers' experiences. Most of the discussions focused around ways in which a seemingly complex idea (e.g., concept map and/or vee diagram) could be modified to suit the level of the targeted students. For example, using an example of a Year 5/6 (Stage 3) activity students were currently engaged with (i.e. *Christmas in Different Countries*), the teachers brainstormed and collaboratively recommended ways a vee diagram could be completed using the appropriate language and vocabulary for a Stage 3 class. The discussions provoked exchanges of ideas between primary and secondary teachers resulting in heightened awareness and appreciation of the importance of Stage appropriate language whilst simultaneously highlighting that cognitive processes such as *observing events*, *making connections between ideas*, *reporting*, *thinking*, and *reasoning* basically underpin and permeate all science and mathematics syllabus outcomes from primary to secondary.

Discussions in the second reflection session continued to revolve around the need to further develop teachers' understanding of, and proficiency modifying, an "abstract" vee diagram for use in primary classrooms. As a result, the primary teachers took the lead in making suggestions and confirming revisions to the displayed language of a *Christmas Activity* vee diagram (prepared by the first author based on the previous session's discussions) to make it more Year 5/6 appropriate. A similar discussion eventuated with a *"Volume Activity"*, namely revising the language of the displayed vee diagram so that it was more appropriate for primary level (see Afamasaga-Fuata'i & McPhan, 2007 for more details). Again, the primary teachers led the discussions and exchanges of ideas.

The rest of the data below, focused on the MPS primary teachers' experiences with the innovation, struggles to come to terms with the innovative strategies, and their eventual adaptation and adoption of concept mapping as part of their classroom teaching-learning-assessment practices. Therefore, the focus question for this chapter is: *"What are the issues and concerns associated with the introduction and practice of an innovative strategy in primary classrooms?"*

On-Going Professional Support for the Teachers

Preparation of Teaching and Learning Resources

Various types of concept maps were constructed by the researcher (hereafter refers to the first author) at the beginning of the 2006 school year based on the primary mathematics syllabus, to provide a readily available set of resources for classroom applications, in anticipation of potential use by the MPS primary teachers who, in previous reflection sessions, indicated that even concept maps were too abstract for their primary students and consequently, from their point of view at the time, may not be applicable as a teaching and/or learning strategy for their school.

Initial Site Visits

At the beginning of the 2006 school year, with the MPS school leader and also the K-Year 1 teacher away on study leave, site visits involved meetings with just the other two primary teachers (i.e., for the Year 2/3/4 and Year 5/6 classes). The meetings discussed at length the basic literacy and numeracy problems of their students and other problems concerning school support services, which impinged on the teachers' capacities to participate more fully in the research project. In recognition of these practical demands on teachers' time, it was mutually agreed that initially, they would trial only concept maps, not vee diagrams, in their classrooms. Subsequently, it was suggested that the Year 5/6 would begin with a class-constructed map whilst the Year 2/3/4 class would begin with individually-constructed maps given the existing diversity of student abilities within this one class. Again, both teachers reiterated their preference to work with concept maps only. The outcome of the discussion was for both teachers to focus on concept maps in the next few weeks and for the researcher to visit their classrooms for observation of the language used in their teaching (i.e., classroom discourse). The latter was important in order to authentically base further suggestions on how concept maps could be developed to closely align with the teachers' existing pedagogical practices and classroom discourse as a first step towards conceptualising a pedagogical bridge to the introduction of the innovation in their classrooms.

Classroom Observation Visit

On the eve of the observation visit, the Year 2/3/4 teacher voluntarily chose to pull out of the research project citing personal reasons, thus the researcher observed only the Year 5/6 class as they solved problems on *Addition of Decimals*. Of particular interest for concept mapping purposes was the language Sue (a pseudonym) used during the lesson, which provided an authentic and appropriate basis for a list of words to begin constructing a concept map. Whilst observing the lesson, the researcher also noted that some of the students struggled when doing some of the problems on their own. In Fig. 5.1 is an example of one of the concept maps constructed by the researcher based on a problem the Year 5/6 class was engaged with during the classroom-observation visit. Figure 5.2 shows the concept map co-constructed by the Year 5/6 class and facilitated by the teacher as evidence of her self-initiated attempt at incorporating a concept map into her classroom activities prior to the observation visit.

Portfolio of Teacher Resources

To provide additional professional support to the MPS teachers, a portfolio of concept maps was prepared. Using Ruiz-Primo's (2004) continuum of directedness, concept maps may be classified on the basis of the amount of information (i.e., concept lists, linking words, and/or hierarchical structure) that is provided as part of the mapping task. Specifically, the continuum ranges from "high directedness" to "low directedness" where the least cognitively demanding task is the *Fill-in-the-Map* (i.e., *Fill-in-Lines* or *Fill-in-Nodes*) type on the left side of the continuum with the *Construct-a-Map (no concepts, linking words or structure provided))* type on the extreme right as the most demanding. In between, from left to right are *Construct-a-Map (Concepts & Linking Phrases Provided); Construct-a-Map (Concepts Provided & Structure Suggested)* and *Construct-a-Map (Concepts Provided)*. More importantly, the researcher's portfolio of semi-structured concept maps, consisted of a variety of *Fill-in-Nodes* and *Fill-in-Lines* maps, based on the observed Year 5/6 lesson on *Addition of Decimals*, and others from the Stages 1 (Years 1–2) to 3 (Years 5–6) of the *NSW K-6 Mathematics Syllabus* (NSWBOS, 2002) such as *Lines*, *Shapes*, and *Geometry* (see Figs. 5.3, 5.4 and 5.5) specifically designed to supplement the *Construct-a-Map* types that were collaboratively developed by the teachers during the PD workshop.

The main objective of the portfolio was to scaffold the teachers' and students' learning trajectories with semi-structured maps (i.e., *Fill-in-the-Map* type instead of *Construct-a-Map (no concepts, linking words or structure provided* type)) until both teachers and students were sufficiently confident in using concept maps. Ultimately, the goal was for each individual (teacher and/or student) to create their own hierarchical structures of interconnecting concepts to reflect their idiosyncratic, existing understanding of a topic and as an alternative means of prompting the teachers to be

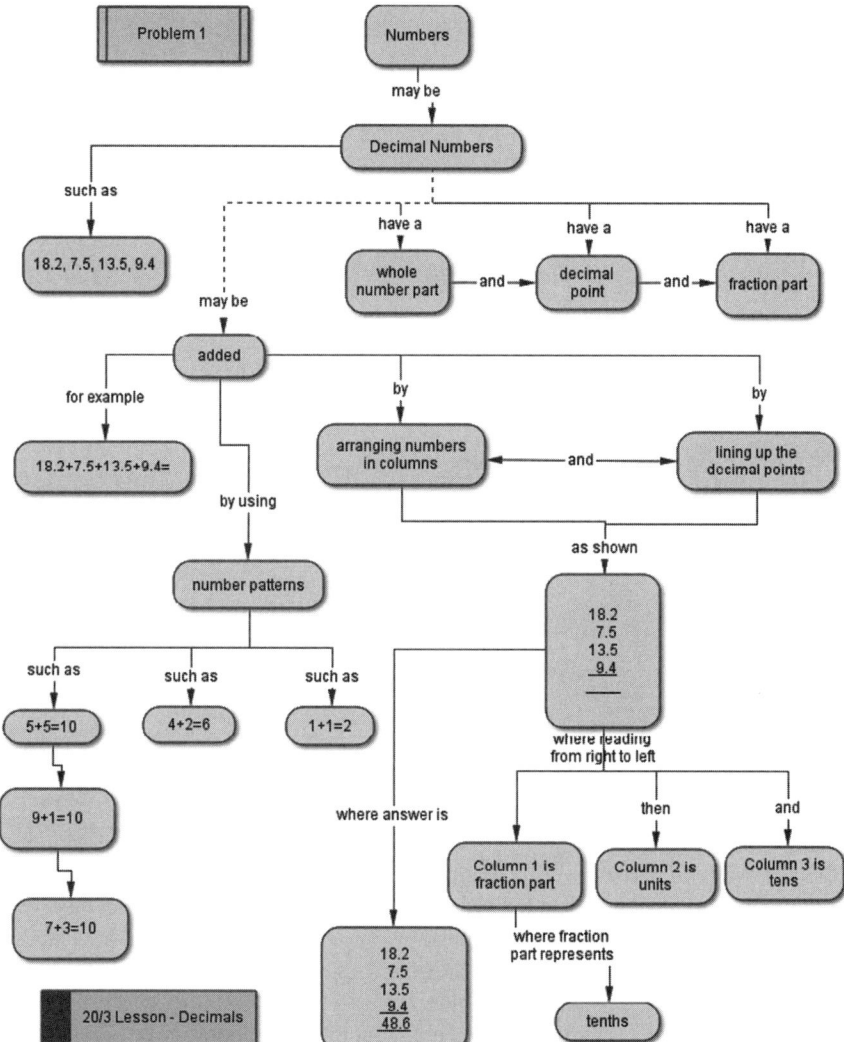

Fig. 5.1 One of the concept maps constructed by the researcher based on an observation of the mathematical language used by the Year 5/6 teacher in a lesson on *Addition of Decimals*

more creative and flexible in their conceptualisation of a classroom application of concept mapping.

More Site Visits and Reflection Sessions

Site meetings with the K-Year 1 teacher (pseudonym Carrie, who had returned from leave) and Sue (Year 5/6 teacher), enabled further collaborative discussions around the portfolio of maps including those based on the Year 5/6 observation visit from the previous term. In addition, the computer software *Inspiration*[TM] was purchased

Fig. 5.2 A brainstorming concept map co-constructed by the Year 5/6 class and teacher to introduce the science topic *Adapting for Life*

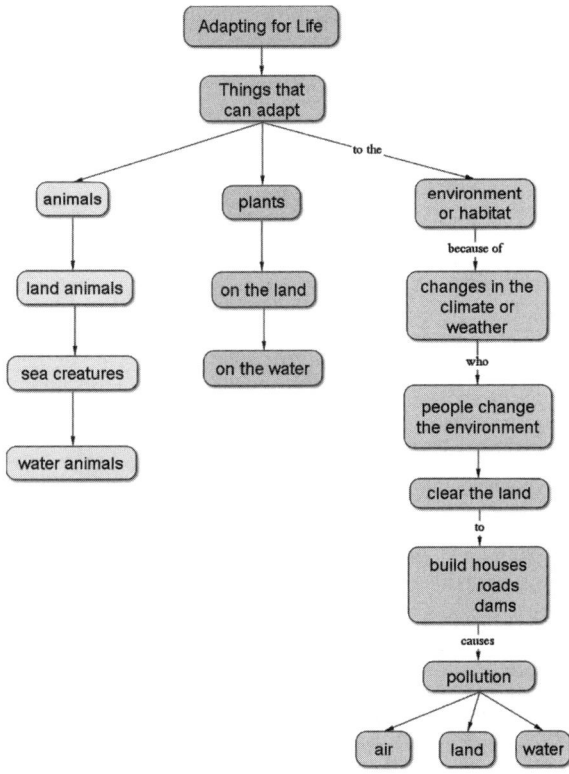

to assist with the classroom preparation of concept maps. The software provided another tool and resource to address the concern that constructing concept maps was messy and took too long. It had a tremendous impact on re-energising both teachers to attempt concept mapping more readily and more frequently making them feel less apprehensive about concept mapping.

Since the two teachers still appeared reluctant to create opportunities for students to begin concept mapping claiming that it would be too difficult for their students, the researcher offered to provide them with some pre-prepared semi-structured concept maps on whatever topic they were currently teaching, which they can further modify/revise as they see fit to start their students off. Accordingly, the K-Year 1 teacher advised she was working with students on describing relative *Positions* of objects by taking the whole class for a walk around the school grounds followed by further review of student understanding by using a wall-size picture of the school grounds. The Year 5/6 teacher, in comparison, had been teaching *Watery Environment* in the previous two weeks and had already started to brainstorm ideas with her students using a concept map (see Fig. 5.2). Outcomes of this consultative meeting were that (1) the researcher would drop off samples of maps the following week and (2) classroom visits would be scheduled soon after.

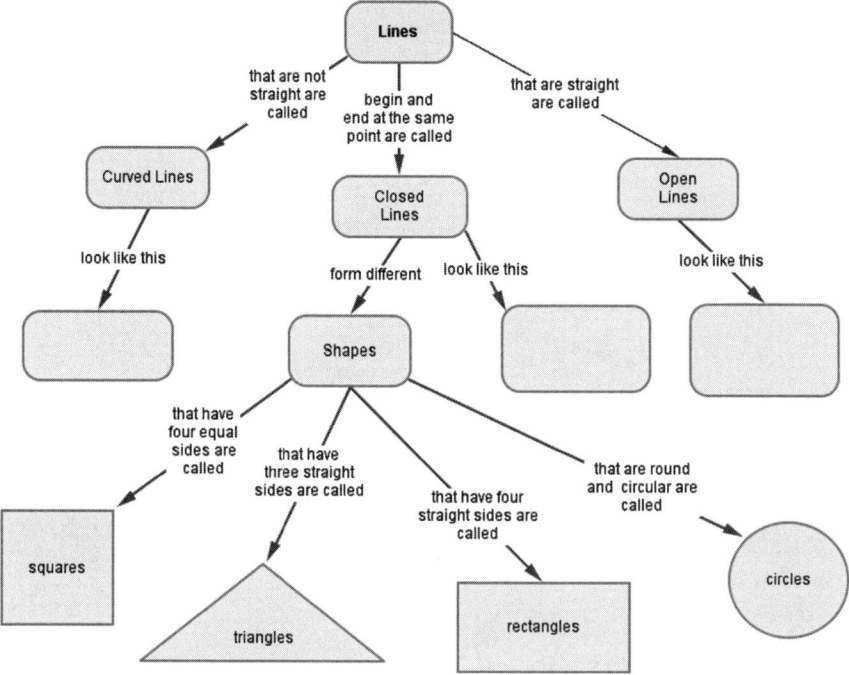

Fig. 5.3 A *Fill-in-Nodes* concept map – *Lines*

Concept Mapping Activities

First Mapping Activities

K-Year 1 Class – Activity 1 – Fruits' Position I

The K-Year 1 teacher began with a class demonstration using a white-board version of the *Fruits' Position* map shown in Fig. 5.6. By working through each link on the map and by posing questions, Carrie invited students to suggest how the links should be described. The class demonstration was followed with individual work in which students were asked to write in "*descriptions of positions*" on the blank links. Assisting students individually and in small groups were two parent-teacher aides, the teacher associate and Carrie. Shown in Fig. 5.6 is an example of a student-completed concept map from this first activity. In her written assessment of the activity, Carrie wrote: "*I was pleasantly surprised to see how well the children coped with this activity and were able to use the language of position*". For further improvement, she recommended that, for subsequent activities, "*As this class is Early Stage 1, most children had some difficulty writing the words onto the lines. I recommend labels be used*".

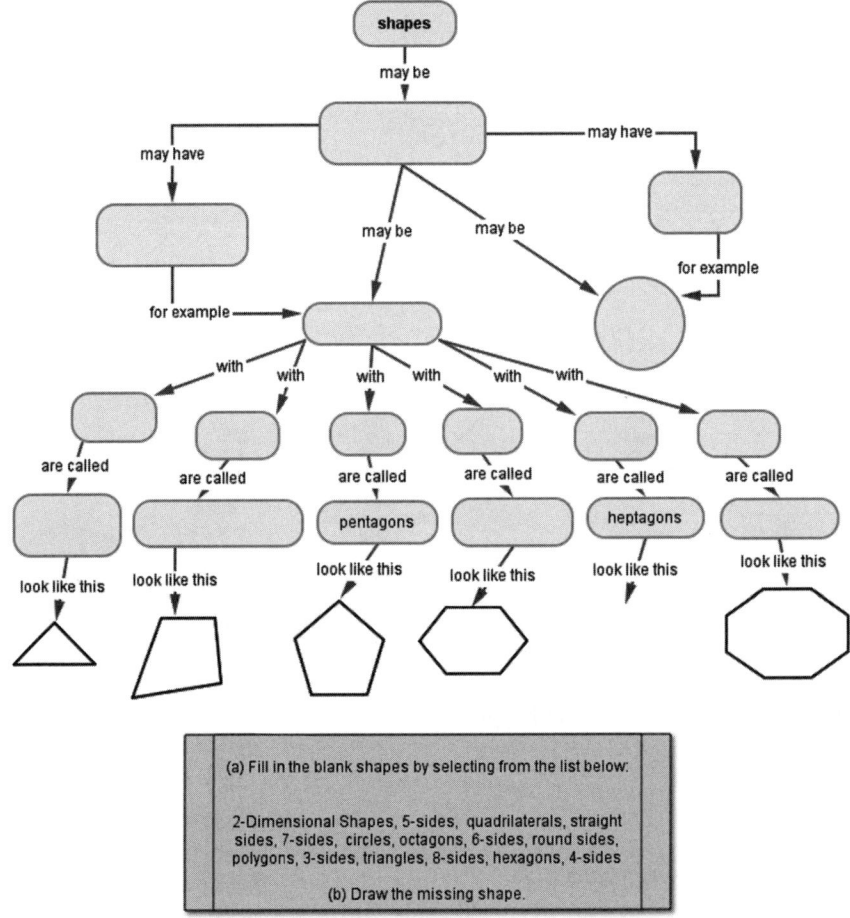

Fig. 5.4 A *Fill-in-Nodes* concept map – *Shapes*

Year 5/6 Class – Activity 1 – Adaptive Features of a Fish

A week before this mapping activity, the Year 5/6 class had already completed a brainstorming concept map as a class (i.e., Fig. 5.2) and had completed a diagram in which students labeled the different parts of a fish using a given list of words. Hence this mapping task included the same fish diagram on one side (i.e., prior knowledge), which the students were asked to complete again to refresh their memories before completing a Fill-in-Nodes concept map, based on the textbook summary of Adaptive Features of a Fish the teacher had used in her teaching. From the given list of "concept names and phrases", students were to complete the blank nodes by selecting the appropriate words/phrases from the list. Figure 5.7 shows an example of a student-completed concept map. The teacher and teacher associate provided

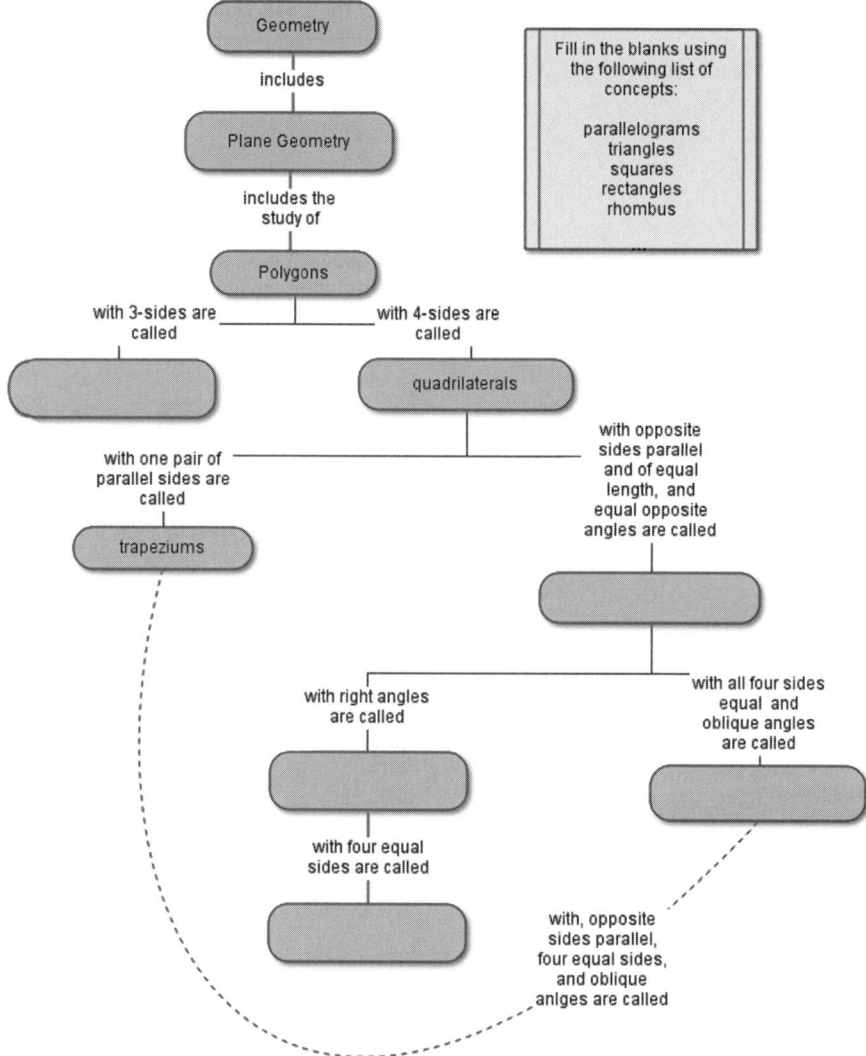

Fig. 5.5 A *Fill-in-Nodes* concept map – *Geometry*

assistance to the students by answering their queries, facilitating, and guiding them in completing their maps.

Second Mapping Activities

The second mapping activities for both classes involved the application of concept maps that were structurally different from those of the first activity.

Fig. 5.6 A *Fill-in-Lines*
concept map – *Fruits'
Position* – K-Year 1 activity 1

Picture of Fruits and Vegetables

Use the <u>Picture</u> above to describe the position of
fruits and vegetables on connecting lines.

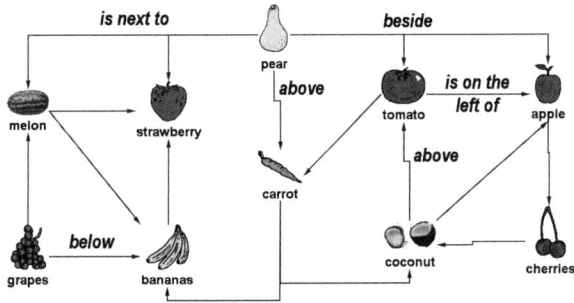

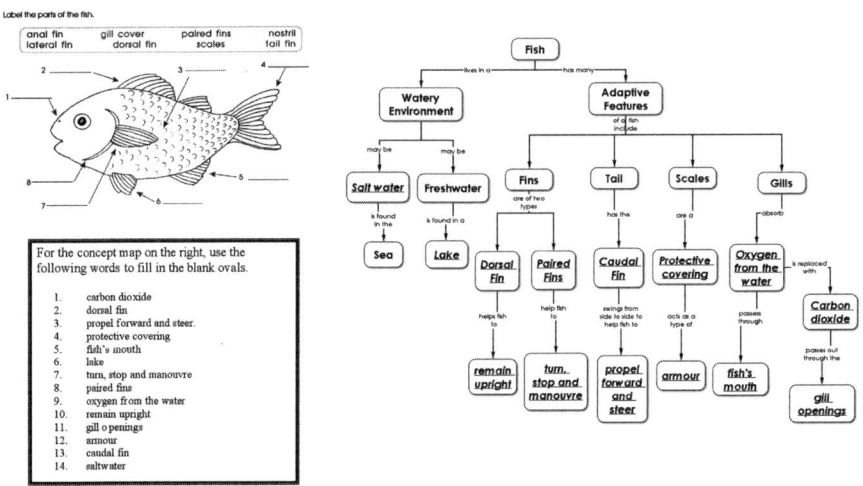

Fig. 5.7 A *Fill-in-Nodes* concept map – *Adaptive Features of a Fish* – Year 5/6 activity 1

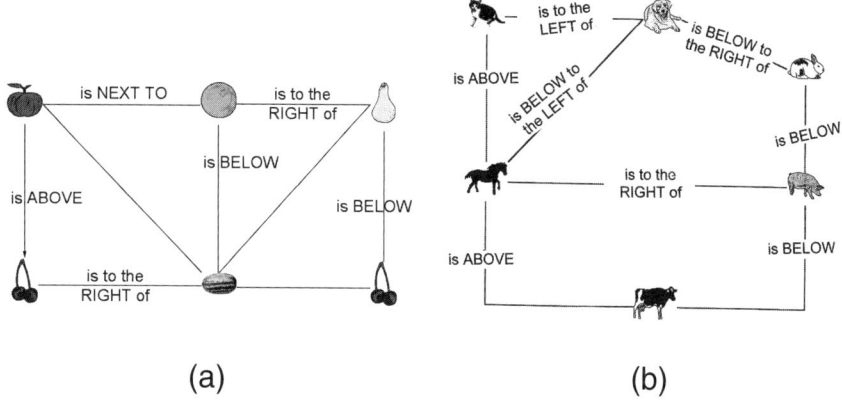

(a)   **(b)**

Fig. 5.8 *Fill-in-Lines* concept maps – *Position* – K-Year 1 activities 2 & 3

K-Year 1 Class – Activity 2 – Fruits' Position II

Another *Fill-in-Lines Fruits' Position* concept map was used in the K-Year 1 class. This time, instead of students writing in the 'linking words', the students had to chose one word from a selection of cardboard-cut-out-words before gluing it on the appropriate link. Unlike the first visit, this time they worked in groups of 4 but individually completed their own maps, facilitated by two parent-teacher aides, teacher associate and the teacher. Figure 5.8a shows an example of a student-completed map.

Year 5/6 Class – Activity 2 – Adaptive Features of a Fish

Unlike their previous mapping activity, the Year 5/6 students were given cardboard-cut-out words that were used in the previous activity but this time they were to "*create their own meaningful hierarchy*" to display their understanding of the fish's adaptive features. That is, using the given cardboard words, the students creatively constructed their own hierarchical structure and linking words to communicate their understanding of the topic. This is an example of a *Construct-a-Map (Concepts Provided)* type according to Ruiz-Primo's continuum. Shown in Fig. 5.9 is an example of a student-completed map for this mapping activity.

Strategies that assisted in progressing to this benchmark of completing two mapping tasks included working closely with the teachers, hearing their concerns and providing support where necessary as the teachers came to terms with a different way of thinking whilst at the same maintaining sight of the aims of the research project, namely, to introduce innovative teaching and learning strategies in the hope of making a difference with students' engagement with learning. The comments from the project's critical friend reflected the spirit and importance of achieving this milestone: "*The participation in this project is quite demanding of teachers because they have to get their heads around to a different way of thinking, as well*

Fig. 5.9 A *Construct-a-Map (Concepts provided)* concept map – *Fish* – Year 5/6 activity 2

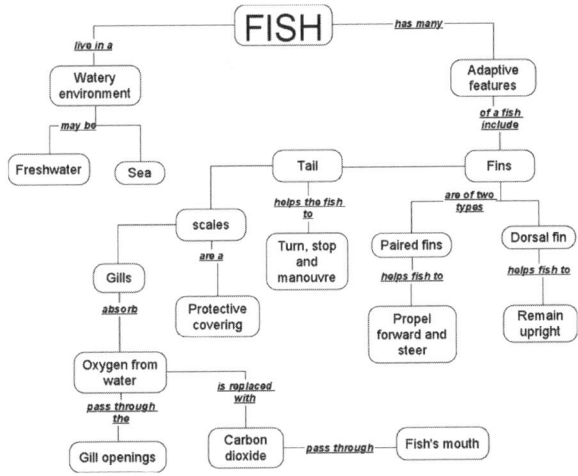

as encouraging their students to do so. This is, however, very worthwhile" (Project Milestone Reporting Tool).

Additional Mapping Activities

By the end of the first two mapping activities, the MPS teachers appeared more confident with incorporating concept mapping into their classroom practices and hence more receptive to discussing more concept mapping activities.

K-Year 1 Class – Activity 3 – Animals' Position III

The teacher used another *Animals* concept map that was left with her after the second classroom visit for this third concept map activity. Interestingly, instead of working individually or in small groups but individually completing their own maps, this time Carrie organised the students into 4 small groups with each student having up to 2 turns in gluing in a cardboard word on the appropriate link thereby completing a group map cooperatively and more efficiently. Indeed, this development demonstrated positive growth in progressively developing class/group arrangements to enhance the efficiency of completing mapping activities. Carrie's increasingly refined execution of the activities reflected her growing enthusiasm and demonstrated her developing confidence with the whole idea of concept mapping. Figure 5.8b shows an example of a collaboratively-constructed concept map for this activity. Carrie wrote in her reflection journal: "*This activity worked much better as each child took it in turns to pick up a card. They all assisted in positioning it onto the map*". In her assessment of the results of the concept map activity, she wrote: "*The children worked well. Cards had to be read to them and children helped each other to position them on map. Final maps were done well*".

Year 5/6 Class – Activities 3 and 4 – Fractions and Equivalent Fractions

Year 5/6 class had been working on *Fractions* since the second visit so the teacher requested some semi-structured concept maps to assess her students' understanding of fractions. Subsequently, beginning with a simple *Fill-in-Map Fraction* map, students individually filled in blank lines and nodes to reflect their understanding of what fractions mean. An example of a student-completed map is in Fig. 5.10. The fourth mapping activity was filling in lines and blank nodes in a bigger and structurally more complex concept map on *Equivalent Fractions*. An example of a student-completed map is in Fig. 5.11.

With both activities, students individually completed their own maps with the teacher, teacher associate and parent-teacher aide moving around interacting with, facilitating and helping students with their queries. Over three visits, this class had evidently progressed through a series of semi-structured concept maps and one open-structured concept map in both science and mathematics. A positive

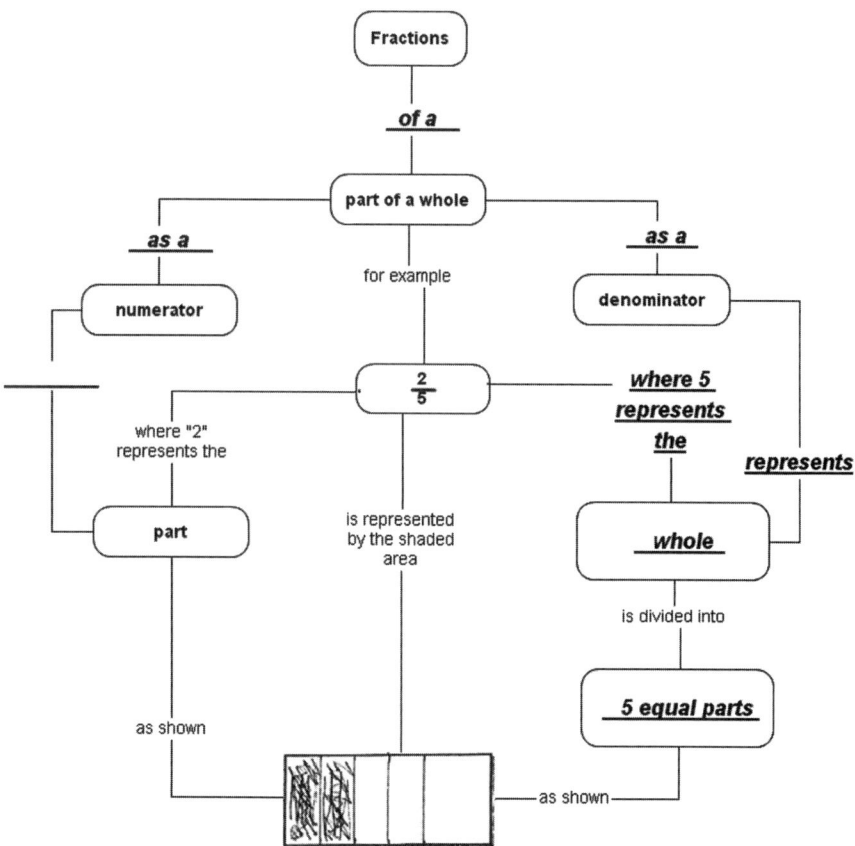

Fig. 5.10 A *Fill-in-Map* concept map – *Fractions* – Year 5/6 activity 3

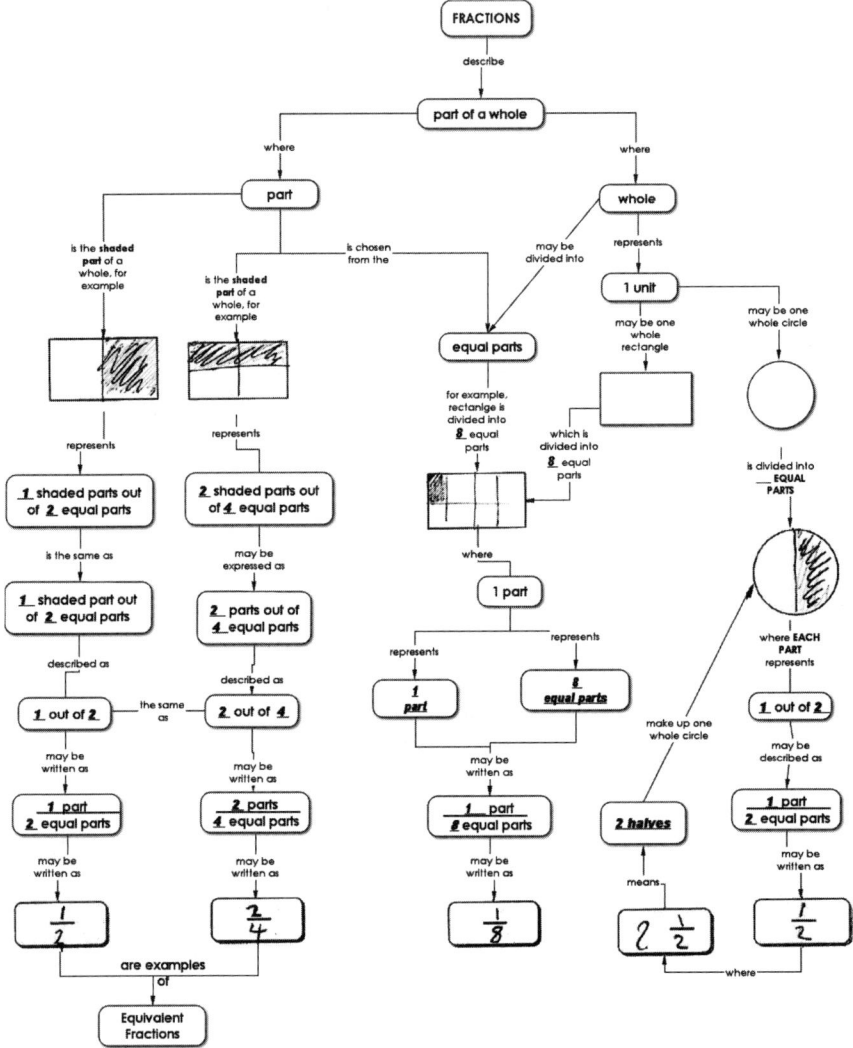

Fig. 5.11 A *Fill-in-Map* concept map – *Fractions* – Year 5/6 activity 4

consequence of this evolving engagement was the emerging confidence exhibited by Sue – who was beginning to critically formulate her own ideas of how she would construct concept maps to align more with her style of teaching and consistent with the type of language practiced in her classroom. This change marked a major step forward and a positive indicator of things to come.

Strategies that contributed to the completion of these mapping tasks were the multiple site visits, discussions and continuous support to the teachers until they gained enough confidence to start introducing and using concept maps in their classrooms.

The MPS parent-teacher aides contributed to the smooth administration of small group activities especially in the K-Year 1 classroom and also the Year 5/6 classroom, both of which had a diverse range of student abilities. Evidently, the persistence and continuous meetings and consultations with the two teachers eventually paid off. The concept maps up to this point had been prepared by the researcher mainly to allow the teachers space and time to focus only on the task of introducing them in the classroom. By the third classroom visit, the two teachers were beginning to appreciate the educational value of the concept maps particularly its application as a means of reviewing and assessing students' summative understanding of a topic. In summary, these additional mapping activities benchmarked a turning point in the attitude and beliefs of the two teachers as evident by their enthusiastic discussion of future possibilities, and in particular, of how they would design their final project activities.

Teachers' Self-Designed Final Mapping Activities

Although the maps the teachers/students had used by this stage, were semi-structured with the option that students filled in the blank lines or nodes by selecting words from a given list or their own words, the teachers had, through the implementation process, realised the value of the maps as a means of assessing students' knowledge and understanding of topics already taught, for example, descriptions of *Position* for the K-Year 1 class and *Watery Environment* and *Fractions* for the Year 5/6 class.

Continuing to meet with the MPS teachers as they progressed their self-designed final project, the researcher provided critical feedback and professional advice as the need arose. Ideas to resolve emerging concerns were brainstormed with suggestions on how they could be practically realised in the classrooms. Subsequent consultative meetings with the MPS teachers indicated steady progress with classroom activities with students using *Inspiration*^{TM} to construct concept maps in the Year 5/6 class after brainstorming and co-constructing a class concept map on *Space* (see Fig. 5.12) while the K-Year 1 class, on the other hand, had started brainstorming and mind mapping their ideas on *Animals*, see Fig. 5.13.

On-going professional support and regular face-to-face meetings with the participating teachers became necessary to manage risks that threatened the successful completion of the teacher-initiated mapping activities (i.e., final projects), and to provide on-going support as the teachers made the professional transition from receptors of ideas to implementers of innovative ideas. Healthy progress and positive achievements were noted with the MPS teachers' final projects. Comments from the critical friend stated: *"This was an interesting meeting – not least because of the diagrams made by students – and the teachers' comments on the outcome and the process. It seems as if teachers are learning alongside their pupils"* (Project Milestone Reporting Tool). Furthermore reflection sessions provided continuous support to the MPS teachers as they progressed the implementation of their final projects with their students. For example, the K-Year 1 teacher and her students were constructing hierarchical concept maps by revising and extending their class-constructed mind map

Fig. 5.12 A *brainstorming* concept map – *Space* – Year 5/6 activity 5

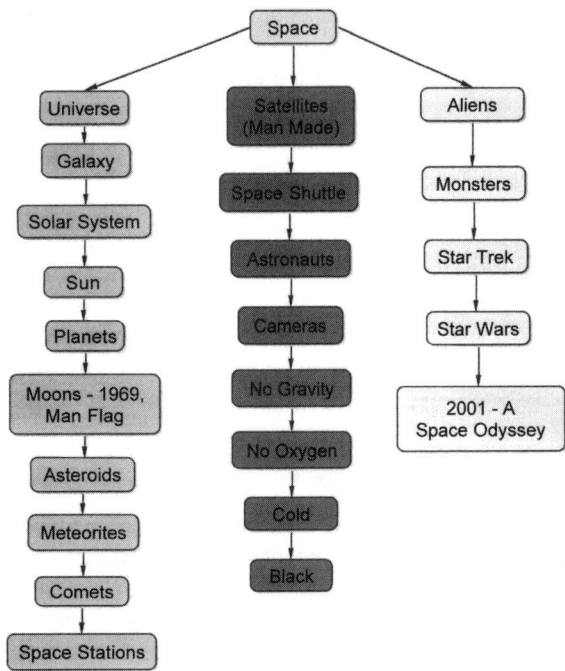

Fig. 5.13 A *brainstorming* mind map – *Animals* – K-Year 1 activity 4

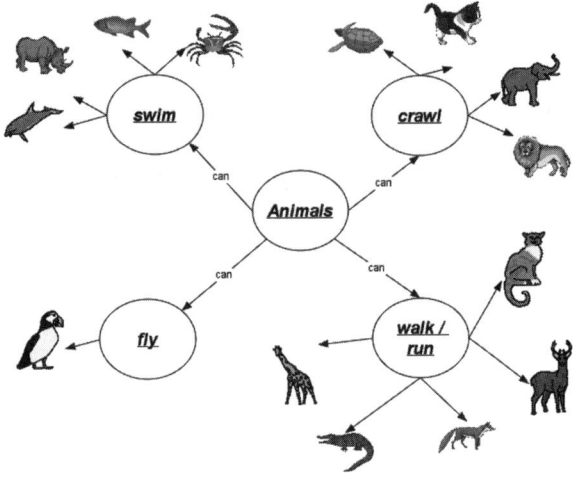

on *Animals* (Fig. 5.13). To facilitate the construction of concept maps, the two primary teachers instigated the coming together of the two classes to enable peer tutoring of K-Year 1 students by the Year 5/6 students in learning how to use the software *Inspiration*[TM]. The Year 5/6 teacher, on the other hand, facilitated her students' work in constructing *Space* concept maps using *Inspiration*[TM]. This individual

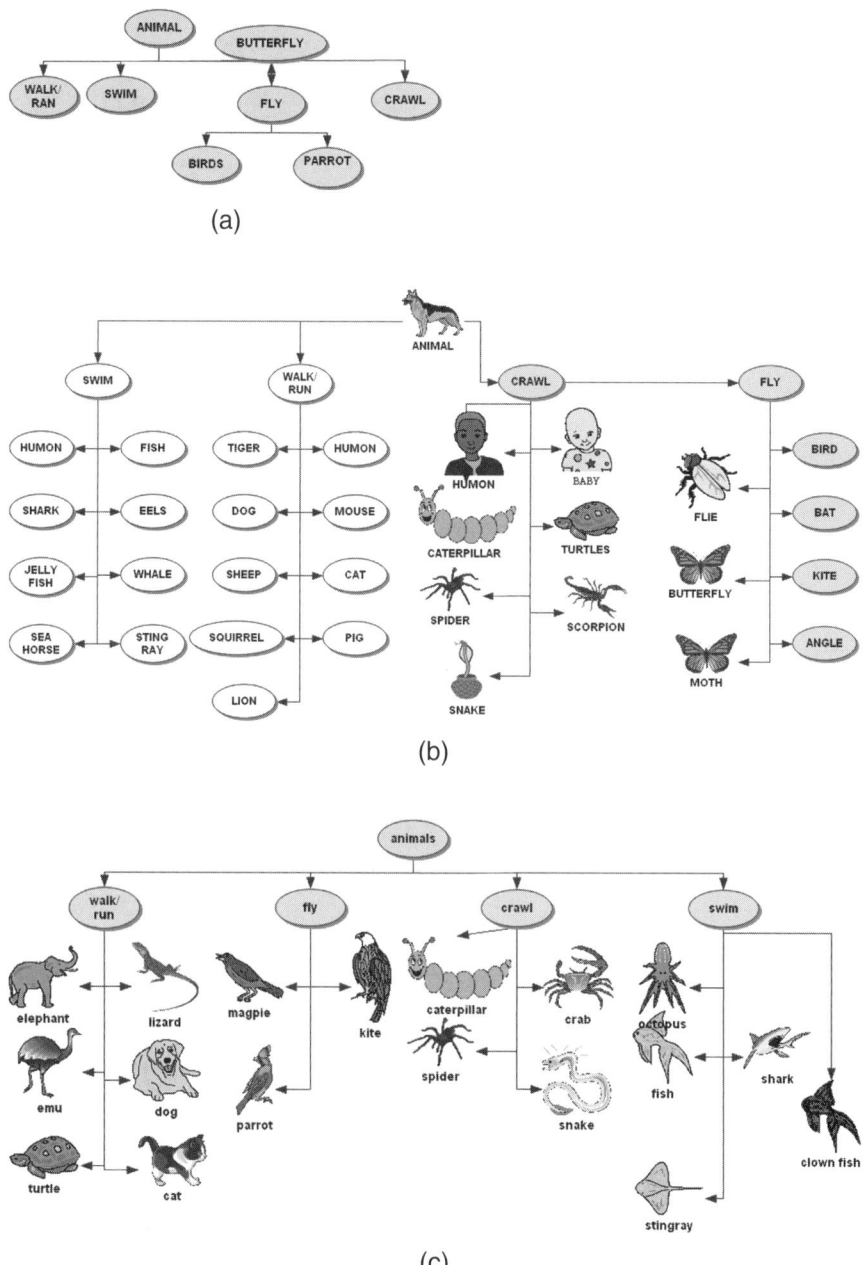

Fig. 5.14 Students' concept maps – *Animals* – K-Year 1 Activity 5

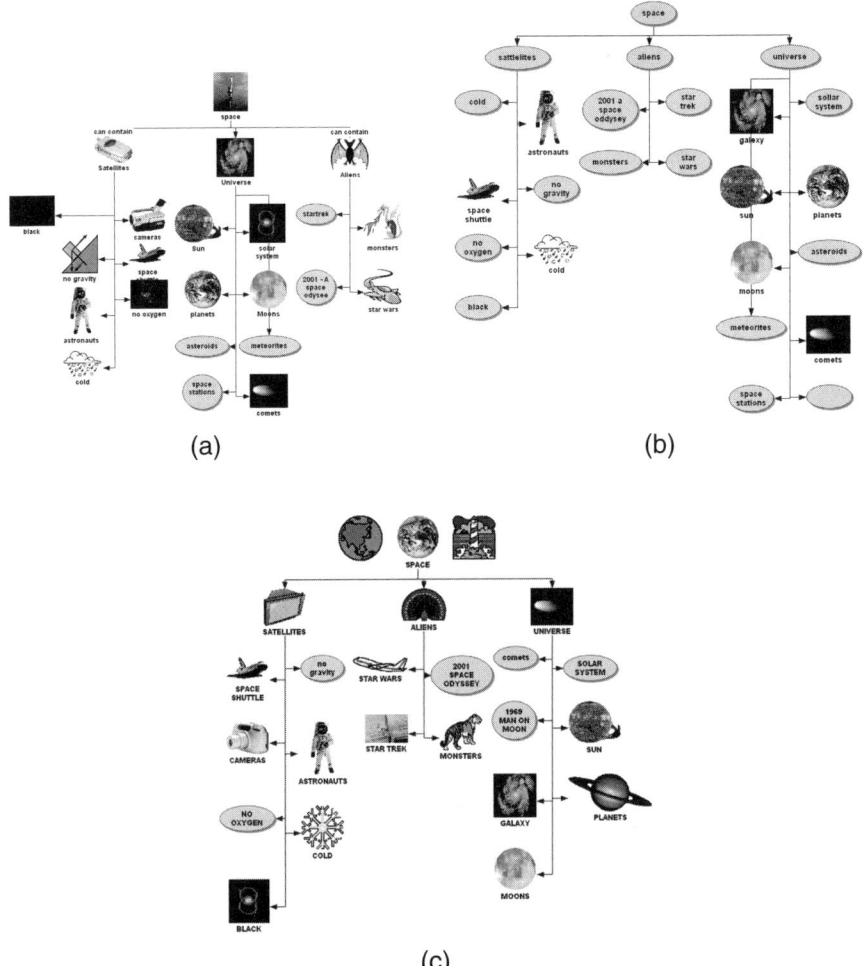

Fig. 5.15 A student's concept map – *Space* – Year 5/6 activity 6

construction of maps followed after their class-constructed concept map in Fig. 5.12. Some examples of K-Year 1 concept maps constructed using *Inspiration*^{TM} are shown in Fig. 5.14a, b and c whilst some student-constructed concept maps from Year 5/6 are shown in Fig. 5.15a, b and c.

In summary, regular on-going professional collaborations between the teachers and the researcher and on-going support while the teachers brainstormed their ideas, developed and finalised their respective projects assisted in achieving this milestone. The availability of the researcher in providing on-going support to the teachers paid off particularly as the MPS teachers conceptualised their projects, developed and implemented them in their respective classrooms.

Post-Activity Reflective Session

The two MPS teachers shared their reflections of their difficult journey up to this point in a final reflection session prior to the completion of their final project. In direct contrast to their initial concerns, they enthusiastically discussed how they would improve mapping activities if they were to do them again. They felt more confident and were in a much stronger and better position to plan and take charge of their own final mapping activity in preparation for the scheduled 1-day conference at the end of the school year to showcase their students' work. The final reflection session was valuable and informative in that it enabled the researcher to share with the two teachers her views of their successful progress in spite of initial difficulties, and the two teachers also shared their own reflections of their experiences and completed an evaluation questionnaire. For example, the teachers commented: "the lesson (using concept maps was) quite successful and the concept map was a very useful tool in enhancing the learning that had already taken place prior to the mapping activity" and "valuable for (her) to see that the (students) could apply (their) knowledge in a more abstract way."

The post-activity session also benchmarked the major progress demonstrated by the two teachers in terms of their increasing proficiency and confidence to incorporate concept mapping as part of their repertoire of teaching and assessment strategies. Activities that assisted the achievement of this milestone included multiple consultations and site visits to encourage and provide support to teachers, and to collaboratively negotiate a realistic pathway forward to ensure progress towards the achievement of the aims of the research project, namely, the incorporation of concept maps as an innovative teaching and learning strategy as part of the teachers' classroom culture and practices.

By working collaboratively and cooperatively with the two teachers to gain their confidence and trust, and providing on-going support through resources and professional advice collectively contributed to the successful implementation of the teacher-designed final projects. As a consequence, over the five terms, the teachers eventually became more comfortable and receptive to having the researcher visit their classrooms whilst engaged with concept mapping activities.

Discussion

The reality of pressures and demands on teachers' time was an influential factor on the implementation of an innovative strategy in a real school, something that is also noted by Loughran and Gunstone (1997) when conducting professional development programs with teachers. It needed to be taken into serious consideration and balanced against what could be realistically achieved in a relatively short period of time. For the MPS primary teachers, it was the struggle to come to grips with the perceived difficulty level of the innovative strategies and the need for more time to grapple with how they could be seamlessly incorporated as part of their normal classroom practices. It became quickly apparent very early in the research project that unless teachers themselves felt personally comfortable with the innovation, it would take awhile before their students can use them. Thus the trial period was important and necessary to enable teachers to think reflectively about the new ideas; this was most important with the MPS primary teachers.

Reflective Sessions

Initial reflection sessions met its objective of providing a forum whereby teachers shared their trialing experiences as evidence of where they were along their learning trajectories, albeit some were more progressed than others. Secondly, the sessions enabled discussions and sharing of ideas amongst professionals on how an innovative strategy that was perceived as being too difficult could be modified and applied to primary level topics. Through the convergence of reflective thinking and professional collaboration and cooperation over two school terms, an incremental approach was eventually determined for the MPS classes, which was more appropriate and less cognitively demanding for primary students particularly in the initial introductory stage. An essential component of the research project was the promotion of collegial interactions and professional exchanges in a non-threatening environment, particularly in the early stages, to support teachers as they struggled to come to terms with the innovative strategies.

Despite the MPS teachers' struggles in the beginning, their leading role in the professional exchanges and sharing of ideas during the first two reflection sessions was significant as it provided them with the opportunity to share their primary teaching expertise with other participating teachers to modify and revise an innovative strategy to be more suitable for primary students. Over time and through on-going professional support and negotiations, the two primary teachers eventually made a professional decision to focus only on one innovative strategy, and thereafter became more willing and open-minded about exploring possibilities in their own classrooms as evident by their progressive engagement and increasing leadership roles with subsequent concept mapping activities up to the final self-designed projects.

School Realities

The realities of school timetabling, normal school activities, unforeseen disruptions and classroom situations were real issues that any innovative research project such as this had to respect and work around. End-of- and beginning-of-school-year busy schedules were real obstacles and there was a need to be cautious and prudent at those times in encouraging the teachers to be more proactive and flexible about applying the innovation. The consultative meetings were essential to minimise risks to the aims of the research project. With sensitivity and persistence and due consideration for the teachers' viewpoints, an amicable solution can be and was negotiated without necessarily alienating the teachers in the process.

On-going Professional Support

Multiple visits to MPS by the researcher provided continuous support and encouragement to the teachers even in the face of transparent reluctance and disinterest in continuing with the research project. This was to be expected especially given the high demands on teachers' times in dealing with their own classes and off-class duties and hence the researcher had to patiently and gently persevere to gain teachers' confidence and trust. The progress made and achieved in terms of the willingness of the teachers to remain in the project, despite one withdrawal, was worthy of time and effort spent.

Since one of the aims of the research project was to cultivate functioning collaborative partnerships between university's school of education and real schools, any problems encountered in realising this effective collaboration were real and when they occurred, had to be sensitively and diplomatically worked through in a professional manner by the parties concerned. Getting teachers to agree to classroom visits was a mammoth task. However, as mentioned earlier, with sensitivity and persistence, and respect for the teachers, a compromise was negotiated without necessarily losing sight of the bigger aims of the research project. Also, the importance of having a school leader, that is prepared to put in the time to drive the innovation within the school, was truly tested in the process of achieving the first few classroom visits.

Summary of Class Contributions

For the K-Year 1 class, they started off using three different versions of *Fill-in-Lines* concept maps to further enhance their learning of relative positions and, from the teacher's perspective, to assess what they have learnt, after doing physical, pictorial and concrete activities on *Positions*. The final project evolved from a class-brainstormed list of words on *Animals* to a class-mind map, before working collaboratively with Year 5/6 students as peer tutors to learn how to use *Inspiration*^{TM}, with individual students, subsequently, constructing their own concept maps using *Inspiration*^{TM}.

For the Year 5/6 students, they started off using a *Fill-in-Nodes* concept map with a *given-list-of-concepts* to further enhance previous learning, and from the teacher's perspective, to assess students' understanding after a unit on *Adaptive Features of a Fish*. A follow-up activity required students to individually construct their own concept map from a given list of concepts on *Adaptive Features of a Fish.* After a unit on fractions, students completed a simple *Fill-in-Map* concept map to illustrate their understanding of fractions, numerators and denominators before continuing onto to complete a more complex *Fill-in-Map* concept map to assess their understanding of "part of a whole"; its multiple representations using diagrams, word descriptions, and notations; and equivalent fractions. The final project evolved from a brainstormed list of words on *Space* and then a class-constructed concept map with students subsequently constructing their own hierarchical concept maps using *Inspiration^{TM}*.

Innovation and Professional Practice

On-going site visits and consultative meetings with the participating teachers to provide support as they designed, developed, finalised and implemented their final projects collectively contributed to the progressive development of pedagogical processes and procedures to ensure students were adequately prepared for, and actively engaged in, completing and/or constructing hierarchical concept maps to reinforce their previous learning and to communicate their summative knowledge and understanding at the end of a unit of work. Overall, a major highlight of the research project, as presented here, was the success with the Matilda Primary School teachers and their K-Year 1 and Years 5/6 students in the way they eventually adapted and adopted concept mapping as a part of their normal classroom practices. This is a particularly significant achievement given the initial difficulties encountered by the two teachers very early in the program. However, as it turned out, with ongoing professional support and collegial collaborations between the researcher and the teachers, over time, they eventually designed and developed their own concept mapping classroom applications.

Through within-school professional collaborations between the two teachers, they cooperatively organised across-Stage peer tutoring and mentoring (between K-Year 1 and Year 5/6 students) with the older students mentoring and assisting the younger ones in using *Inspiration^{TM}* so that the latter can construct their own concept maps on *Animals*. This across-level peer tutoring would not have been possible if the two teachers themselves had not made a conscious professional decision to collaborate, by initiating professional dialogues and negotiations to enable this innovative change in their classroom cultures and practices resulting in their respective students co-learning and co-assisting each other with concept mapping using a computer software.

This exemplary case study confirms a previous research finding on teacher change, namely, that implementing innovation in schools and incorporating it into classroom practices is dependent on *teacher change*. The latter is a long-term process with the most significant changes in teacher attitudes and beliefs occurring *after* teachers begin implementing a new practice successfully and observed changes in learning (Guskey, 1985). As Grimbeek and Nisbet (2006) pointed out, professional development for a particular purpose (in this case, innovation) is an ongoing and cyclical process that focuses on developing teachers' knowledge, attitudes and beliefs about the innovation, and including as well *teacher change,* which would not come about unless teachers themselves experienced demonstrable success of the innovation on student learning. In line with this perspective, there was evidence in the data presented, which demonstrated that the MPS teacher were sufficiently convinced of the value of concept mapping on their students' learning for them to have gone as far as they did and designed, along with their students, final projects to demonstrate their own growing proficiency and competence with concept mapping. For example, the MPS teachers became confident and proficient to the extent that they designed, developed and implemented final project activities, which culminated in a peer tutoring arrangement between the two classes of primary students, with the older students assisting the younger ones with the use of the computer software *Inspiration^{TM}* enabling the younger ones to control the mouse to construct their own concept maps.

The biggest challenge to the research project and innovation in particular, was sustaining the interest and commitment of the teachers to the project amidst practical realities and competing realities of teaching in schools.

The data presented demonstrated that the effective implementation of an innovation in classrooms requires a much longer period of ongoing professional development and collaboration to support teachers as they get used to the innovation, trial the innovation with sufficient time to design and implement their self-initiated activities. It is, and was a long-term and cyclic process as the teachers made the transition from receptors of innovation to implementers of innovation and ultimately even innovators themselves as they adapted and extended the innovation to suit their own particular classroom practices and teaching programs. As evidenced by outputs from the research project, the MPS teachers constructed concept maps of mathematics and science topics in the PD workshop; they became proficient, alongside their students, in constructing hierarchical concept maps to align with classroom practices; and used concept maps to further enhance student learning and also as assessment tools while MPS students had their own individually or collaboratively constructed concept maps, which illustrated their understanding of previously taught topics. Overall, the MPS teachers incorporated concept maps into the teaching, learning and/or assessment of student learning for some topics throughout the school year as a result of their participation in the research project. Finally, this case study contributes findings to the literature on the kinds of issues and concerns that are encountered when research-based innovative ideas are applied in actual practice in primary classrooms.

Implications

There are two main implications in adapting concept maps for use by students as learning tools, and for use by teachers as assessment tools, both of which are based on continuous reflection and negotiations. Firstly, a focus on the conceptual interconnections provides students with the opportunity to clarify their own understanding of the links and integration between concepts. These maps then represent a snapshot of student understanding at a particular point in time. Assessing the displayed connections or systems of interconnecting concepts can reveal the extent to which meaningful learning has taken place leading to further reflection and discourse, not only with the individual student, but with other students as well. For example, assessing the entries on the concept map provides an opportunity for the teacher to negotiate meaning with the individual student, and/or an opportunity for the teacher to provide further clarifications and/or pose challenging questions to prompt student thinking and reasoning as a class or in small groups.

Secondly, the students and teachers involved in the project have been exposed to and have experienced concept mapping to different levels. It is now up to the teachers to go forward with the strategies in ways that are suitable for their students as learning and/or assessment tools. The MPS teachers indicated that they are continuing with concept mapping as a strategy to brainstorm ideas about a new topic and then as an assessment tool after a unit by having students construct concept maps to indicate their understanding. Peer tutoring was also highly recommended as a means of working collaboratively with new students and/or new schools as a way forward. These suggestions, based on the teachers' own experiences, provide viable directions for further research to identify the most efficient ways of transforming research-based ideas into actual practice in more real classrooms for a wider impact.

Acknowledgment This project was funded by a grant from the Australian School Innovation in Science, Technology and Mathematics (ASISTM) Project, which is part of the Australian Government's Boosting Innovation, Science, Technology and Mathematics Teaching (BISTMT) Programme.

References

Afamasaga-Fuata'i, K. (1998). *Learning to solve mathematics problems through concept mapping & vee mapping.* Samoa: National University of Samoa.

Afamasaga-Fuata'i, K. (2004). Concept maps and vee diagrams as tools for learning new mathematics topics. In A. J. Canãs, J. D. Novak, & F. M. Gonázales (Eds.), *Concept maps: Theory, methodology, technology. Proceedings of the first international conference on concept mapping,* (Vol. 1, pp. 13–20). Navarra, Spain: Dirección de Publicaciones de la Universidad Pública de Navarra.

Afamasaga-Fuata'i, K. (2005). Students' conceptual understanding and critical thinking? A case for concept maps and vee diagrams in mathematics problem solving. In M. Coupland, J. Anderson, & T. Spencer (Eds.), *Making mathematics vital. Proceedings of the twentieth biennial conference of the Australian Association of Mathematics Teachers (AAMT)* (Vol. 1, pp. 43–52). Sydney, Australia: University of Technology.

Afamasaga-Fuata'i, K., & McPhan, G. (2007). *Vee diagrams as a tool for teacher professional development: Learning, reflecting and planning. Proceedings of the New Zealand Association for Research in Education (NZARE).* National Conference. CD-ROM.

Ausubel, D. P. (2000). *The acquisition and retention of knowledge: A cognitive view.* Dordrecht, Boston: Kluwer Academic Publishers.

Grimbeek, P., & Nisbet, S. (2006). Surveying primary teachers about compulsory numeracy testing: Combining factor analysis with Rasch analysis. *Mathematics Education Research Journal, 18*(2), 27–39.

Guskey, T. (1985). Staff development and teacher change. *Educational Leadership, 42,* 57–60.

Loughran, J., & Gunstone, R. (1997). Professional development in residence: Developing reflection on science teaching and learning. *Journal of Education for Teaching, 23*(2), 159–178.

New South Wales Board of Studies (NSWBOS). (2002). *K-6 Mathematics syllabus.* Sydney, Australia: NSWBOS.

Novak, J. D. (2002). Meaningful learning: The essential factor for conceptual change in limited or appropriate propositional hierarchies (LIPHs) leading to empowerment of learners. *Science Education, 86*(4), 548–571.

Novak, J. D., & Gowin, D. B. (1984). *Learning how to learn*. Cambridge, UK: Cambridge University Press.

Ruiz-Primo, M. (2004, September 14–17). Examining concept maps as an assessment tool. In A. J. Canãs, J. D. Novak, & Gonázales (Eds)., *Concept maps: Theory, methodology, technology. Proceedings of the first international conference on concept mapping* (pp. 555–562). Dirección de Publicaciones de la Universidad Pública de Navarra, Spain.

Part III
Secondary Mathematics Teaching and Learning

Chapter 6
Evidence of Meaningful Learning in the Topic of 'Proportionality' in Second Grade Secondary Education

Edurne Pozueta and Fermín M. González

This chapter describes an experiment using concept maps to teach a mathematics topic. The main goal was to detect and evaluate signs of meaningful learning in the students through the analysis of their concept maps, in a setting in which second grade students worked through an innovative instructional module on the topic of mathematical proportionalities. The conceptually transparent instructional module was designed to include introductory, focus and round up activities following the LEAP (Learning about Ecology, Animals and Plants) Model developed at Cornell University (USA).

The study was based on the comparative analysis of the concept maps drawn by the students before and after the implementation of the instructional module, using a model presented by Guruceaga and González (2004) that enabled us to identify various features in students' concept maps, providing a tool to monitor meaningful learning and detect any tendency towards rote learning.

The results show that the implementation of a theoretically grounded instructional module enabled a group of students to learn about mathematical proportionalities more meaningfully. An additional finding, providing extra added value to the study, was the emergence of three clear patterns of Concept Maps. As shown in the study, through meaningful learning, original map structures indicating the possibility of undesired outcomes can be appropriately adjusted to contribute towards an integral education (involving the heart, mind and body) for our students.

Theoretical Background

Over the course of the last 30 years profound changes have taken place in the teaching of mathematics. The international community of experts in the didactics of mathematics continues to strive to find appropriate models. It is clear therefore that we are currently undergoing a period of experimentation and change (Guzmán & Gil, 1993). On the one hand, the increasing scientific and technological development of

E. Pozueta (✉)
Public University of Navarra, Pamplona, Spain
e-mail: epozueta@unavarra.es

K. Afamasaga-Fuata'i (ed.), *Concept Mapping in Mathematics*,
DOI 10.1007/978-0-387-89194-1_6, © Springer Science+Business Media, LLC 2009

our Western societies has created a demand for very high levels of knowledge in the area of mathematics, widely considered to be one of the most important, if not the most important, areas of the school curriculum. On the other hand, there is also growing awareness, evidenced by findings from various surveys such as the PISA report 2003 (Programme for International Student Assessment), of the fact that our schoolchildren are seriously lacking in numeracy skills (Rico, 2005).

Meanwhile, the research on human learning processes has moved forward with new series that have had a strong impact in educational circles. One of these is Novak's educational theory, Novak (1982), which proposes that, for children to learn more actively and effectively and modify their understanding of mathematical activity, it is necessary to relate what is already known about the nature of knowledge and human learning to the teaching of mathematics. Novak claims that meaningful learning (ML) is the result of the constructive integration of thinking, feeling and acting leading to the empowerment for commitment and responsibility. From his constructivist approach, Novak (1988) stresses the concept of ML and the way it is treated by Ausubel, Novak, and Hanesian (1976) in his assimilation theory.

One of the main problems in the learning of mathematics, according to Novak (1998), is that most of the instructional material is conceptually confusing. In other words, there is a failure to present either the concepts or the conceptual relationships that schoolchildren require in order to understand the meaning of the mathematical ideas in question. When it comes to designing and implementing classroom instructions that satisfy the conditions for ML, there is a need for tools to facilitate learning; the concept map (CM) is one such tool.

The ML requires conceptually transparent curricular and instructional material design (Ausubel et al., 1976), hierarchised, so the new information can be integrated in the previously conditioned student's cognitive structure. This condition for the ML must fill every single knowledge area, even being valid for the mathematics. So, to make this process operative, the design of an instructional module (IM) will be necessary, an innovative IM based on the use of CM as a tool to facilitate ML around a mathematic topic.

Some of the papers presented at the First International Conference on Concept Mapping (Pamplona, 2004), and later published in the conference report, referred to the use of CMs in mathematics topics. One example is that of Afamasaga-Fuata'i (pp. 13–20), which stresses the usefulness of CMs as tools to help learners achieve a better and deeper understanding of certain selected mathematics topics. Another is that of Serradó, Cardeñoso, and Azcárate (2004, pp. 595–602) where CMs are highlighted as tools to help in the diagnostic assessment of the obstacles that arise when imparting mathematical knowledge, and described as a source of information to promote the professional development of teachers.

The Second International Conference on Concept Mapping (San José, Costa Rica, 2006) included a special session on Concept Mapping in mathematics, moderated by leading researchers: Nancy R. Romance from Florida Atlantic University (USA), Jean Schmittau the State University of New York at Binghamton (USA) and Karoline Afamasaga-Fuata'i the University of New England (Australia). The proceedings of that session list a number of case studies illustrating the use of concept maps in the teaching of mathematics. These include the use of concept maps to help students to grasp the concept of the positional system in a study conducted by Jean Schmittau and James J. Vagliardo (2006, pp. 590–597); the development of the concept mapping approach to the teaching of mathematics in secondary schools in an article by William H. Caldwell, Faiz Al-Rubaee, Leonard Lipkin, Della F. Caldwell, and Matthew Campese (2006, pp. 170–176); a study presented by Rafael Pérez Flores from the Universidad Autónoma Metropolitana de México (pp. 407–414); the assessment of multidimensional concept maps by M. Pedro Huerta from the Universitat de València (2006, pp. 319–326); concept maps in the learning stages of Van Hiele's educational model in a study (2006, pp. 383–390) conducted by Pedro Vicente Esteban Duarte, Edison Darío Vasco Agudelo, and Jorge Alberto Bedoya Beltrán.

Through the mention of such communications and studies directed to achieve a deeper understanding of mathematics teaching, we try to underline the importance and the validity of CMs in topics related to school mathematics. This way, we are focusing on a new field for the application of CMs as an unbeatable tool to foment ML not rote learning (RL) in students and to identify different patterns that might have some predictive value.

The Topic of the Proportionality

From our experience as teachers, one of the most intriguing topics in the teaching of mathematics is that of proportionality. From the earliest times, proportionality has played a part in human understanding of the surrounding world. Thus, for example, faced with the impossibility of directly measuring distances, resort was made to comparative methods. The idea first arose in astronomy and later in science as a whole, both for defining new magnitudes and for expressing numerical relationships, working with indices, constants or rates. Proportionality is therefore a basic concept in mathematics and is of major importance in the school curriculum (Fiol & Fortuny, 1990), since it is associated with most mathematical content and the content of other subjects such as *physics, biology, chemistry, etc*. It is not a simple concept, however. Rapetti (2003) noted that it is not easy to acquire the notion of proportion because it challenges learners with a range of situations of varying levels of numerical complexity and different types of magnitudes. The need to consider quantities in relation to one another, in addition to approaching them in absolute terms, presents a problem to some learners and creates an obstacle to their understanding of the mathematical content surrounding the notion of proportionality.

According to Azcárate and Deulofeu (1990), in order to deal with proportionality as a function model, it is first necessary to explore concepts such as ratio and proportion, and ways of solving proportionality tasks. Although proportions played a fundamental role for the Pythagoreans, historically, this proved to be a major obstacle in the development of the general concept of function. When working with proportions, it is hard to identify the relationship between two different magnitudes, since compared magnitudes are not always of the same nature. Proportions therefore conceal the underlying interdependence of different magnitudes. The Greeks, however, always made their proportions homogeneous by means of ratios, each of which was formed by two magnitudes of the same type. This apparently came about as a result of the geometric significance of the magnitudes. In other words, length was compared with length, or area with the area, with the result that a ratio between different magnitudes became meaningless.

The prestigious journal, *Mathematics Teaching in the Middle School* (MTMS), an official publication of the National Council of Teachers of Mathematics (founded in 1920 with members from the United States and Canada) is a widely acknowledged resource for students and teachers. It contains articles and seminars presenting activities, ideas, strategies and problems, and, since the year 2000, has devoted over 20 articles to proportionality and the teaching/learning issues this involves.

This is a patent illustration of the current interest in proportionality as a key teaching/learning topic in the first years of secondary education. Ben-Chaim, Fey, Fitzgerald, Benedetto, and Miller (1998) specifically claim that proportional reasoning is the core mathematical topic during the last years of primary and first years of secondary education.

Furthermore, in the search for reliable references on the analysis of proportionality in a teaching/learning context, it is important to consider the existing research into the definition of ratio and/or proportion. Lesh, Post, and Behr (1988), for example, describe and compare the views of different authors such as Vergnaud, Schwartz and Kaput, on the nature of ratios.

One characterisation of cognitive development is to be found in Piaget's theory (Piaget & Inhelder, 1972), which identifies the ability to reason proportionally as a primary indicator of formal operational thought, one of Piaget's levels of cognitive development. Piaget approached the subject primarily in his studies of probability, and the laws of physics and spatial relationships. He claimed that the notion of proportion belongs to the level of formal relationships, in other words, that operations are not performed directly on concrete objects; they are operations on operations. In his research into the development of child thought, the author mentions having faced the problem of explaining the process involved in learning to understand proportions and found proportional reasoning to manifest itself around the beginning of secondary education.

Further key contributions have been made by authors such as Freudenthal (1983), who showed that proportional reasoning tasks can be divided into two types: those involving internal ratios and those involving external ratios. Internal ratios are ratios between terms belonging to the same magnitude, while external ratios are comparisons between values of two different magnitudes.

In a brief review of previous research into the concepts of ratio and proportion, Nesher and Sukenik (1989) report that the standard procedure in many studies is to test subjects' understanding of ratio and proportion by asking them to solve ratio problems (with or without illustrations; written or oral) and then analyse the problem-solving strategies used in their answers. One of the main strategic errors identified by them in students of different ages is the use of the additive strategy, where the relationship between ratios is viewed as the difference between terms and students fail to perceive its multiplicative nature.

The above theoretical considerations and the importance of the subject of proportional reasoning revealed in them suggested to us the need for an IM designed to teach this mathematical topic to second grade secondary schoolchildren, incorporating the ideas presented in Novak's theory of education. Thus, the aim was to create conceptually transparent material, based on the use of CMs as a means to facilitate meaningful learning and enhance understanding of the concept of proportionality and the ability to solve related problems, with particular emphasis on two points; one, ratios between different values of the same magnitude and, two, the potential proportional relationship between two different magnitudes and the various forms in which such a relationship may be expressed.

Research Planning and Design

The study presented here was made possible thanks to the researchers' participation in one of the subgroups of the research team that conducted the GONCA project (González & Cañas, 2003), which was financed by the Government of Navarra Education Department.

One of the aims of the GONCA project was to assess the effectiveness of CMs versus rote learning, as evaluation, teaching and learning tools. Another of its proposals was that the students should use CmapTools software to construct their concept maps. This software was developed at the Institute for Human and Machine Cognition to allow users to construct CMs using a very simple interface and thereby make the maps easy to save on a universally accessible Internet resource (González & Cañas, 2003).

In the present study, we wish to highlight various features of concept mapping as a means towards ML.

It plays an extremely useful role in an innovative IM to teach proportionality, making it easier, as recommended by Ausubel et al. (1976), to take into account pupils' prior knowledge of the topic to be dealt with in the classroom. The maps produced by the students prior to instruction can be used to establish each individual's point of departure in the learning process. Ausubel also proposed that the more inclusive concepts relating to the topic be presented at the start of the instruction period, and that the more specific concepts be dealt with later. Hence there is a need to clarify which concepts are to be included in the instruction, define their meaning, identify the hierarchical relationships and reconciliations between them and establish how this frame of reference relates to what the pupils already know. Novak recommends teachers to create a reference map, to arrange the relevant concepts, both inclusive and specific, relating to the chosen topic; in this case, proportionality. Once the reference map has been created (see Fig. 6.1), one can proceed to identifying the most significant conceptual nodes around which to plan and schedule classroom activities.

Concept maps provide students with a learning tool to help them cover all the content items of the module.

They can be used to evaluate students' prior knowledge of this topic and monitor their learning progress. We will therefore base our study on the comparative analysis

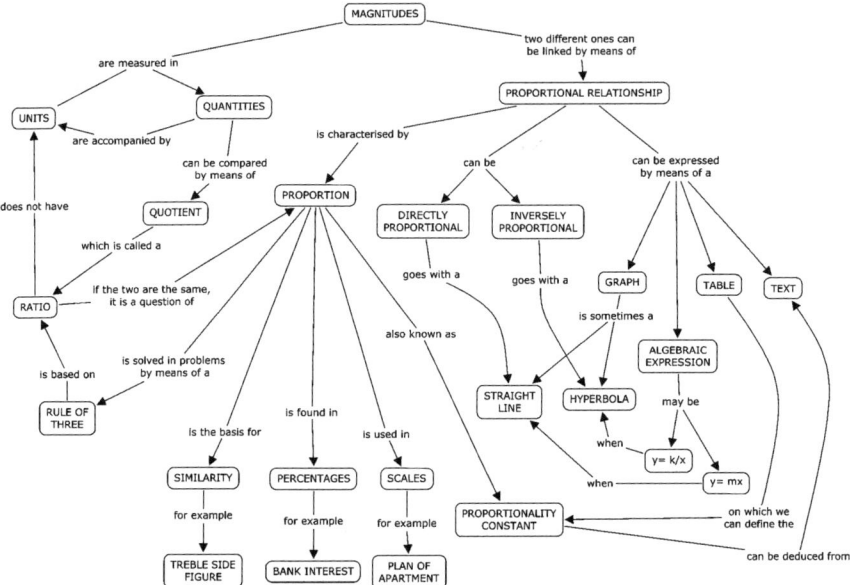

Fig. 6.1 Reference map (Pozueta, 2003)

of the children's maps before and after instruction, following the model presented by Guruceaga and González (2004). This will allow us to focus on various features of concept maps (see Table 6.1). Thus, our pupils' concept maps will become the means by which we will be able to detect the level of meaningful learning they have attained, and will also serve to alert us if the process has involved rote or mechanical learning rather than meaningful learning. The following table (Table 6.1) shows the indicators that were checked. After analysis, the pupils' maps can be classified into groups showing similar characteristics and tendencies.

The Setting in Which the Concept Maps were Used

The fieldwork took place throughout the 2002–2003 academic years, in the state-aided school, San Fermin Ikastola, in the outskirts of Pamplona (Spain). Classes are conducted in the Basque language and the school caters for infants, primary, secondary and high school pupils. The research, which was conducted by a highly experienced secondary mathematics teacher, comprised several stages.

The first was to create the above-mentioned reference map (see Fig. 6.1), which was geared to the teaching of second-grade secondary students (13 to 14 year olds). Twenty five (25) concepts were selected and the main relationships between them were identified, taking into account the didactic aims to be covered in the instructional module.

Table 6.1 Learning indicators

Indicators of rote/mechanical learning	Indicators of meaningful learning
• No clear differentiation between concepts and linking phrases; direction of the relationships between concepts not shown	• Clear differentiation between concepts and linking phrases; shows the direction of the relationships between concepts
• A minor number of concepts are used	• Most of concepts are used
• A high frequency of erroneous propositions: illogical conceptual hierarchies	• A decreasing trend in erroneous propositions
• An incorrect hierarchical ordering of concepts in terms of their inclusivity	• There is coherence in the hierarchical organisation of the concepts in terms of their inclusivity
• The most inclusive concepts are not identified	• The most inclusive concept is identified
• Shows long linear relationships, chaining of concepts	• Examples of super-ordination of an inclusive concept
	• Progressive differentiation between inclusive concepts
	• Linear relationships between concepts are fewer or totally absent
• Crossed links are few in number and erroneous: a sign of weak integrative reconciliations	• There are numerous crossed links revealing high-level integrative reconciliations

At this point it should be emphasised that the module was intended to present and develop the topic of proportionality using mathematical situations involving ratio, such as similarity, percentages and scales. In the final level of the hierarchical map, therefore, we can see specific examples of such situations. The map does not reflect the necessary condition of the equality of corresponding angles when defining similar figures; it simply reflects the fact that proportional reasoning is involved in the definition. The map was meant, moreover, to clearly illustrate the difference between ratio defined as a relationship between different values of the same magnitude and the proportional relationship that may exist between two different magnitudes, as well as the different ways in which this proportional relationship can be expressed. Thus, the four ways of expressing a relationship between two magnitudes appear on the right-hand side of the map, indicating in each case the type of proportional relationship involved, that is, whether it is directly or inversely proportional. The concept of *ratio* is defined as in the vast majority of lower secondary school mathematics textbooks, despite the fact that Freudenthal (1983) considers such a definition to be a violation of its meaning.

Sixty three (63) students from two second-grade secondary school classes each drew a concept map prior to instruction. They were asked to use the same list of 25 concepts used in the reference map.

These maps enabled the teachers to check the point of departure for each individual student and they served as a guide for the design of the innovative instructional module on proportionality. The instruction largely followed the second-year secondary education mathematics syllabus, but it should be noted that in the textbooks for this level the concepts involved usually appear under different topics: *proportionality, similarity, scales, percentages, linear functions, etc.* hence there was a need to relate them within the context defined above in order to create the reference map. The instructional sequence was adapted from Project LEAP (Learning about Ecology, Animals and Plants, 1995). This was a plan that grouped activities into three stages: introduction, focusing and summary. The process begins with the presentation of the most inclusive concepts, after which progressive differentiations and the more significant reconciliations are made, before finally applying the information discussed throughout the instruction period. When designing activities for this module, which was written in the Basque language ready for presentation in the classroom, ideas were taken from several published texts.

The teacher carried out the instruction using the same approach in both classes during the second term of the school year. In other words, there was no control group; on the one hand, there was no need for altering the Mathematics Department's previously defined teaching planning and, on the other hand, it was important to respect the school's philosophy about not creating any type of distinction among the students. From the methodological point of view, pupils were expected to work individually during the presentation and summary stages, and in small groups of five during the focusing stage. Although group work in the classroom was not standard practice during mathematics lessons, the second grade students on the whole showed a positive attitude towards working in groups, except for a few isolated cases of low contributors. During the instruction period, the pupils also created additional concept maps using CmapTools software, already installed in the computers in the school's computer room. Concept mapping did not form part of the standard mathematics program at the school but had been adopted by the natural science department. The pupils in this age group were therefore already trained in the creation of concept maps relating to the content of the area of natural sciences, and thus had no difficulty in using the technique. In the last activity of the instructional module, each student was asked to produce a final concept map using the 25 concepts that had featured in the reference map prior to instruction.

The concept maps created before and after instruction by 32 of the 63 students who participated in the experiment were then subjected to a comparative analysis.

Description and Discussion of the Findings

In this section, we will discuss the indicators described above (see Table 6.1) with a view to performing a comparative analysis of the maps produced by the pupils before and after having worked through the instructional module. The analysis will focus on the following issues:

The Way in Which the Pupils Differentiate Concepts from Links

On the whole, their differentiation of concepts and links was very similar in both maps; in other words, most of the pupils were able to make the correct distinction between the concepts and the linking phrases on the maps created both before and after instruction. Of the 32 pupils, 25 were found to differentiate correctly between concepts and links, remembering to use arrows and explaining all the links, on both maps.

On the whole, the quality of the links was also similar in both maps by all pupils. Their propositions tended to be rather poor, but half the pupils showed a tendency towards a gradual enrichment of their propositions.

Utilisation of Concepts

As shown in Table 6.2, only two of the 32 pupils (numbers 10 and 19) used all 25 of the concepts mentioned on the list they were given during the creation of the pre-instruction concept map. The same two were also the only ones of the whole group who reduced the number of concepts used from 25 to 24 in the second map. That is, they used a greater number of concepts in the map prior to instruction than in the one they created after instruction was complete. The majority of the pupils were found to make more use of the concepts in the second map than in the first, 11 pupils having managed to use all 25 concepts (higher frequency of use). It is important to stress that 24 out of the 32 pupils did not use the concept *ratio* in the map drawn prior to instruction. Of these, 21 introduced it in the second map; pupils numbers 24, 25 and 26 failed to use it in either of their maps, while pupil number two used it in the first but not in the second. With regard to the three examples featured on the list of concepts: *bank interest*, *plan of an apartment* and *treble the side of the figure,* it should be noted that, of the whole group, 27 pupils failed to use one, two or three of these examples in the map prior to instruction.

Propositions Formed from Links Between Concepts

From the information provided in Table 6.3, we can see that in the map constructed prior to instruction, 19 pupils out of the whole group were able to identify up to 15 relationships, and none identified fewer than 7. In the post instruction map, all the pupils were able to make at least 11 propositions, 29 of them made between 21 and 29, and six managed to make 30 or more links.

With regard to the share of erroneous or inaccurate propositions relative to the total included in the maps, there is clear evidence of a decreasing trend in all the pupils.

It is worth mentioning the case of three pupils, numbers 3, 16 and 27, who, out of the total number of propositions in their post-instruction maps, (totals of 26, 29 and 39, respectively) had no incorrect propositions whatever.

Table 6.2 Use of concepts relating to proportionality

Pupil	Number of concepts used (out of 25) in first map	Percentage of total number of concepts used in first map	Number of concepts used (out of 25) in second map	Percentage of total number of concepts used in second map
1	17	68	23	92
2	22	88	22	88
3	13	52	21	84
4	13	52	21	84
5	17	68	25	100
6	10	40	20	80
7	13	52	25	100
8	10	40	17	68
9	22	88	25	100
10	25	100	24	96
11	12	48	23	92
12	16	64	22	88
13	15	60	25	100
14	16	64	24	96
15	18	72	23	92
16	14	56	25	100
17	18	72	24	96
18	17	68	25	100
19	25	100	24	96
20	12	48	23	92
21	9	36	21	84
22	15	60	18	72
23	15	60	18	72
24	18	72	19	76
25	8	32	10	40
26	14	56	14	56
27	19	76	25	100
28	11	44	25	100
29	18	72	25	100
30	15	60	18	72
31	17	68	25	100
32	16	64	25	100

Specifically, 28% failed to introduce one of the three examples; *treble the side of the figure* was the one least used. Another 28% failed to introduce two examples; but more than half of these introduced all three examples after instruction. A further 28% failed to incorporate any of the examples; but more than half of this group incorporated all three after receiving instruction. One concept that was frequently omitted from the map produced after instruction was *proportionality constant*, which 22% of the pupils failed to use.

In terms of the percentage of erroneous or inaccurate propositions in the totals, 15 of the 32 pupils made up to 50% inaccurate propositions in the first map, while in the subsequent map only two made more than 50% and 16 made 10% or less.

Table 6.3 Propositions in relation to proportionality

Pupil	In first map: Erroneous propositions out of total	Percentage in first map	In second map: Erroneous propositions out of total	Percentage in second map
1	8/20	40	2/32	9.4
2	20/24	83.3	14/24	58.3
3	10/13	76.9	0/26	0
4	8/14	57.1	4/24	16.7
5	10/17	58.8	4/30	13.3
6	4/9	44.4	5/27	18.5
7	5/12	41.6	2/26	7.7
8	2/11	18.2	4/11	36.7
9	5/23	21.7	1/32	3.1
10	19/36	52.8	4/28	14.3
11	2/13	15.4	2/26	7.7
12	9/17	52.9	8/26	30.8
13	6/14	42.8	2/29	6.9
14	5/15	33.3	6/27	22.2
15	4/17	23.5	1/22	4.5
16	3/13	23.1	0/29	0
17	10/18	55.5	1/26	3.8
18	8/16	50	5/28	17.8
19	8/25	32	7/26	27
20	5/11	45.5	2/27	7.4
21	5/8	62.5	10/19	52.6
22	8/14	57.1	6/18	33.3
23	4/18	22.2	2/21	9.5
24	5/8	62.5	5/24	20.8
25	4/7	57.1	1/11	9.1
26	7/13	53.8	1/14	7.1
27	11/18	61.1	1/27	3.7
28	3/12	25	0/39	0
29	9/17	52.9	5/30	16.7
30	8/14	57.1	6/21	28.6
31	11/15	73.3	6/24	25
32	9/15	60	3/30	10

Levels of Hierarchy

In relation to this point, it is worth mentioning the differences between the pre-instruction and post-instruction maps. The pre-instruction maps vary considerably in structure from one pupil to another, while bearing little resemblance to the reference map but displaying certain patterns as regards the grouping of concepts.

The concept of *magnitudes* is directly linked to *quantities* or *units* and a linear sequence is made from there to *proportion,* which includes four types: *proportional relationship, directly proportional, inversely proportional and constant of proportionality.*

The concept of *magnitudes* is linked with different forms of expression: *scales, tables, graphs, percentages and text.*

The concept of *quantities* is associated with the *rule of three*, which is identified as a *formula* while other examples of formulas are on $y = mx$ and $y = k/x$.

The concept of *scales* is associated with the concept of *plan of an apartment* and the concept of *percentages* with *graphs.*

Generally speaking, the hierarchical levels are more clearly defined in the post-instruction maps, especially in the upper part of the map. Thus, 34% of the 32 pupils were able to present the hierarchical levels correctly throughout the whole map.

All the maps begin at the top with a similar structure to that of the reference map, with a more inclusive concept clearly differentiated. It is worth noting that, in many of them, the concept of *proportion* is not associated with the concept *proportional relationship*, showing that the students have failed to make reconciliation between these two concepts. Another point worth mentioning is that in some of these maps the four forms of expression of a proportional relationship are not placed on the same hierarchical level.

Crossed Links

Although we can report no crossed links on the pre-instruction maps, they did appear on 50% of the post-instruction maps. The crossed links on all maps referred to the same concepts, in other words, transversal propositions were made linking the four different forms of expression with each of the two types of proportional relationship.

This description illustrates the fact that, overall, the cognitive structure of the pupils can be described as more organised and hierarchically ordered after instruction. There is a considerable increase in the number of concepts used to construct the post-instruction maps and the pupils' cognitive structure appeared to have been enriched with the incorporation of the concept of *ratio*, the inclusive nature of which was correctly shown in many cases. Judging from the pattern that can be observed when the two maps of each pupil are compared for the percentage of erroneous or inaccurate propositions out of the total, the pupil's cognitive structure can also be said to have become more logical. In the post- instruction maps, it is possible to detect an increasing ability to establish clearer and more accurate hierarchical levels, especially those that come highest on the map.

In many cases a clear differentiation was made between concepts such as *proportional relationship* and *proportion.* Fifty six percent (56%) of the pupils succeeded in differentiating one or other of these concepts, which were inclusive on the reference map.

The crossed links that appear on the post-instruction maps suggest that the majority of the pupils had succeeded in making an integrative reconciliation between the different forms of expression of the two types of proportional relationship. Furthermore, we are able to report that the three pupils who were able to differentiate the concepts of *proportional relationship* and *proportion* also managed to reconcile them, since they established a correct link between the two.

As we discuss this, it is interesting to note that there were three clear tendencies in the concept maps constructed by the pupils after instruction. This same observation has already been highlighted in a study by González (1997), describing three groups of pupil-constructed concept maps, each sharing common characteristics. In the case that concerns us, it was possible to observe similar groupings of maps with similar characteristics. Thus, one of the first tendencies to be observed was the obvious restructuring in the post-instruction map made by three of our pupils, who were able to produce something very close to the reference map they had been shown. The pre-and post-instruction maps made by pupil, I.I., (see Figs. 6.2 and 6.3) are included as an illustration of the progress made by this group of students.

This particular pupil used 14 concepts in the first map, but increased this to 25 in the final map. It is interesting to observe how the concept of *ratio* is incorporated with its full meaning, shown in its linkage with *proportion* and *proportional relationship*. The concept of proportional relationship is assigned the correct hierarchical level, while the four forms of expression, that is, text, table, formula and graph, for both types of proportionality presented and handled in the instructional module are perfectly defined on a lower level. The crossed links between these types of expression are also shown. Furthermore, the three mathematical contexts treated

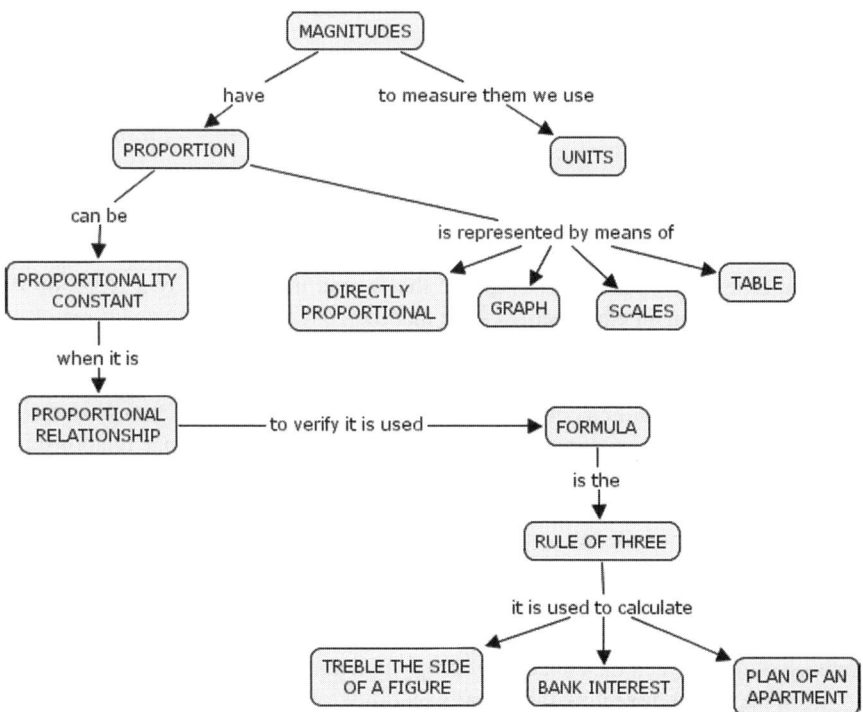

Fig. 6.2 I.I.'s first map

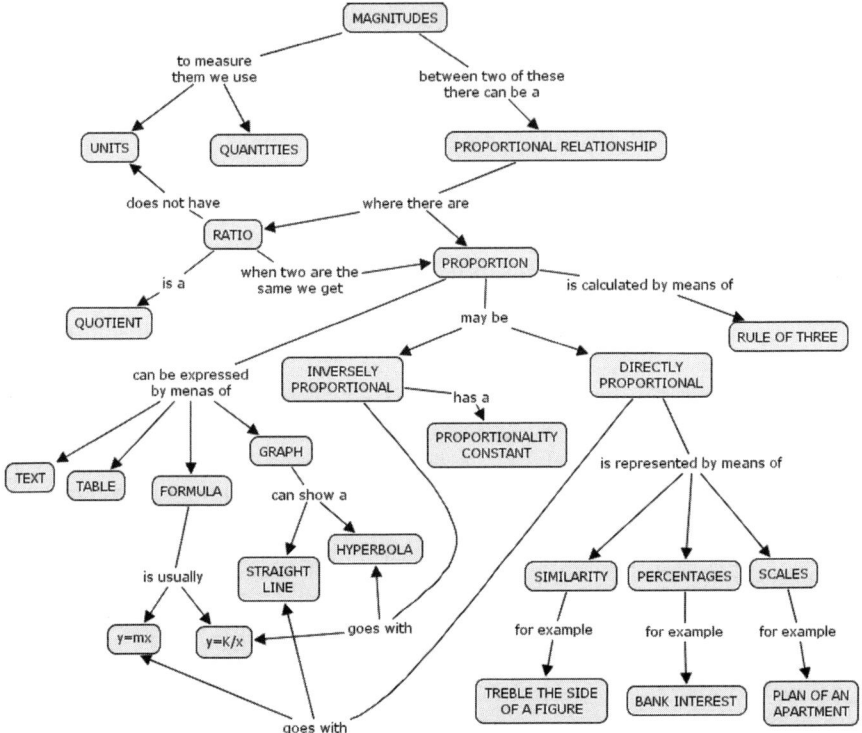

Fig. 6.3 I.I.'s final map

in the module as examples of direct proportionality are shown on another hierarchical level.

Another tendency can be observed on the second maps made by a group of 19 pupils. Here, the main characteristics are that the concept of *proportional relationship* (or, in its absence, the concept of *proportion*) is differentiated, there are few incorrect links, no linear sequences, the hierarchical levels are largely well defined, the logic of the discipline is observed and, in some cases, reconciliations are made between some forms of expression and both types of proportional relationship. This tendency is reflected in the way pupil, I.M., has modified the second map (see Figs. 6.4 and 6.5).

Note that the 19 concepts used in the first map increased to 25 in the second. There are also clear signs that this pupil has begun to give a new meaning to the concepts of *ratio* and *proportion*. However, despite having clearly differentiated the concept of *proportional relationship* and correctly placed all the subordinate concepts, the link between the concepts of proportion and proportional relationship has not been identified, revealing this pupil's inability to reconcile these two concepts. Crossed links between the different forms of expression of the two types of proportionality were also missing in this map. Finally, we can report on a third group of

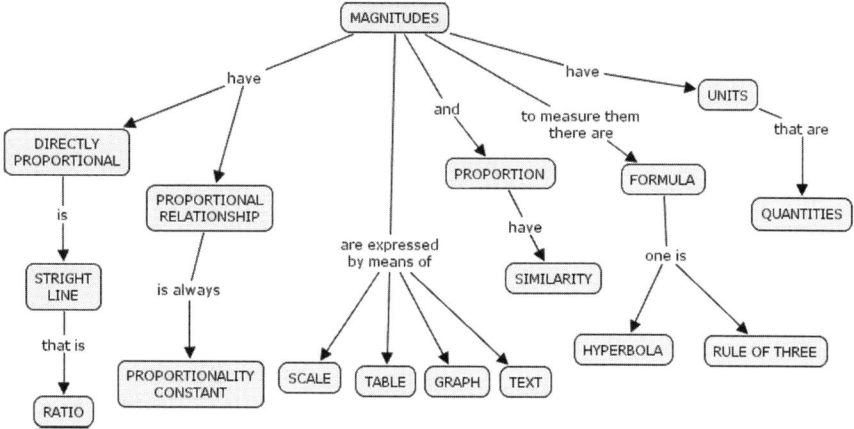

Fig. 6.4 I.M.'s first map

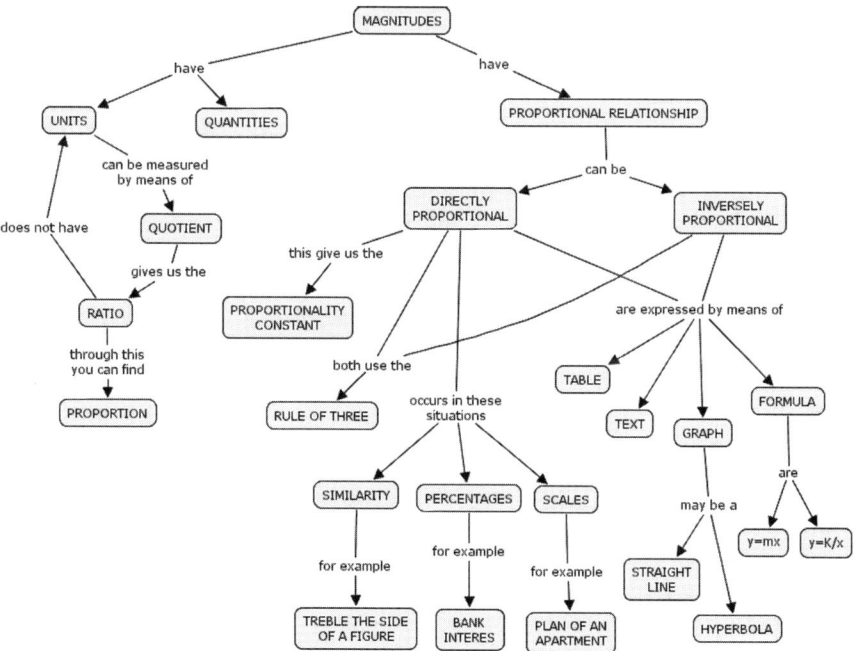

Fig. 6.5 I.M.'s final map

10 pupils, whose final maps showed many features that were more characteristic of rote learning than meaningful learning, given the total absence of differentiations and the very few examples of reconciliation.

The frequency of erroneous links is generally higher and there is barely any evidence of adherence to the logic of the discipline. Some of these maps feature some

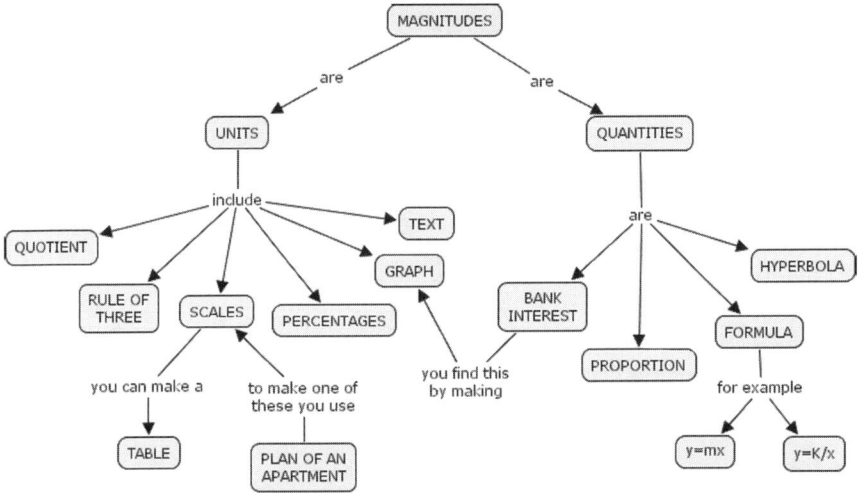

Fig. 6.6 H.Z.'s first map

rather confused nodes. If we take the maps drawn by pupil, H.Z., (see Figs. 6.6 and 6.7) we are able to observe that although all 25 concepts are present in the second concept map, 6 of the 24 links are incorrect. What is more, the concepts are not placed on the appropriate level in the hierarchy, nor does the structure of the map reveal a progression from the more inclusive to the more specific.

There is an abundance of linear sequences and no kind of reconciliation. What does appear is a series of concepts, particularly in relation to the three mathematical contexts used in the instructional module as examples of direct proportionality, but they are linked arbitrarily and meaninglessly in the upper levels of the hierarchy.

Conclusion

The above results give us reason to believe that the implementation of the innovative instructional module for the topic of proportionality was successful in promoting more meaningful learning in the classroom.

We base this judgment on the criteria used to identify indicators of meaningful learning in the concept maps created by our pupils. The indicators include a considerable increase in the number of concepts used in the post-instruction maps, a significant reduction in errors or inaccurate propositions, increased clarity in the levels of hierarchy and coherence with the inclusivity of the concepts, a decrease in the number of linear chains and confused nodes between concepts and an increase in the progressive differentiations reconciled integratively, all of which clearly reveal the pupils' greater ability to achieve meaningful learning.

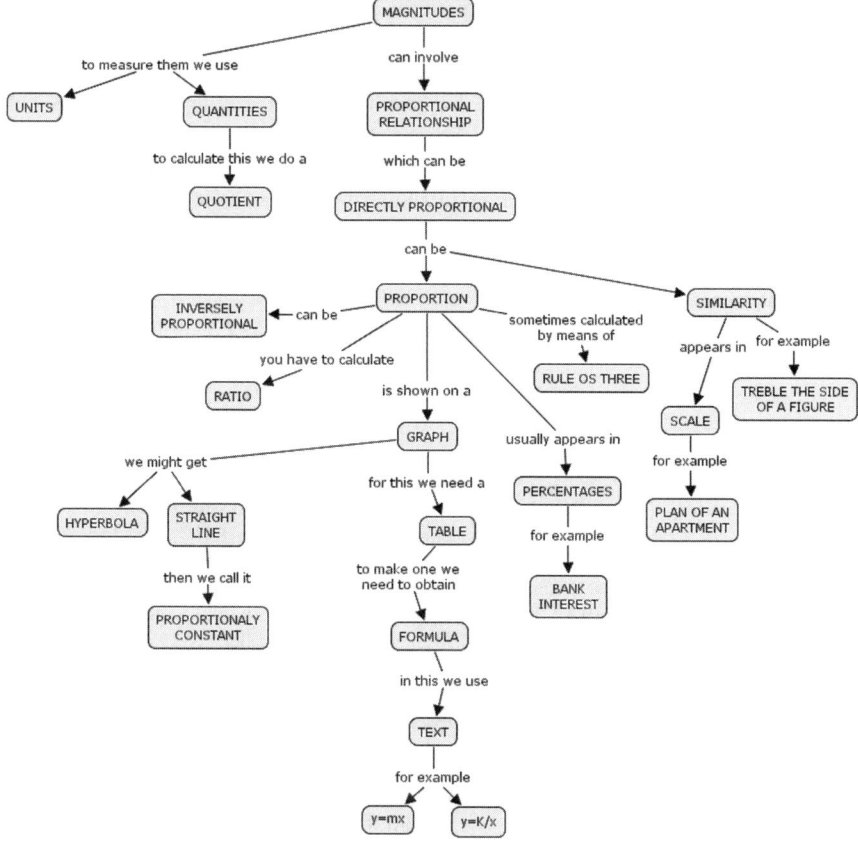

Fig. 6.7 H.Z's final map

The research also revealed problems in terms of the pupils' lack of ability to reconcile *ratio* or *proportion* with that of *proportional relationship*. They clearly had serious difficulty in identifying the most important inclusive concepts, and this obviously hindered their capacity to establish richer and more reconciled progressive differentiations. In our view, however, it is important to note that the implementation of a theoretically grounded instructional module in a standard classroom setting gave a group of pupils the opportunity to learn about the topic of proportionality in a more meaningful manner.

Concept mapping revealed itself as a useful tool enabling us to design an innovative and conceptually more transparent instructional module on a complex mathematical topic, and also to check the pupils' prior knowledge of the topic and track their learning process.

In addition, three clear patterns emerged in the concept maps made by the pupils, revealing three different tendencies among them. In future research, it would be interesting to try to identify different patterns in the concept maps constructed by our pupils. An operational definition of the maps and their correlation with the pupils' academic performance and certain aspects of their personality, attitudes and behaviour could turn them into valuable predictors of learning capacity. As our research has patently shown, it might be possible through meaningful learning to modify the patterns that emerge in the original maps, which are predictive of undesired effects, and thus achieve better results in terms of a more integrative education (in cognitive, emotional and psychomotor terms) for our pupils.

Acknowledgements The above research, which was carried out within the framework of the GONCA project, was made possible thanks to funding provided by the Government of Navarra Education Department (Grant 294/2001, 27 December).

References

Afamasaga-Fuata'i, K. (2004). Concept maps and vee diagrams as tools for learning new mathematics topics. *Concept maps: Theory, methodology, technology, CMC 2004* (pp. 13–20). Pamplona, Spain.

Ausubel, D. P., Novak, J. D., & Hanesian, H. (1976). *Psicología educativa. Un punto de vista*.

Azcárate, C., & Deulofeu, J. (1990). Funciones y gráficas. Madrid: Editorial Síntesis.

Ben-Chaim, D., Fey, J. T., Fitzgerald, W. M., Benedetto, C., & Miller, J. (1998). Proportional reasoning among 7th grade students with different curricular experiences. *Educational Studies in Mathematics, 36*, 247–273.

Caldwell, W.H., Al-Rubaee, F., Lipkin, L., Caldwell, D. F., & Campese, M. (2006). Developing a concept mapping approach to mathematics achievement in middle school. In A. J. Cañas & J. D. Novak (Eds.), *Concept maps: Theory, methodology, technology, CMC 2006* (pp. 170–176). San José, Costa Rica: Universidad de Costa Rica.

Esteban, P. V., Vasco, E. D., & Bedoya, J. A. (2006). Los Mapas Conceptuales en las Fases de Aprendizaje del Modelo Educativo de Van Hiele. In A. J. Cañas & J. D. Novak (Eds.), *Concept maps: Theory, methodology, technology, CMC 2006* (pp. 383–390). San José, Costa Rica: Universidad de Costa Rica.

Fiol, M. L., & Fortuny, J. M. (1990). *Proporcionalidad directa. La forma y el número*. Madrid: Editorial Síntesis.

Freudenthal, H. (1983). *Didactical phenomenology of mathematical structures*. Dordrecht: Reidel Publishing Company.

González, F. M. (1997). Evidence of rote learning of science by Spanish university students. *School Science and Mathematics, 97*(8), 419–428.

González, F. M., & Cañas, A. (2003). GONCA Project: Meaningful learning using CMapTools. *Advances in technology-based education: Toward a knowledge-based society* (pp. 747–750). Badajoz, Spain.

Guruceaga, A., & González, F. M. (2004). Aprendizaje significativo y Educación Ambiental: análisis de los resultados de una práctica fundamentada teóricamente. *Enseñanza de las Ciencias, 22*(1), 115–136.

Guzmán, M., & Gil, D. (1993). *Enseñanza de las Ciencias y la Matemática*. Madrid: Editorial Popular S.A.

Huerta, M. P. (2006). La evaluación de Mapas Conceptuales Multidimensionales de Matemáticas: Aspectos Metodológicos. In A. J. Cañas & J. D. Novak (Eds.), *Concept maps: Theory, methodology, technology, CMC 2006* (pp. 319–326). San José, Costa Rica: Universidad de Costa Rica.

Lesh, R., Post, T., & Behr, M. (1988). Proportional reasoning. *Numbers concepts and operations in the middle grades* (pp. 93–118). NCTM. Reston, VA: Erlbaum.

Nesher, P., & Sukenik, M. (1989). Intuitive and formal learning of ratio concepts. *Proceedings of the 13th. annual conference of the international group for the psychology of mathema tics education* (pp. 33–40). Paris, France.

Novak, J. D. (1982). *Teoría y práctica de la educación*. Madrid: Alianza Editorial S.A.

Novak, J. D. (1988). Constructivismo humano: un consenso emergente. *Enseñanza de las Ciencias*, 6(3), 213–223.

Novak, J. D. (1998). *Conocimiento y aprendizaje*. Madrid: Alianza Editorial S.A.

Perez, R. (2006). Mapas conceptuales y Aprendizaje de Matemáticas. *Concept maps: Theory, methodology, technology, CMC 2006* (pp. 407–414). San José, Costa Rica: Universidad de Costa Rica.

Piaget, I., & Inhelder, B. (1972). *De la lógica del niño a la lógica del adolescente*. Buenos Aires: Paidós.

Rapetti, M. V. (2003, Noviembre). Proporcionalidad. Razones internas y razones externas. *Suma*, 44, 65–70.

Rico, L. (2005). *Competencias matemáticas e instrumentos de evaluación en el estudio PISA 2003*. PISA 2003. Madrid: INECSE. Ministerio de Educación y Ciencia.

Schmittau, J., & Vagliardo, J. J. (2006). Using concept mapping in the development of the concept of positional system. *Concept maps: Theory, methodology, technology, CMC 2006* (pp. 590–597). San José, Costa Rica: Universidad de Costa Rica.

Serradó, A., Cardeñoso, J. M., & Azcárate, P. (2004). Los mapas conceptuales y el desarrollo profesional del docente. *Concept maps: Theory, methodology, technology, CMC 2004* (pp. 595–602). San José, Costa Rica: Universidad de Costa Rica.

Chapter 7
Concept Mapping as a Means to Develop and Assess Conceptual Understanding in Secondary Mathematics Teacher Education

Jean Schmittau

A case study of the concept maps of two pre-service teachers illustrates the potential of concept mapping to the teacher educator. The maps reveal much about whether future secondary teachers grasp the nature of mathematics as a conceptual system, understand the conceptual content of mathematical procedures, and possess the requisite pedagogical content knowledge to mediate such understandings to future learners. The map of one of the two teachers reveals that she possesses these understandings. The map of the other shows a formalistic understanding of mathematics. Concept mapping also functions as an epistemological heuristic for pre- and in-service teachers.

Case Study

The relatively small number of mathematics presentations at the First and Second International Conferences on Concept Mapping held in 2004 and 2006, in Pamplona, Spain and San Jose, Costa Rica, respectively, suggests that across much of the world the use of concept mapping in mathematics lags far behind its applications in the sciences. This is clearly the case in the US where, despite the emphasis on conceptual understanding that characterizes the reform standards of the National Council of Teachers of Mathematics (NCTM, 1989, 2000), concept mapping in mathematics continues to be under utilized. This is unfortunate, since it has the potential to begin to counteract the superficial treatment of concepts occasioned by the failure to develop a coherent curriculum that identifies essential concepts and probes them in sufficient depth (Schmidt, Houang, & Cogan, 2002). Indeed one of the outcomes of the TIMSS study was the characterization of the US curriculum as "a mile wide and an inch deep". The standards developed by the fifty states continue to reflect the prevailing national tendency toward a multiplicity of topics and consequent superficial coverage of mathematics concepts.

J. Schmittau (✉)
State University of New York at Binghamton, Binghamton, New York, USA
e-mail: jschmitt@binghamton.edu

K. Afamasaga-Fuata'i (ed.), *Concept Mapping in Mathematics*,
DOI 10.1007/978-0-387-89194-1_7, © Springer Science+Business Media, LLC 2009

137

In October 2006, citing the excessive number of topics required in the standards of the various states (as high as 80 in one state's fourth grade curriculum), NCTM promulgated a set of "focal points" which target a small number of both concepts and procedures to be taught and mastered at each grade level. The inclusion of procedures in the focal points is significant, since although they were not excluded in the NCTM Standards, procedures have clearly been de-emphasized in much of the published literature. In addition, reform mathematics curricula in the US have all but abandoned the teaching of algorithms in recent years, preferring to consign computation to calculators instead (Morrow, 1998). This practice has not only been widely criticized (Wu, 1999; Schmittau, 2004), but has, in fact, been a major factor in fueling the US "math wars".

At the root of this false dichotomy between mathematical procedures and mathematics concepts is the notion that algorithms can only be taught mechanically, as they so often were in the past, and are therefore, incompatible with the teaching of concepts. So accepted has the presumed dichotomy between concepts and procedures become, that the nation's leading newspapers reported that the NCTM focal points represented a move "back to basics", a reference to the rote learning of mechanical procedures that had often characterized mathematics learning prior to the reforms heralded by the NCTM Standards (1989). These media reports prompted the NCTM president to write letters to the editors of two of the most prominent US newspapers, the New York Times and the Washington Post, in an attempt to correct this misperception and reaffirm NCTM's commitment to conceptual understanding as well as procedural competence in mathematics. Indeed, rather than a mechanical "back to basics" approach to the teaching of mathematics procedures, the focal points appropriately require that students understand the meaning behind the mathematical algorithms in which they are to demonstrate fluency. It is important, therefore, that algorithms be understood and taught not as divorced from concepts, but as historical and epistemological analysis reveals them to be – namely, fully conceptual cultural historical products. Concept mapping can serve as a useful tool to enable the linking of algorithms with their conceptual content.

It is also imperative that prospective teachers master the pedagogical content knowledge that will enable them to teach algorithms conceptually, and concept mapping can be a valuable means of both promoting and assessing this important understanding in teacher education. The case study that is the focus of this chapter illustrates the manner in which concept mapping can function as an assessment tool for the teacher educator.

The background/classroom setting for the case study is as follows. The construction of a concept map of multiplication was assigned to pre-service teachers approximately six weeks into a graduate mathematics education course. The students were enrolled in a Master of Arts in Teaching program leading to certification to teach secondary mathematics in grades 7 through 12. Questions concerning the meaning of multiplication in the various numerical domains in which it functions were threaded throughout the next few class sessions and became the focus of intense thinking and discussion by the students. The Vygotskian tenet – that in order to understand a concept it must be viewed in its full developmental history – marked the discussions, mandating student engagement in both conceptual and historical analysis. In this chapter, the concept maps of two of the graduate student pre-service teachers are presented by way of illustrating the contrast between a formalistic and a truly conceptual understanding of mathematical content that the concept map has the power to reveal.

The case study presented here provides an example of the manner in which a concept map can alert the teacher educator to whether or not students are understanding mathematics as a conceptual system (in which procedures are fully integrated), or grasping it at the level of mere formalism (Schmittau, 2003). Both students drew maps of their understanding of the concept of multiplication, subsequent to class discussion of the concept as noted above. Pseudonyms are used for student names, and since the maps were quite large, space considerations limit presentation to partial sections of each.

In Fig. 7.1A, a section of Janet's map reveals her understanding of what multiplication is, viz., "a change in units in order to take an indirect measure" (cf. Davydov, 1992). Stephen, however, sees multiplication as an "operation which is composed of" an "operator" and "operands" (Fig. 7.1B). Janet's definition can be meaningfully taught to children in the early elementary school years (Davydov, 1992), while Stephen's would make very little sense to students prior to college mathematics.

Janet's definition reflects the development of multiplication from the need to take a measurement or count of a quantity of objects or units sufficiently numerous to render a direct count tedious and subject to error. In such a circumstance, it is advantageous to construct a larger unit (or multiple) of the unit of interest, and use it to take an indirect count. This is done in the case of the area of a rectangle, for example, where the number of unit squares in a row are counted, and then the number of rows are counted in order to indirectly obtain the area as the number of square units.

Such an understanding of multiplication underlies the algorithm for obtaining the product of multi-digit numbers. It does, however, require that the products of single digit numbers (multiplication "facts") be committed to memory. It is not sufficient that the calculator can call up these "facts"; they must be stored in the human memory if they are to be recognized in subsequent mathematical studies in the myriad of conceptual interrelationships into which they enter. Janet refers to these in her concept map, indicating that their commitment to memory is important for mathematical understanding, and should not be by-passed as is often now occurring in the wake of the reform movement. All of these points were emphasized during our extensive class discussions of the concept of multiplication and the various numerical domains in which it is defined. During these we noted the inadequacy of the "repeated addition" definition that is ubiquitous in US textbooks (Schmittau, 2003). Accordingly, neither Janet nor Stephen invoked this notion in their concept maps of multiplication.

In Fig. 7.2, a portion of Janet's map reflects the centrality of the concept of area to her understanding of multiplication.

In the case of rectangular area, the unit changes from a unit square to a row of such squares, which can then be counted to obtain the area in square units. Janet

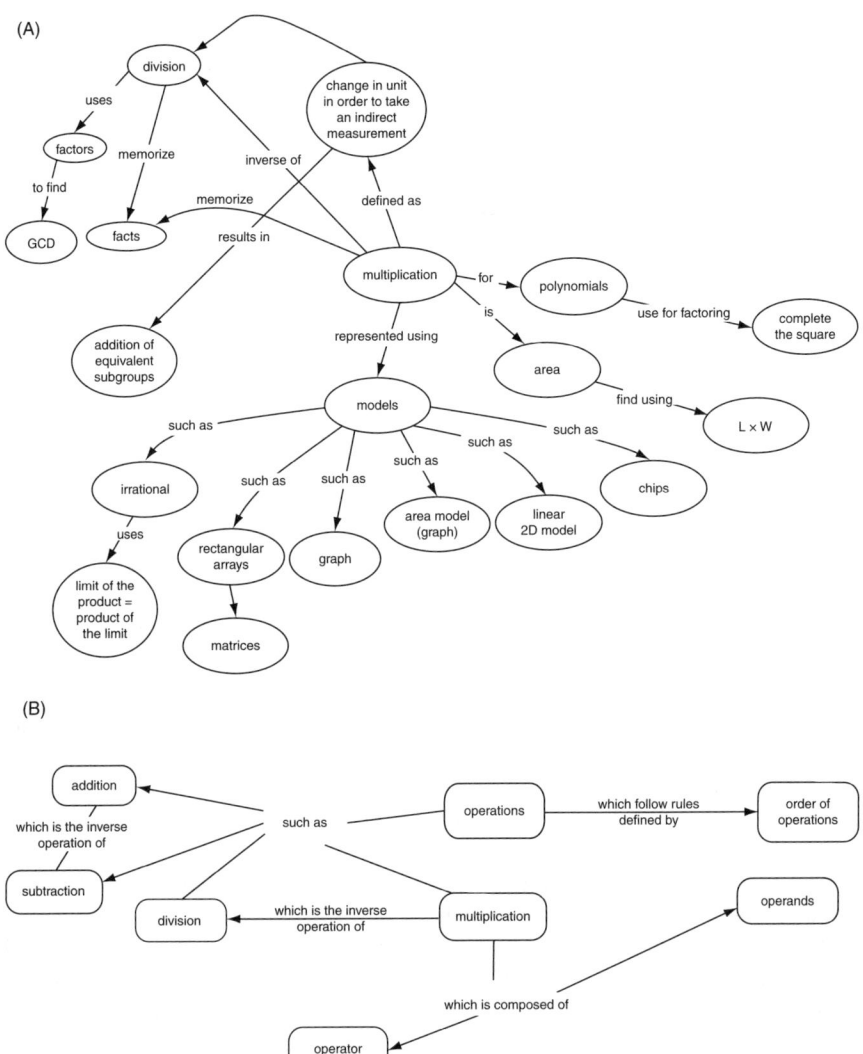

Fig. 7.1 Janet's (**A**) and Stephen's (**B**) maps of the concept of multiplication

also presents area models for the conversion of a sum of terms (polynomial) to
a product (factoring) following the methods of Al-Khowarizmi (Karpinski, 1915).
Such area models are the conceptual foundation for the completion of the square.
The upper model for the solution of the quadratic equation "$x^2 + 10x = 39$" shows
Al-Khowarizmi's method, while the lower model displays the solution using algebra
tiles. In his development of algebra 1,000 years ago, Al-Khowarizmi solved this
equation by first drawing a square (having dimensions x by x), then dividing 10 by
4, and using the result to add a rectangle of dimensions "x" by 2.5 to each side of the

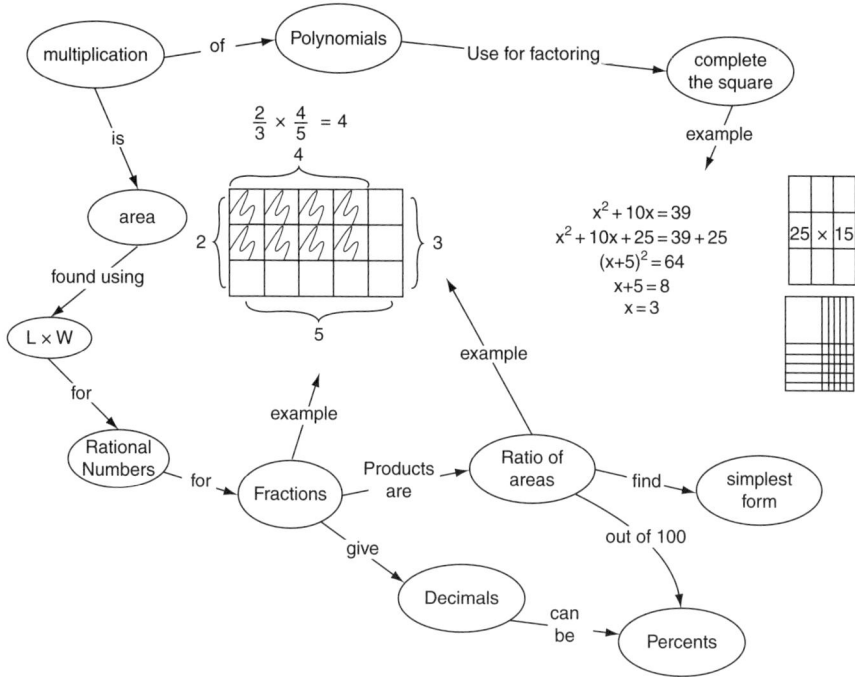

Fig. 7.2 Janet's map showing the historic role of the concept of area in the development of algebra and the factoring of polynomials, and the concept of area underlying the algorithm for the multiplication of fractions

square. There were as yet no algebraic symbols, but since it is cumbersome to use word labels for variables, in my presentation to Janet and Stephen's class, I labeled the length of these four rectangles "x". Al-Khowarizmi then literally completed the square (the new larger square) by adding the four small squares in each corner. Each of these had an area of $2.5 \times 2.5 = 6.25$, and since there were four of them, their total area was 25 square units. Now the original equation (again we are using symbols that Al-Khowarizmi lacked) becomes $x^2 + 10x + 25 = 39 + 25 = 64$. Since the length of a side of the square is $x + 5$ and the area of the square is 64, the square has dimensions 8 by 8, and $x + 5 = 8$. Hence, x is 3. (Al-Khowarizmi's geometric method did not permit negative solutions.)

Janet's lower model shows an algebra tile model for the same problem. Algebra tiles are a modern manipulative designed to geometrically model polynomials. Although the ten "x by 1" rods cannot be broken (to obtain rods having a dimension of 2.5 as in Al-Khowarizmi's solution), they can be arranged to produce a new larger square by placing five of them along the side and five along the bottom of the x by x square and completing the new square with 25 unit squares.

Janet's inclusion of these models in her map is important for several reasons. First, the map reveals area as the central antecedent concept necessary for an understanding of factoring. Many secondary students fail to grasp factoring despite the use of algebra tiles, and teachers typically do not understand why a model that is so transparent to them is not equally so for their students. If teachers realize that area is insufficiently understood by many students (Schmittau, 2003), they will understand why area models sometimes fall short of their anticipated effect. Second, Janet's map reveals her knowledge of the cultural historical development of the solution of the quadratic equation by factoring, as well as its current rendering by the use of a popular manipulative. Her map reveals that she has internalized from the class discussions, the relevant content and pedagogical content knowledge necessary to teach this concept meaningfully.

Some teachers resist the use of algebra tiles, believing that a solution method that is purely "algebraic" (i.e., symbolic), is just as effective. However, the fact that Al-Khowarizmi tested his invented algebraic methods against geometric models and that geometric models were used up to the nineteenth century, should caution against the tendency to omit this important step in concept development. Further, Al-Khowarizmi invented this method one thousand years ago, some 500 years before the creation of algebraic symbols. So immediately engaging students at the level of symbolic expression omits from their ontogenetic experience the equivalent of hundreds of years in the phylogenetic development of this concept. Such an approach violates Vygotsky's insistence on the necessity for tracing the full developmental history of a concept, and can scarcely be considered a recipe for adequate conceptual understanding.

If we explore this portion of Janet's map further (Fig. 7.2), we see an area model for the multiplication of fractions also, in which the algorithm for the product of two fractions is identified by Janet as a "ratio of areas". Janet now understands what virtually none of my graduate students in mathematics understand prior to our class discussions, viz., that the reason why the algorithm for the product of two fractions "works", is because the product of the numerators and the product of the denominators are *areas* whose ratio is the product of the fractions. In the example used by Janet, viz., "2/3 × 4/5", the shaded area of the rectangle represents the product of the numerators "2 × 4", while the area of the larger rectangle is the product of the denominators "3 × 5". Their ratio, 8/15, is four-fifths of two-thirds, the product of the two fractions.

Stephen (Fig. 7.3) states simply that "integer fractions" . . . "have the multiplication formula a/b × c/d = ab/cd which is a ratio of area". Without further elaboration or a model representation, it is difficult to discern from his map alone whether he grasps the precise nature of this ratio. This propositional acknowledgement alone, however, is more than is commonly perceived by the typical pre-service or in-service teacher.

Finally, the current reform movement in mathematics often advocates that teaching go no further than the use of manipulatives, and then allow students to "construct" their own algorithms (Morrow & Kenney, 1998). There is, however, no certainty that such student constructed algorithms will be correct. It is essential, therefore, that the powerful *culturally and historically constructed* algorithms be connected with their conceptual content and these understandings be mediated to learners. Abandoning the teaching of such general methods in favor of dealing only with "concepts" relegates procedural knowledge further to the rote end of the

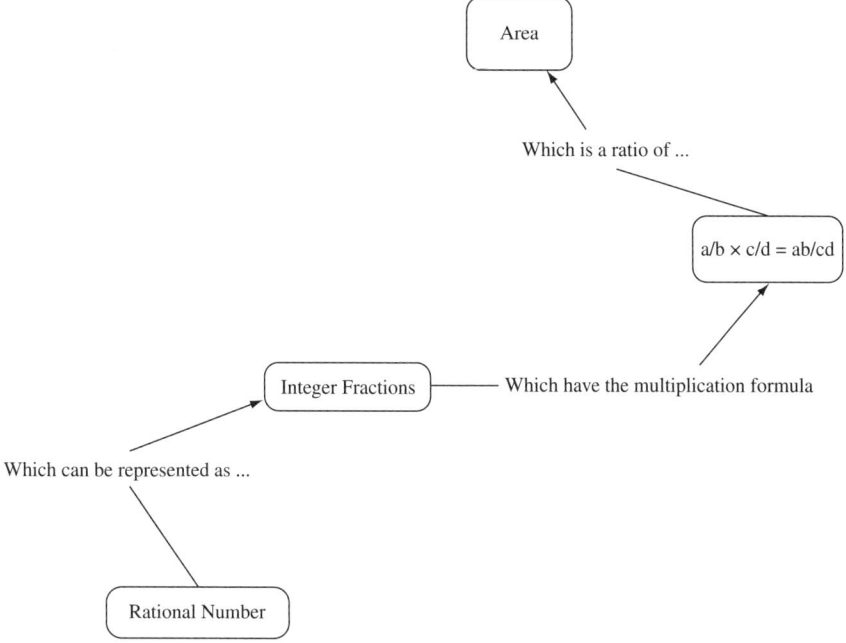

Fig. 7.3 Stephen's map showing a formalistic understanding of fraction multiplication

meaningful-rote learning continuum (Novak & Gowin, 1984), reducing it to little more than a memorized sequence of calculator keys. Janet's map, however, reveals that she is aware of the conceptual and procedural development of multiplication in the historical progression of mathematical knowledge, and furthermore, possesses the pedagogical content knowledge necessary to teach it effectively to students.

Another aspect of multiplication students find incomprehensible concerns the product of two negative numbers. In Fig. 7.4A, Janet's map indicates that multiplication of negative numbers can be modeled using chips to represent positive and negative charges. Such a model makes use of Ausubel's concept of an advance organizer, which is often necessary because of the early grades (frequently 5th or 6th) at which this topic is introduced. Here, the notion of charged particles serves this function.

In the model showing containers of charged particles, Janet first shows +2(−3). Beginning with 3 positive and 3 negative charges in the container on the left (for a net charge of zero), two groups consisting of 3 negative charges in each, are added, resulting in a net charge of −6. From the middle container (starting with 6 positive and 6 negative charges), two groups of 3 positive charges are removed, representing −2(+3) and leaving a net charge of −6. From the container on the right (again starting with 6 positive and 6 negative charges), two groups of 3 negative charges are removed, leaving a net charge of +6. Use of such an advance organizer or other model appropriate for this age group is imperative, because the conceptual integration of the product of two negatives is to be found in the system of complex numbers, which is not studied until high school, long after this topic has been taught.

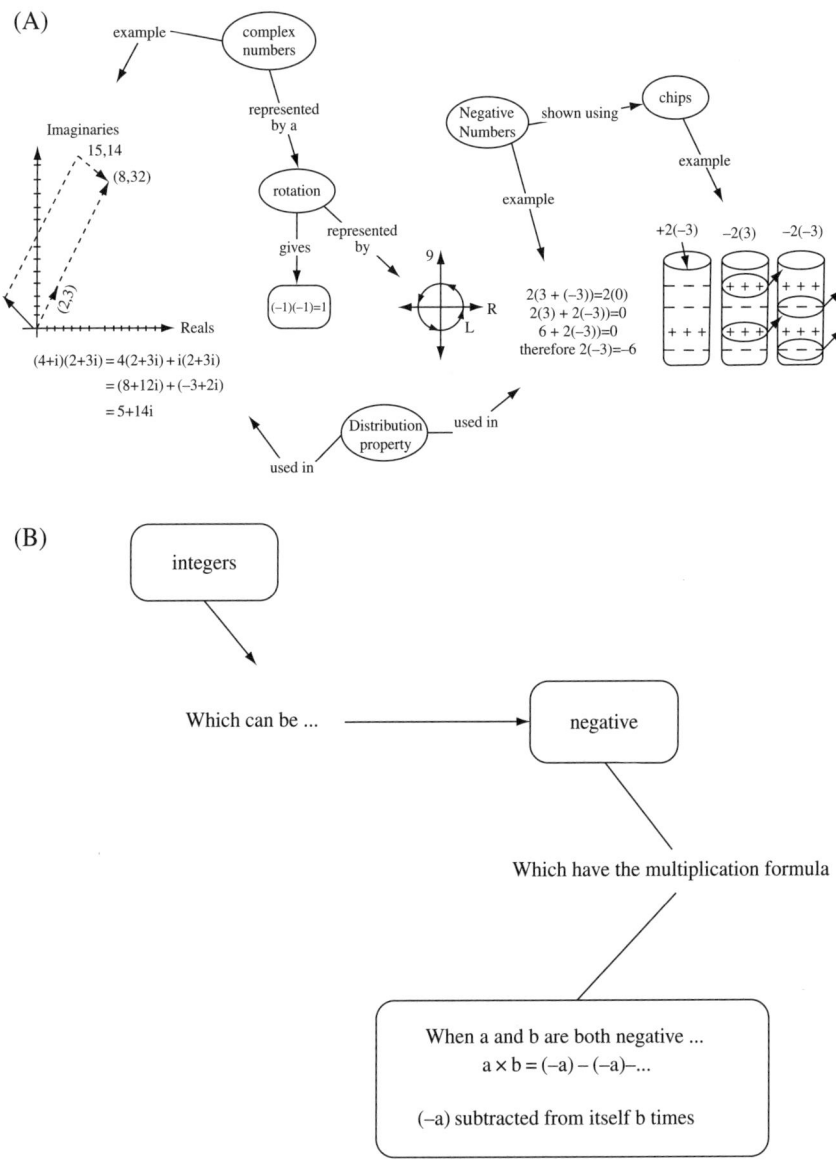

Fig. 7.4 Janet's (**A**) and Stephen's (**B**) differing understandings of the multiplication of two negatives

Accordingly, Janet's map shows a real understanding of the conceptual connections I presented to her class, linking the product of two negatives to the complex number system. Her map displays a linkage from negative numbers to the product "$(-1)(-1) = 1$" which is linked to multiplication of "complex numbers" and "represented by a rotation" shown in the complex plane. This is the actual inception of the concept of multiplication by a negative number, and Janet's representation of $(4 + i)(2 + 3i)$ in the complex plane together with her assertion that the "distributive property" is "used in" obtaining this product, suggests an understanding that multiplication by the scalar quantity "4" repeats the vector $(2,3)$ four times, quadrupling its norm (length) and resulting in the vector $(8,12)$.

Multiplication of $(2,3)$ by "i", however, in a decided break with the meaning of multiplication for real numbers, produces a 90° counterclockwise rotation of this vector, resulting in the vector $(-3,2)$. In class I showed students that the vector $(1,0)$ when multiplied by "i" rotates to the vector $(0,1)$. Multiplying by "i" again results in the vector $(-1,0)$. Hence, the fact that $(-1)(-1) = 1$ is due to the fact that -1 multiplied by itself reflects two further 90° counterclockwise rotations, i.e., the vector $(-1,0)$ is rotated 180° to $(1,0)$. Janet's model shows four 90° counterclockwise rotations to produce the real number "1", which she states is "$(-1)(-1)$". This, together with linkages to the graph of the product of two complex numbers obtained using the "distributive" property, connected to "complex number" that are "represented by a rotation" modeled in the complex plane, suggests that she has internalized what was taught in my graduate course, and has both the relevant content knowledge and pedagogical content knowledge to teach this concept with meaning. A teacher who possesses these understandings may be expected to point out the connection to multiplication of negative numbers when multiplication of complex numbers is taught. None of my graduate students in mathematics have ever made this connection. They learn it for the first time in my graduate course.

Stephen's map deals with this topic very minimally, without models. His map states that for complex numbers "multiplication is defined by ... $(a+bi)(c+di) = (ac - bd) + (cb + ad)i$". This section of his map is unconnected to the section dealing with "negative numbers" (Fig. 7.4B). Here he states that "When a and b are both negative ... $a \times b = (-a) - (-a) - ...$", that is, "$-a$ subtracted from itself b times". If that is the case, then $(-3)(-2) = (-3) - (-3) = 0$, rather than $+6$. Hence, despite the fact that several meaningful ways of modeling this concept were presented during our class discussions, Stephen's map suggests that he views multiplication of two negatives as repeated subtraction, which is an inaccurate conceptualization.

Stephen and Janet were both present in the same classes in which the conceptual content of multiplication that is reflected in Janet's map above was taught and discussed. However, the evidence from their maps is that one internalized the concept in its systemic interconnections, while the other continued to see it through a formalistic lens. Janet's map gives evidence that she possesses the requisite conceptual understanding, and historical and pedagogical content knowledge, to mediate the concept of multiplication meaningfully to students, without separating its so-called "procedural" from its "conceptual" content. Indeed, it appears that she can move rather seamlessly between the two. Stephen's map, in its entirety, has considerable extension, encompassing multiplication of matrices, determinants, and the cross products of vectors. But the connections are consistently formalist and give no evidence that his teaching will go beyond a formalistic approach. Both students are nearing completion of the masters' degree, but their maps reveal very different understandings of this fundamental concept.

Epistemological Value

While Ausubelian theory emphasizes the conceptual connections that are requisite for meaningful learning (Novak & Gowin, 1984), Vygotskian theory points to the need to unfold the historically developed conceptual content from its encapsulation in symbolic expression in order to pedagogically mediate the full restructuring of the concept (Davydov, 1990). In the examples above, the concept maps produced were made subsequent to class discussions and presentations on the conceptual and historical analyses of multiplication with masters' level students preparing to be high school teachers. I typically require that doctoral students conduct such analyses on their own and frame pedagogical recommendations for the improvement of instruction based upon their findings. James Vagliardo's analysis of the concept of logarithm is illustrative of the role of concept mapping in this process (cf. Chapter 9).

In addition, concept mapping may be used to reveal to pre-service and in-service teachers why it is occasionally imperative that they teach topics not contained in their textbooks or the state mathematics curriculum. In mapping mathematical concepts taught in middle school and searching for their conceptual roots, it becomes clear that the concept of positional system, for example, is a central antecedent concept with the power to render more effectively the meaning of such concepts as decimals, fractions, and polynomials (cf. Chapter 3). But the concept of positional system cannot be attained by studying only base ten (Vygotsky, 1986); for adequate conceptualization of such a superordinate concept the study of multiple bases is required. The folly of superficially covering many topics is simultaneously revealed; only by establishing a conceptual base of concepts central to the future development of mathematics, can students begin to grasp the nature of mathematics as a conceptual system. Yet two decades into the reform movement, US curricula continue to cover too many topics each year, to repeat the same topics year after year, and with little increase in depth (Schmidt, Houang, & Cogan, 2002). The NCTM focal points if properly implemented may serve as a much needed corrective to this situation. It would seem that their implementation and the reform movement in mathematics as well could benefit from the pedagogical potential of concept mapping.

Acknowledgements My thanks to James J. Vagliardo for his expert assistance in digitizing the concept mapping sections, and to the two pre-service teachers who graciously provided the concept maps discussed in this chapter.

References

Davydov, V. V. (1990). *Types of generalization in instruction: Logical and psychological problems in the structuring of school curricula*. Reston, VA: National Council of Teachers of Mathematics.

Davydov, V. V. (1992). The psychological analysis of multiplication procedures. *Focus on Learning Problems in Mathematics, 14*(1), 3–67.

Karpinski, L. C. (1915). *Robert of Chester's Latin translation of the algebra of Al-Khowarizmi*. New York: Macmillan.

Morrow, L. J. (1998). Whither algorithms? Mathematics educators express their views. In L. J. Morrow & M. J. Kenney (Eds.), *The teaching and learning of algorithms in school mathematics* (pp. 1–6). Reston, VA: National Council of Teachers of Mathematics.

Morrow, L. J., & Kenney, M. J. (Eds.) (1998). *The teaching and learning of algorithms in school mathematics*. Reston, VA: National Council of Teachers of Mathematics.

National Council of Teachers of Mathematics. (1989). *Curriculum and evaluation standards for school mathematics*. Reston, VA: Author.

National Council of Teachers of Mathematics. (2000). *Principles and standards for school mathematics*. Reston, VA: Author.

Novak, J. D., & Gowin, D. B. (1984). *Learning how to learn*. New York: Cambridge University Press.

Schmidt, W., Houang, R., & Cogan, L. (2002). A coherent curriculum: The case of mathematics. *The American Educator, 26*(2), 10–26.

Schmittau, J. (2003). Cultural-historical theory and mathematics education. In A. Kozulin, B. Gindis, S. Miller, & V. Ageyev (Eds.), *Vygotsky's educational theory in cultural context* (pp. 225–245). New York: Cambridge University Press.

Schmittau, J. (2004). Vygotskian theory and mathematics education: Resolving the conceptual-procedural dichotomy. *European Journal of Psychology of Education XIX*(1), 19–43.

Vygotsky, L. S. (1986). *Thought and language*. Cambridge, MA: MIT Press.

Wu, H. (1999). Basic skills versus conceptual understanding: A bogus dichotomy in mathematics education. *American Educator,* Fall Issue, 1–7.

Chapter 8
Concept Mapping a Teaching Sequence and Lesson Plan for "Derivatives"

Karoline Afamasaga-Fuata'i

The chapter presents a student teacher's work from a study, which investigated secondary preservice teachers' use of concept maps and vee diagrams as pedagogical tools to (i) guide the critical analysis of the content of a mathematics syllabus, and (ii) develop their skills in designing activities that promote working mathematically. Through in-class presentations and critiques of concept maps, student teachers engaged in the processes of reasoning, justifying, verifying, and validating to ensure that visually displayed interconnections effectively reflected their intended meanings. Bobby's concept maps presented here, illustrate the conceptual structure underpinning a teaching sequence, a lesson and an assessment plan as part of a required course assignment, to communicate his perceptions of what it means to developmentally and conceptually teach "Derivatives" in contrast to simply compiling a sequential list of sub-topics. Main insights from the findings suggested that constructing concept maps (a) prompted Bobby to reflect more deeply about his own mathematics knowledge beyond the assignment topic and (b) challenged him to strategically organize his conceptual analysis results into hierarchical displays of concept networks to parsimoniously and meaningfully illustrate the interconnectedness between key and subsidiary concepts as his pedagogical planning progresses from a 2-year curriculum and topic syllabus notes to a teaching sequence, lessons and an assessment plan.

Introduction

Whilst syllabus outcomes and key ideas are useful to guide the planning of teaching sequences, "they are only 'frameworks' – teachers need in-depth knowledge of mathematical concepts and processes so as to enrich them" (Bobis, Mulligan, & Lowrie, 2004, p. 25). Given the prevailing curricular emphasis on encouraging students to think mathematically (New South Wales Board of Studies (NSW BOS), 2002), there is a need to conduct research into innovative ways of supporting student teachers' pedagogical and mathematical thinking and reasoning in deeper and more

K. Afamasaga-Fuata'i (✉)
University of New England, Armidale, Australia
e-mail: kafamasa@une.edu.au

K. Afamasaga-Fuata'i (ed.), *Concept Mapping in Mathematics*,
DOI 10.1007/978-0-387-89194-1_8, © Springer Science+Business Media, LLC 2009

conceptually based ways. Hence, the main study explored ways in which growth in understanding and pedagogical mediation of meaning of mathematical concepts and processes could be supported, by investigating secondary student teachers' use of concept maps and vee diagrams as they (i) critically analyse the content of the junior and senior secondary mathematics syllabus (NSW BOS, 2002), (ii) illustrate and communicate their conceptual understanding of syllabus outcomes, activities and problems, and (iii) develop requisite skills in designing conceptually rich activities to promote working and communicating mathematically.

The study was guided by Ausubel's theory of meaningful learning which proposes that learners' cognitive structures are hierarchically organized with more general, superordinate concepts subsuming less general and more specific concepts (Ausubel, 2000; Novak, 2004). By constructing concept maps and vee diagrams (maps/diagrams), students illustrate publicly their interpretation and understanding of a topic/problem in terms of interconnections between concepts, principles and methods. Concept maps are hierarchical graphs of interconnecting concept nodes with links connecting relevant concepts. Descriptive words on the links describe the meaning of the relationship between the connected concepts. Examples of concept maps are provided later. A vee diagram, on the other hand, is a vee structure situated in a problem with its lefthand side depicting the conceptual information underpinning the methods of solving a problem displayed on the righthand side. Vee diagrams are not presented in this chapter but some examples are found in Chapters 2 and 4.

Recent research (Afamasaga-Fuata'i, 2007, 2006, 2005, 2004a, 2004b) with Samoan undergraduate mathematics students demonstrated the usefulness of maps/diagrams as valuable meta-cognitive tools to scaffold students' thinking and reasoning, to illustrate their developmental and conceptual understanding of mathematics topics, and to enhance efficiency in communicating mathematically as they learnt new mathematics topics and/or solved mathematics problems in their university mathematics courses. Through participation in social critiques over the semester, students received constructive feedback to further improve their individually-constructed maps/diagrams. Their final topic-concept-maps were structurally more complex and differentiated than initial maps as a result of thinking and reflecting about their own understanding, interacting with others and concept mapping. Whilst these studies focused on undergraduate students' use of maps/diagrams as learning tools, the main study that is reported here was with student teachers at an Australian regional university; it focused on the applications of maps/diagrams as pedagogical tools to analyse syllabus documentations and to plan learning activities.

The following sections briefly describe the Australian study's methodology before presenting data from one student teacher's concept mapping work when developing a teaching sequence and lesson plans as part of a required course assignment.

Methodology

The main study was a design experiment in which student teachers critically analysed syllabus outcomes, problems and activities (i.e. *critical analysis*) for underlying concepts and principles (i.e. *conceptual structure*) before illustrating the results on maps/diagrams followed by an examination of (a) the kinds of *discourse* that emerged during *critiques* of presented maps/diagrams; (b) the nature of *student reflections* on how their construction and mapping experiences impacted on the way they planned, thought and viewed the teaching of mathematics topics; (c) the types of participation norms (i.e. *socio-mathematical norms*) established and practiced for the development and critique of maps/diagrams during weekly workshops; and (d) the types of *practical means* by which the researcher "orchestrated relations among these elements" (Cobb, Confrey, diSessa, Lehrer, & Schauble, 2003, p. 9).

The sample included ten internal students enrolled in two secondary mathematics education courses (i.e. junior and senior secondary) who agreed to participate. The lecturer-researcher introduced and used maps/diagrams in her presentations of materials during weekly workshops and student teachers practised constructing maps/diagrams individually and collaboratively followed by presentations and critiques in-class. Required course assessments, in parts, required students to prepare unit plans and lesson plans for various content areas of the *NSW 7–12 Mathematics Syllabus* (NSW BOS, 2002).

The study was in two phases. First, as learners, student teachers practised and constructed maps/diagrams to illustrate and communicate their conceptual and methodological understanding of the mathematics content, of syllabus topics, and that embedded in activities and problems. Second, as student teachers preparing for teaching practicum, they developed lesson plans and activities using maps/diagrams to guide instruction. Required course assignments included some questions on the applications of maps/diagrams to planning teaching sequences, lesson plans and/or learning activities.

Data Collected

Data collected, over the semester, included maps/diagrams constructed and presented in workshops and final maps/diagrams constructed as part of required course assignments, students' reflection journals, and researcher's field notes. This paper presents the case study of a student teacher (Bobby, a pseudonym), who used concept maps to prepare a teaching sequence and lessons on the topic "Derivatives" based on the syllabus notes: *Section 8. The Tangent and the Derivative of a Function* for the *Higher School Certificate (HSC) Mathematics 2/3 Unit – Years 11–12* (NSW BOS, 2002, pp. 50–53). The task of developing a teaching sequence and two consecutive lessons to introduce the formal definition of derivatives constituted part of Assignment 1 for the senior secondary mathematics education course.

The next sections present data from Bobby's concept mapping experiences during the early part of the semester leading up to the completion of the first assignment. Actual quotes from Bobby's reflection journal are italicised and enclosed in quotes.

Data Analysis

Learning to Concept Map

The key characteristics of concept maps, namely the (i) hierarchical organization of key and subsidiary concepts and (ii) inclusion of linking words on connecting lines to form propositions from chains of "node –linking-words → node" triads, were illustrated and demonstrated through a number of pre-prepared concept maps. During group/individual work in weekly workshops, student teachers practiced concept mapping selected topics/problems/activities and concerns were addressed as they emerged during these activities. For Bobby, he identified, selected and ranked key and subsidiary concepts of the selected topic/problem/activity before organising them hierarchically from most general to more specific concepts.

While constructing a hierarchy of nodes, reflecting upon the emerging network of interconnections, selecting linking words, critically evaluating and assessing the map's overall validity in terms of the discipline knowledge, Bobby inevitably realized that he was thinking more deeply and intensively about possible variations of underlying conceptual structures and cognitively deliberating between alternatives. Not surprisingly, Bobby called this preparatory stage *"the verb-type"* by which he meant *"the act of doing the map"* and is *"represented by a pseudo-algorithm to draw the concept map – such as choosing key concepts and possible links"*. Explaining his experiences, he said: *"There are in fact 2 'knowledge constructs' gained from doing concept maps. Firstly the 'verb-type', by which I mean the act of doing the map, even if it ends up in the bin at the end. And secondly, the 'noun-type' by which I mean the end product, the actual map of the conceptual structure"*.

It seems that whilst learning to concept map, Bobby realized for himself that *"there are actually 2 types of maps . . . there is the pre-existing one that is embedded in the mapper's brain, and then there is the map that actually best describes the (unit/problem/activity) for the mapper"*. Basically conjecturing that these *"2 maps and their differences could be described by Vygotsky's zone of proximal development"*, he explained that, *"I was confused as to whether I was mapping my 'prior knowledge construct map' or the 'map of best description' and so I struggled with the concept maps"*. These distinctions (or confusions) between the likely nature and focus of maps were perceived and defined by Bobby as *"dimensions"* of a concept map; see Table 8.1 for his schematic representations of dimensions.

Elaborating further, Bobby proposed yet another dimension namely the *"focus"* of the map. He wrote: *"By this term I mean the 'qualitative nature' of the map – Is it concrete in nature so that it's usefulness lies in teaching a student to solve a particular mathematics problem; or is it descriptive in nature providing an abstract summary of a topic?"* However, sharing and discussing these reflections later on in-class clarified further for Bobby the need to explicate the intended purpose and specific focus of a map, often a common point of confusion when learning to concept map for the first time. That is, explicating the purpose and focus of map first ensures the appropriate selection of concepts, hierarchical organization and suitable linking of relevant nodes to enhance the map's overall cohesiveness and

Table 8.1 Student's perceptions of "dimensions" of a concept map

Knowledge construction	Prior-knowledge construct map[a]	Best description map[b]
Verb-type	(1) Represented by *pseudo-algorithm* to draw concept map – such as choosing key concepts and possible links.	(2) Represented by a *plan* to re-arrange the prior-knowledge construct map to best solve current problem.
Noun-type	(2) *Final copy of concept map* that accurately represents "what is in mapper's head".	(3) Final copy of concept map, which may represent a solution to a mathematics problem or a teacher's unit/lesson plan.

[a] Already existing and may be primitive or erudite but exists and must be discovered.
[b] Varies depending on the nature of the problem; i.e. is it a mathematics problem to be solved; or a content summary of a topic of study by a teacher?

meaningfulness. For example, a concept map of a mathematics problem (*Type 1*) illustrates the conceptual structure embodied by the problem and underpinning its solution whilst a topic concept map (*Type 2*) illustrates the conceptual and epistemological structure of the key ideas (i.e. mathematics concepts and principles) relevant to the topic. As a consequence of such qualitative distinctions, Type 1 map would be more contextualised and situated in contrast to the more general overview and abstract Type 2 ones.

It was becoming apparent from his reflections and schematic representations in Table 8.1 that, for Bobby, preparing and constructing a concept map demanded much deliberation and decision-making, cognitive and analytical processing beyond the mere recall of formal definitions and general formulas. As a consequence, by the time the first assignment was due, Bobby had become increasingly more proficient in selecting key and subsidiary concepts with strengthened skills in hierarchically organizing concepts into cohesive groups and more confident in constructing viable networks of propositional links to communicate his understanding of the task's conceptual structure. Explaining this growth in understanding, Bobby wrote: "*I realized that these 2 types of maps ['prior knowledge construct map' and 'map of best description'], need to be well-defined before mapping begins*".

Through his 2-dimensional schema in Table 8.1, Bobby posed two viable pathways for the construction of a "best description map". Firstly, by progressing vertically down the "prior-knowledge construct map" (Column 2 of Table 8.1) from (1) a pseudo-algorithm through (2) a final copy of what is in the mapper's head and then horizontally across to (3) a final copy representing a solution to a mathematics problem or teacher's unit/lesson plan. Secondly, by progressing horizontally along the "verb-type knowledge construction" (Row 1 of Table 8.1) from (1) a pseudo-algorithm on to (2) a plan to re-arrange the prior-knowledge construct map and down to (3) a final copy of concept map to represent a solution to a mathematics problem or teacher's unit/lesson plan. The choice of pathways appears dependent on whether the focus is a problem or a unit/lesson. Irrespective of the pathway taken, each seems to represent a progressive or developmental trajectory from an initial preliminary version to a finalized "best description map". Presented below are Bobby's final "best description" maps obtained through the first pathway for the purpose of illustrating a teaching sequence and lessons as requested in Assignment 1.

Overview Concept Maps

Instead of designing a teaching sequence directly from syllabus notes, Bobby first of all, situated the topic of "Derivatives" amongst those required for Years 11–12 in the *Mathematics 2/3 Unit* (corresponding to the HSC Mathematics and Mathematics Extension 1 courses, NSW BOS (2002)) to provide a better overview of topics to be taught prior to introducing "Derivatives". Proceeding by identifying the main ideas from syllabus notes, Bobby went beyond the requirements of the assignment and constructed 14 overview concept maps (only *two* are shown here), which covered a range of Years 11–12 prescribed topics. He commented that: "*For me it seems that I must firstly define the entire space of the (unit) before attempting to define the (unit) itself*".

Shown in Figs. 8.1 and 8.2 are Bobby's first two overview concept maps illustrating some of his organisational hierarchies to depict differentiating levels of generality (i.e. Level #) from the most general concepts to progressively more specific concepts towards the bottom of map.

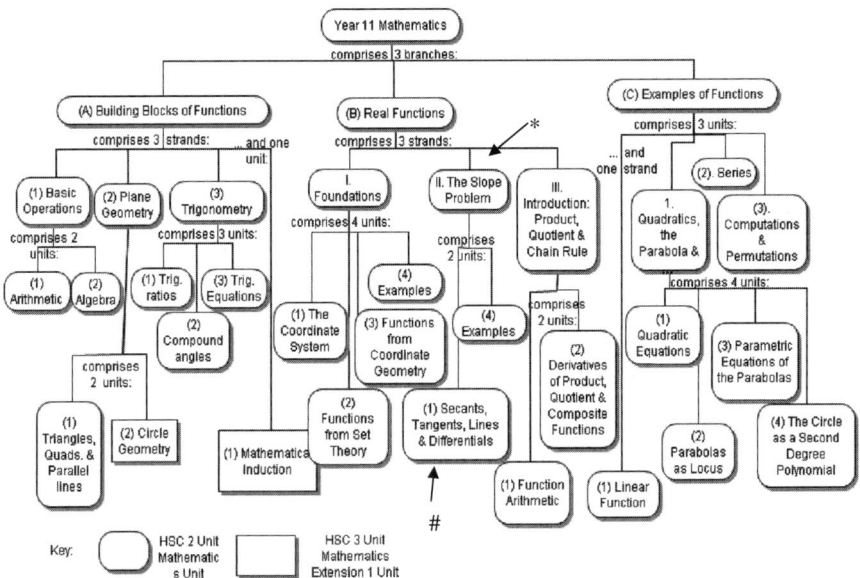

Fig. 8.1 Year 11 overview concept map

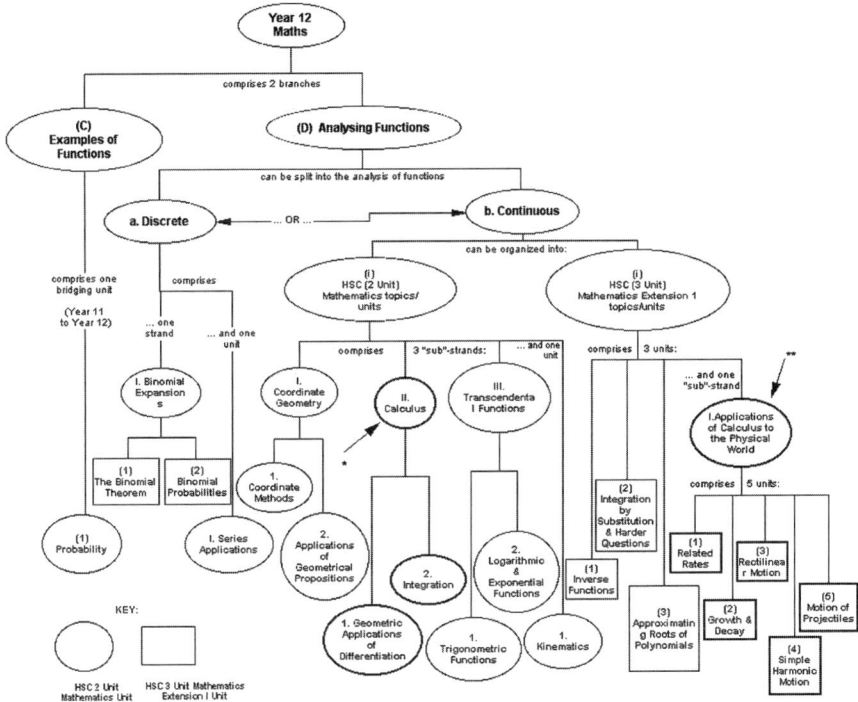

Fig. 8.2 Year 12 overview concept map

For example, Fig. 8.1 is an overview of Year 11 Mathematics (at Level 1) that is subsumed under 3 main concepts namely "(A) Building Blocks of Functions", "(B) Real Functions", and "(C) Examples of Functions" at Level 2, with the order A, B, and C indicating a preferred teaching sequence. Relevant to the topic "Derivatives" is the middle branch subsumed under the Level 2 node: "(B) Real Functions" with a triple-branching link connecting to 3 less general concepts (at Level 3) namely "I. Foundations", "II. The Slope Problem" and "III. Introduction: Product, Quotient and Chain Rule". Again, the ordering I, II, and III suggests that (I) is the required prior knowledge to the topic "Derivatives" embodied by the middle "II. The Slope Problem" sub-branch (marked *). Similarly, the adjacent "(A) Building Blocks of Functions" branch on the left and the adjacent "(C) Examples of Functions" branch to the right, could be likewise read from top to bottom.

In comparison to Figs. 8.1 and 8.2 on Year 12 Mathematics (at Level 1) shows the relevant information in relation to the topic: "Derivatives" such as nodes subsumed under the Level 4 nodes: "HSC (2 Unit) Mathematics topics/units" and "HSC (3 Unit) Mathematics Extension I topics/units" namely "II. Calculus" (marked *) and "I. Applications of Calculus to the Physical World" (marked **). Reading from top-to-bottom, the relevant proposition P1 is: "HSC (2 Unit) Mathematics topics/units *comprises 3 sub-strands*I. Coordinate Geometry, II. Calculus and III. Transcendental Functions *and one unit* 1. Kinematics".

From the "II. Calculus" node is a progressive differentiation double-link to connect to the two terminal nodes "1. Geometric Applications of Differentiation" and "2. Integration" but with no linking words. Situated within the other calculus-related sub-branch (marked **) subsumed under the node: "HSC (3 Unit) Mathematics Extension I topics/units" is the proposition (P2): "1. Applications of Calculus to the Physical World *comprises 5 units:* (1) Related Rates, (2) Growth and Decay, (3) Rectilinear Motion, (4) Simple Harmonic Motion, and (5) Motion of Projectiles". In fact, Fig. 8.2 clearly depicts the nested structure of HSC 2 Unit Mathematics topics within HSC 3 Unit Mathematics and showing that HSC 3 Unit extends topics initially encountered in HSC 2 Unit Mathematics. This inter-relationship is schematically shown by the left-to-right order of the Level 6 concept hierarchies in Fig. 8.2.

Collectively reading from the two maps, Fig. 8.1's middle branch, from left-to-right illustrates the syllabus' expectation and Bobby's plan that the topic "Derivatives" would be introduced via "II. The Slope Problem" (marked *) through secants, tangents, limits and differentials (marked *).

In comparison, Fig. 8.2 provides a more general overview of this sequencing of topics but situated within Year 12 HSC 2 Unit Mathematics (i.e. "1. Coordinate Geometry" to be covered prior to "II. Calculus") and including clear distinctions of topics covered as applications of calculus to the physical world within HSC (3 Unit) Mathematics (marked **). Following on from this general overview of Years 11–12 Mathematics courses, Bobby developed a detailed concept map to illustrate a more developmental approach to "Derivatives" which explicitly builds upon students' prior knowledge of gradients of linear graphs by elaborating further the meaning of the terminal node: "Secants, Limits, Tangents and Derivatives" of Fig. 8.1 (marked #). This process is briefly described next.

Teaching Sequence Concept Map

Bobby's critical and conceptual analysis of *Section 8: The Tangent and the Derivative of a Function* (NSW BOS, 2002, pp. 50–53) yielded 19 main groups of sub-topics of which 5 was identified to be the most relevant for introducing derivatives; see Fig. 8.3 for the 8.5 syllabus referenced sub-topics.

Section 8.3 Gradient of a secant to the curve $y = f(x)$.
Section 8.4a Tangent as the limiting position of a secant.
Section 8.4b The gradient of the tangent.
Section 8.5a Formal definition of the gradient of $y = f(x)$ at the point where $x = c$.
Section 8.6a The gradient or derivative as a function.

Fig. 8.3 Sub-topics relevant to "derivatives" (NSW BOS, 2002)

These sub-topics (Fig. 8.3) eventually formed the basis of Bobby's topic concept map for the introduction of derivatives shown in Fig. 8.4. Selecting the node: "Secants, Limits, Tangents and Derivatives" (from Fig. 8.1, marked [#]) as the titular node at Level 1 of Fig. 8.4, the next hierarchical level shows progressive differentiating triple-links to three main concepts: "(1) The 2-Point Method", "(2) The Limiting Process" and "(3) Derivative Functions" at Level 2. Furthermore, the resulting 3 branches and concept hierarchies appear organized around the three types of knowledge namely (i) prior knowledge, (ii) new knowledge (i.e. derivatives) and (iii) extensions, reflective of the philosophy of preparing learning activities promoted by the mathematics education unit Bobby was enrolled in.

Specifically, the leftmost branch indicates the prior knowledge ("(1) The 2-Point Method" branch) described in the syllabus students require before being introduced to the derivative concept. Emanating from the "(1) The 2-Point Method" node is

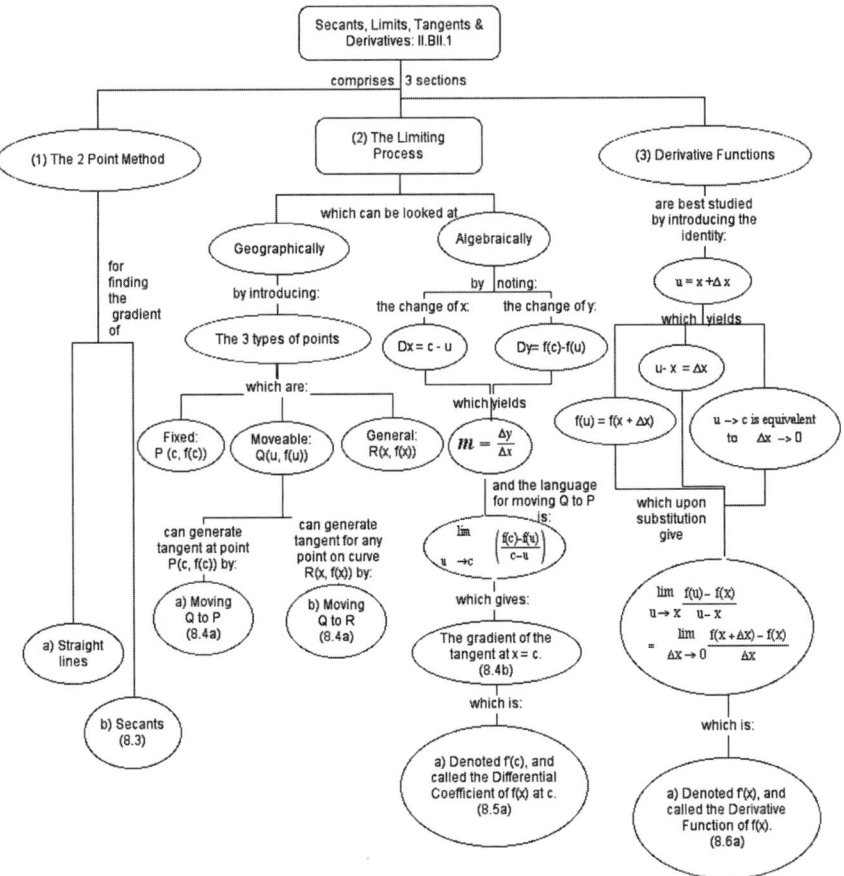

Fig. 8.4 Topic "formal definition of derivatives" concept map

a split-link that generates propositions: (P3): "(1) The 2-Point Method *for finding the gradient of:* (a) straight line*s*" and (P4): "The 2-Point Method*for finding the gradient of:* (b) Secants (8.3)" where 8.3 is a reference to syllabus notes, Section 8.3 (NSW BOS, 2002, p. 50) and the first of the 5 sub-topics listed in Fig. 8.3.

From the middle Level 2 node: "(2) The Limiting Process" are two progressive differentiating split-links to Level 3 nodes: "Geographically" and "Algebraically" which form the extended proposition P5: "(2) The Limiting Process *which can be looked at*: Geographically*by introducing*the 3 types of points *which are* Fixed: P(c, f(c)), Moveable: Q(u, f(u)), and General R(x, f(x))". Emanating from the "Moveable: Q(u, f(u))" node is a split-link that forms propositions P6: "Moveable: Q(u, f(u)) *can generate tangent at point P(c, f(c)) by:* (a) Moving Q to P (8.4a)" and P7: "Moveable: Q(u, f(u)) *can generate tangent for any point on curve R(x, f(x)) by:* (b) Moving Q to R (8.4a)".

On the other hand at the Level 3 node: "Algebraically" of the middle branch, are two differentiating links which formulate an extended proposition P8: "The Limiting Process *which can be looked at* Algebraically *by noting the change of* x: Δx = c–u and *by noting the change of* y: Δy = f(c)–f(u)". The subsequent merging of cross-links (i.e. integrative reconciliation) from the two Level 4 nodes " Δx = c–u" and " Δy = f(c)–f(u)" formulates proposition P9: " Δx = c–u, Δ y = f(c)–f(u) *which yields* m = $\frac{\Delta y}{\Delta x}$" with a single link to the Level 6 node to form the extended proposition P10: "m = Δy/Δx" *and the language for moving Q to P is*: $\lim\limits_{u \to c} \left(\frac{f(c)-f(u)}{c-u} \right)$ *which gives:* The gradient of the tangent at x = c. (8.4b) *which is:* (a) Denoted f'(*c*), and called the differential coefficient of f(x) at c. (8.5a) The middle branch evidently focuses on the geometric (or graphical) introduction of a tangent and the algebraic representation of the limiting gradient as a differential coefficient.

In contrast to the middle branch, the rightmost branch depicts the progressive development (or extension) of the concept "differential coefficient f '(c)" (marked *) to the more general concept "Derivative Functions" (Level 2). Specifically, the first proposition (P11) is: "(3) Derivative Functions *are best studied by introducing the identity*: u =x+Δx" followed by the triple-pronged proposition (P12) "*u* =x+Δx *which yields:* f(u) = f(x+Δx), u–x = Δx, u$\to$x $\equiv$ Δx$\to$0" (Level 4). Cross links from the latter nodes merged to form the proposition P13: "f(u) = f(x+Δx), u–x = Δx, u$\to$x $\equiv$ Δx$\to$0 *which upon substitution give* $\lim\limits_{u \to x} \left(\frac{f(u)-f(x)}{u-x} \right)$ = $\lim\limits_{\Delta x \to 0} \left(\frac{f(x+Dx)-f(x)}{\Delta x} \right)$" *which is* "(a) Denoted f'(x), and called the derivative function of f(x). (8.6a)."

Overall, Fig. 8.4 shows a topic concept map with an explicit organization into 3 main branches, which implicitly suggests a teaching sequence from left-to-right. Furthermore, within each concept hierarchy, there is a logical development of ideas implied by reading from the top to the bottom levels and from left-to-right. Similarly, when reading from the terminal node of a (sub-)branch up to the top level of the adjacent concept hierarchy to the right as described above. The advantage of the visual and more informative display of the interconnectedness of key ideas with respect to each sub-topic (i.e. 8.4a, 8.4b and 8.5a) is clearly depicted by comparing each of the three sub-branches subsumed under the "(2) The Limiting Process" middle branch to the linear sequential list in Fig. 8.3. Of additional interest are the explicit connections between concept maps such as the link between the titular node: "Secants, Limits, Tangents and Derivatives" of Fig. 8.4 and the same-named node in Fig. 8.1 (marked [#]). Taken together, Figs. 8.1, 8.2 and 8.4 clearly illustrate a visual trend from the macro view of main topics in a 2-year mathematics curriculum (Figs. 8.1 and 8.2) to a micro-view of key and subsidiary concepts within a sub-topic (Fig. 8.4); that is, there is an apparent increasingly more detailed elaboration of conceptual interconnections most relevant to "Derivatives" when moving from Figs. 8.2 to 8.1 and 8.4.

The next section presents Bobby's pedagogical concept maps for one of his two lessons to introduce derivatives including his assessment plan for the lessons. His concept maps for the administrative details of the lessons (i.e. time-frame, class, resources etc.) and second lesson are not included here.

Lesson Plan Concept Map

To design the required consecutive lessons to introduce the formal definition of derivatives as requested in Assignment 1, Bobby selected the leftmost and middle branches of the topic concept map in Fig. 8.4 as starting points.

By basically organising his teaching ideas around the 9 lesson-components taught in the mathematics education units, namely, (i) short questions, (ii) homework checking, (iii) introduction, (iv) explication, (v) worked example, (vi) activity, (vii) written record, (viii) homework setting and (ix) conclusion, Bobby re-grouped the components by subsuming them under the three main advance organizers; "OPENING", "BODY" and "CLOSING" as provided in Figs. 8.5, 8.6, 8.7 and 8.8, where the left-to-right positioning of the nodes within a concept hierarchy imply the instructional sequence of the lesson.

The "OPENING" part of the lesson-concept-map (Fig. 8.5) progressively differentiates into three components, namely, "Short Questions", "Homework Checking" and "Introduction" (Level 3). The first node: "Short Questions" links to "Real Functions Foundations" (Level 4) followed by progressively differentiating links, detailing the relevant background knowledge (Level 5) for the introduction of the new topic. Each Levels 4 and 5 node of the "Short Questions" sub-branch represents a separate overview concept map (of the 14 Bobby constructed, not shown here), which further details the relevant key and subsidiary concepts within each sub-topic. Included also on the lesson-concept-map (Fig. 8.5) are examples of short questions Bobby plans on using to review the prior knowledge that is essential as a platform to begin the development of the new ideas.

Subsumed under the "Introduction" node is the same titular node ("Secants, limits, tangents and derivatives", marked **), seen before in the teaching-sequence-concept-map (Fig. 8.4), which signifies the focus of the lessons. Overall, the

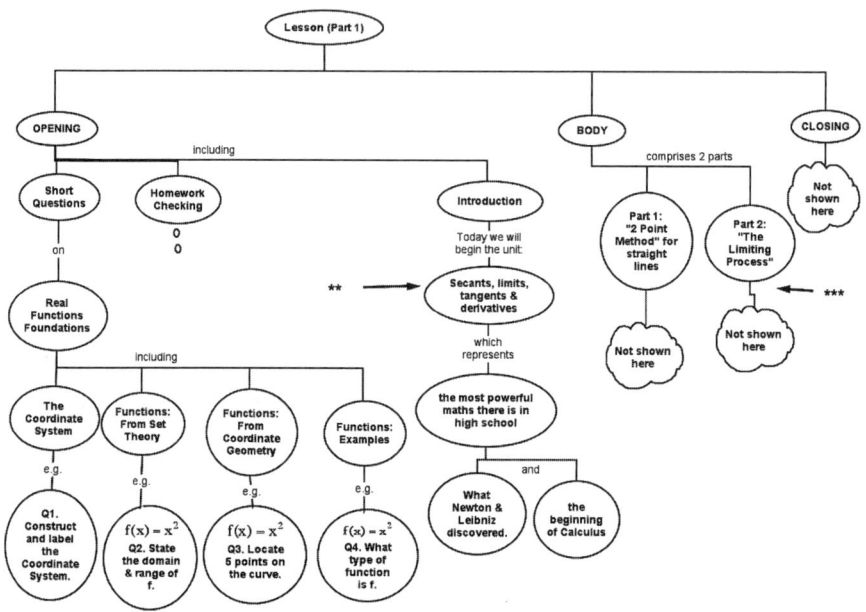

Fig. 8.5 Partial lesson 1 concept map (OPENING)

"OPENING" branch illustrates Bobby's pedagogical intention for the first part of Lesson 1, which begins with short questions followed by homework checking before the introduction. As for the "BODY" branch to the right, the double link indicates "Part 1: 2 Point Method" and "Part 2: The Limiting Process" nodes. Again, the sequence from left-to-right echoes that displayed on the teaching-sequence-concept map (Fig. 8.4), in terms of the intended instructional order. Further details of the two sub-branches are in Figs. 8.6 and 8.7.

Fig. 8.6 details "Part 1: '2-Point Method' for straight lines" of the "BODY" branch with a teaching sequence defined by the left-to-right order of the lesson components at Level 4, namely, "Explication", "Worked Example", "Written Record", and "Activity". Subsequent differentiating links to Level 5 (and beyond) illustrate examples of ideas and questions Bobby planned to utilise to introduce, develop and consolidate students' understanding of the 2-Point Method before progressing to the next stage of the lesson as explicated by the "Part 2: The Limiting Process" sub-branch shown more fully in Fig. 8.7.

A comparison between the 2-Point Method branches of Figs. 8.4 and 8.6 shows that the latter, unlike the more macro view of Fig. 8.4, is more contextualised, as expected, with actual examples at the lesson level. For example, displayed in Fig. 8.6 are examples of points, functions and gradients and a description of Bobby's expectations of students' written record. These represent example questions Bobby planned to utilise to consolidate students' understanding of gradients of straight lines and the transfer of this understanding to a secant of a curve. Overall, this

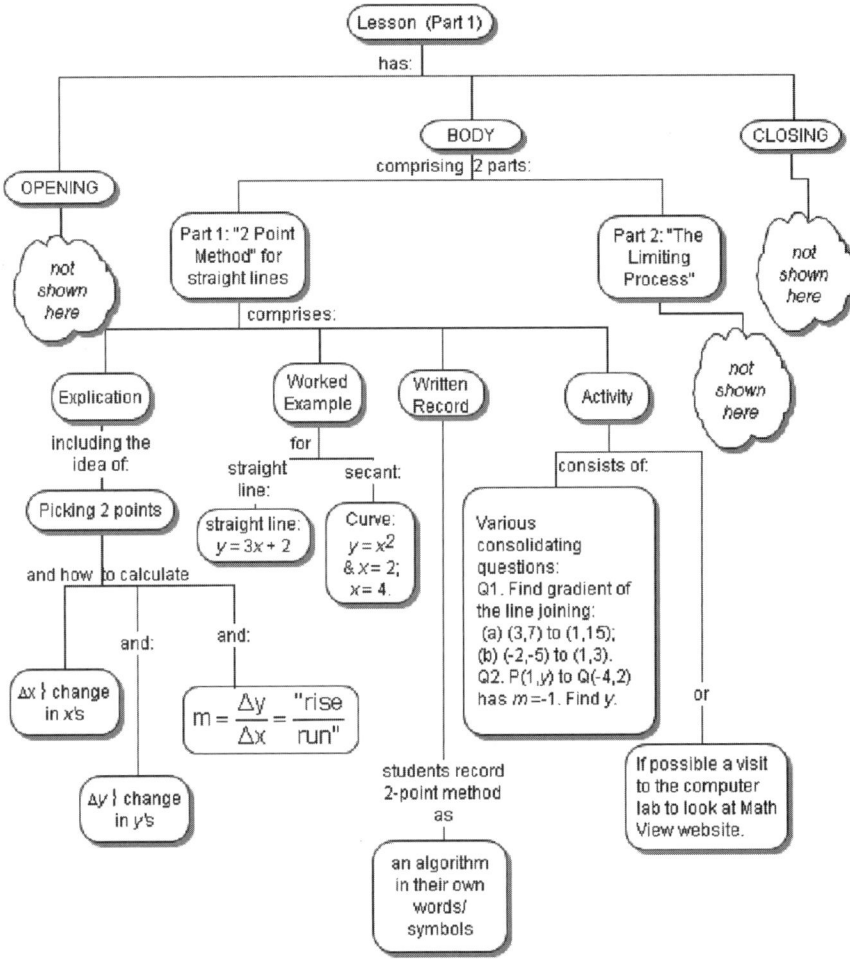

Fig. 8.6 Partial lesson 1 concept map (BODY Part 1)

sub-branch focussed on the development and establishment of students' understanding of the gradients of secants of curves.

Elaborated further in Fig. 8.7 is the "Part 2: The Limiting Process" branch, which in addition to providing the two views (geometric and algebraic) similar to that displayed in the topic-sequence-concept-map (Fig. 8.4), it also includes the same component-sequence and various examples (similar to Fig. 8.6) and details of an investigative activity on finding the numerical limit of the gradient of a secant. Overall, the "geometric" sub-branch expanded the concept of a single secant to include a parade of them to illustrate geometrically the two cases of: Q → P and Q → R.

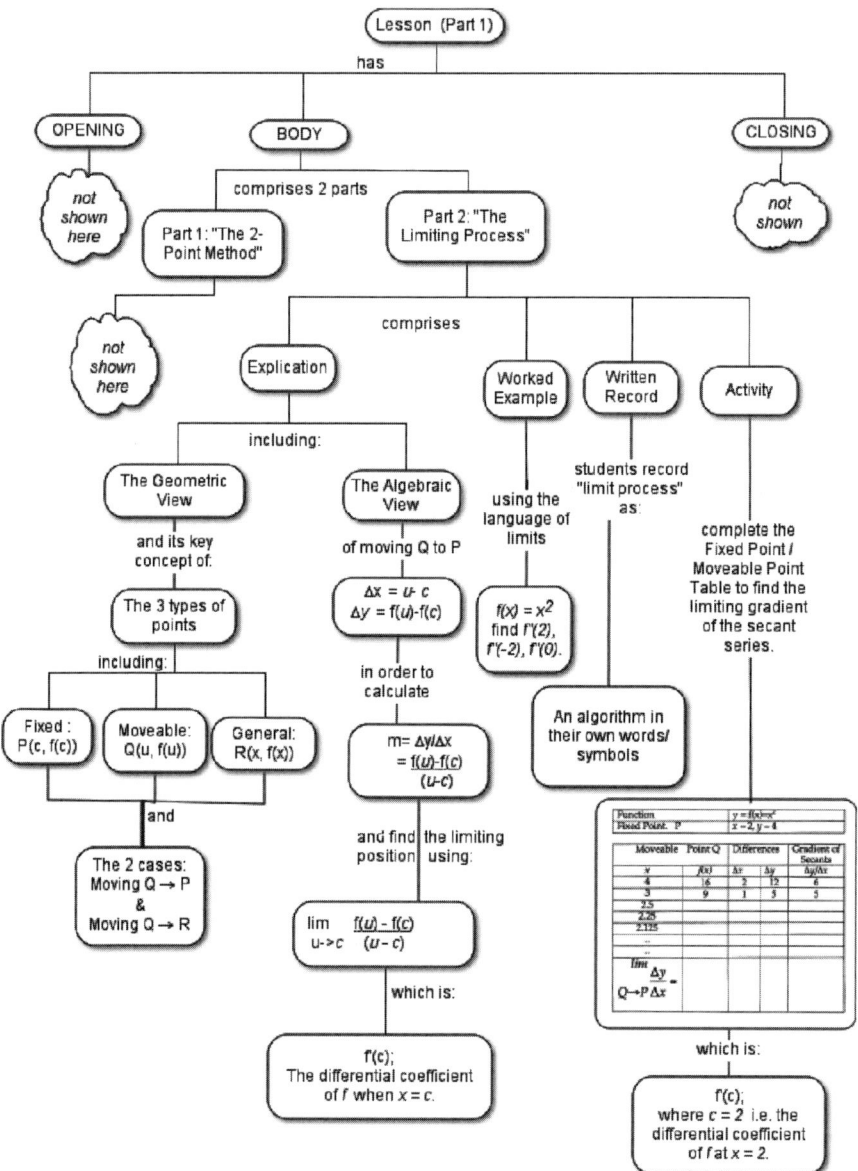

Fig. 8.7 Partial lesson plan 1 concept map (BODY Part 2)

The adjacent "algebraic" sub-branch, in contrast, displays Bobby's intention to develop the concept of the numerical limit of a secant gradient as the secant moves to its limiting position (i.e. as $Q \rightarrow P$) as well as introduce the concept of a differential coefficient at $x = c$. The adjacent lesson-components: "Worked Example" and

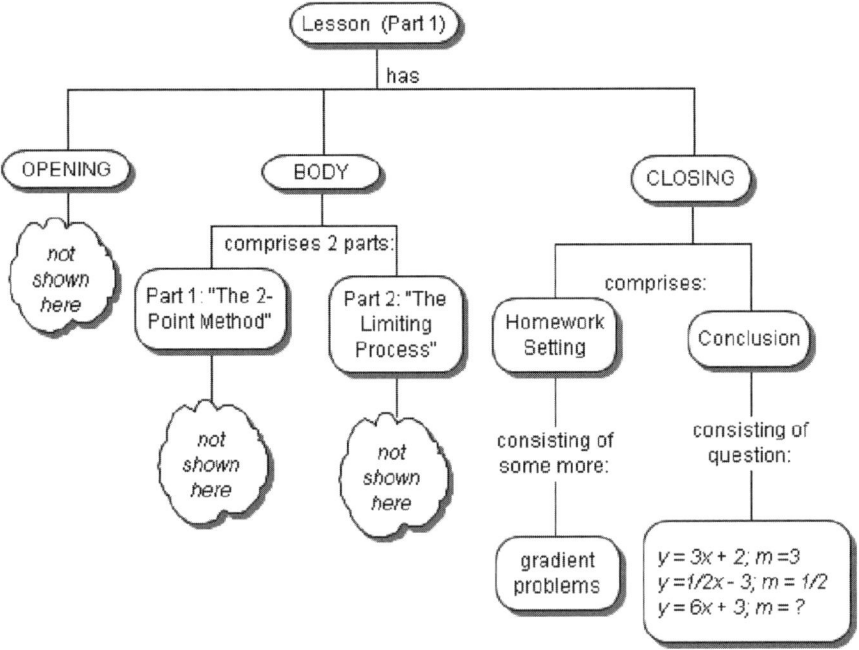

Fig. 8.8 Partial Lesson 1 concept map (CLOSING)

"Written Record" provide the opportunity for students to engage with a worked example and then record their own understanding of the "limit process" before engaging with the investigative activity shown in Fig. 8.7.

The "CLOSING" part of the lesson-concept-map (Fig. 8.8) illustrates Bobby's intentions for the homework and a set of questions to end the lesson. The latter appears to review, examine and/or establish students' understanding of the relationship between a linear function equation and its gradient m; a concept (i.e. m) they had been investigating up to this point of the lesson as the gradient of a line joining two points of a linear function or secant (of a curve) and as a numerical limit of the secant gradient as the secant approaches its limiting position on the curve at $x = c$. Overall, this "CLOSING" sub-branch describes the type of problems intended for homework and subsumed under the "Conclusion" node, is a question set to reinforce students' cumulative understanding of linear functions and the connection between its algebraic form and m.

In summary, the complete concept map for Lesson 1 (Figs. 8.5, 8.6, 8.7 and 8.8) illustrates Bobby's pedagogical intentions to develop students' understanding of the gradients of straight lines (linear function and secant of a curve) and then varying the positions of the second point and observing the numerical limit of the secant gradient as one point approaches the other (e.g. as $Q \rightarrow P$).

Comparing the teaching-sequence-overview-concept-map (Fig. 8.4) and the lesson-concept-map (Figs. 8.5, 8.6, 8.7 and 8.8), it appears that Fig. 8.4 is more of Type 2 (i.e. general overview and abstract) while the lesson-concept-map (Figs. 8.5, 8.6, 8.7 and 8.8) is of Type 1 (more contextualized and situated) following on from Bobby's schema.

Whilst the concept map for Lesson 2 is not shown here, that for Bobby's plan to assess students' understanding of derivatives at the end of the two lessons is in Fig. 8.9.

The map illustrates both the objectives and outcomes and his general plan for assessment. The "Objectives" branch illustrates that students should feel confident about the underlying concepts of the formal definition of derivatives and Bobby's plan to have students construct a concept map to demonstrate their understanding of the key and subsidiary concepts as briefly described in the terminal node of the branch. The "Outcomes" branch to the right, on the other hand, shows four nodes at Level 3, which briefly describe the four key ideas, covered in the two lessons from left-to-right. Whereas the first two nodes display the key ideas developed in Lesson 1 (Figs. 8.5, 8.6, 8.7 and 8.8), the last two nodes represent the focus of Lesson 2 (not shown). Subsequent links to Level 4 connect to the nodes: "2 Point method", "The limiting process" and "The identity $u = x + \Delta x$ and the three corresponding substitutions" before cross-linking to the Level 5 node: "Half-hour test", which indicates the second means of assessing students' understanding of derivatives.

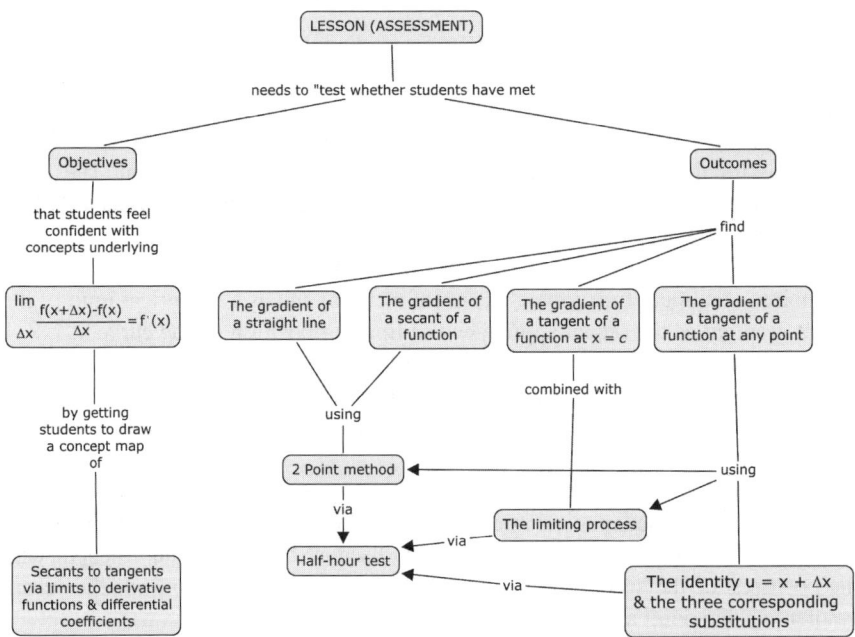

Fig. 8.9 Lesson assessment concept map

Discussion

The discussion of findings are organized around four main points namely (1) concept maps of critical and conceptual analyses, (2) workshop discourse, (3) sociomathematical norms, and (4) practical management of the learning ecology within weekly workshops. Each issue is briefly discussed next.

Concept Maps of Critical Analysis

Bobby's overview concept maps in Figs. 8.1 and 8.2 provided a big picture view of the Years 11 and 12 topics within which the topic "Derivatives" is situated and Fig. 8.9 shows his overview assessment plan for the two consecutive lessons. In contrast, his more situated lesson-concept-map is as illustrated in Figs. 8.5, 8.6, 8.7 and 8.8. Each map represented what Bobby had categorized as "final copy" of a concept map to accurately represent a teacher's unit, lesson or assessment plan.

The visual positioning of concepts within hierarchies and the overall grouping of relevant hierarchies, not only suggested potential teaching sequences (at the macro and micro levels) when the map is read from left-to-right, but it also depicted the level of generality of ideas and/or concepts when read from top-to-bottom. Together, they defined a unique position for a node/hierarchy, roughly paralleling that of a point on the Cartesian plane, whilst simultaneously denoting a relative position amongst a network of nodes/hierarchies highlighting the interrelatedness of ideas.

Findings demonstrated that concept maps provided a parsimonious, visual organization of interconnecting ideas, not only at the macro-level (Figs. 8.1, 8.2 and 8.9), but also at a progressively more in-depth micro-level from a teaching sequence (Fig. 8.4) to a lesson (Figs. 8.5, 8.6, 8.7 and 8.8), which collectively enriched the design of a teaching sequence on "Derivatives".

The cognitive processes of identifying key and subsidiary concepts, hierarchically organising them, constructing and finalising concept maps necessarily required that the student teacher reflected deeply upon his own knowledge of mathematical concepts and processes whilst determining the most viable, visual hierarchical organizations of interconnections he anticipated would promote his future students' conceptual understanding of derivatives. Labelling the preparatory version his "verb-type" or "prior knowledge" map, he proposed that this was a necessary step before finalizing a "noun-type" or "map of best descriptions". His reflective practice when mapping subsequently led him to develop a two-dimensional schema of "verb-type → noun-type" by "prior-knowledge → best-description" to illustrate qualitative differences between types of maps.

Although cognitive demands on the student teacher to critically analyse syllabus documentation, whether or not a concept map is used prior to developing a teaching sequence and lessons would probably be very similar, the significant difference however, is in the extra cognitive and meta-cognitive skills to meaningfully and visually organise ideas into hierarchies of propositional links to display and highlight the "interconnectedness" of concepts across different levels of generality and specificity. Hierarchically organizing concepts evidently challenged Bobby to clarify his thinking as he sought out mathematical principles to provide underlying frameworks that enhanced the cohesiveness and meaningfulness of nested hierarchies (branch). This cognitive exercise appeared to demand reflective, lateral and deeper thinking about mathematics concepts and processes in order to construct visual and schematic representations of meaningful and cohesive knowledge systems (e.g. Figs. 8.1, 8.2, 8.4, 8.5, 8.6, 8.7, 8.8 and 8.9) in contrast to a sequential and linear view of topics from reading notes (e.g. Fig. 8.3).

Workshop Discourse

The kinds of discourse that emerged during critiques of presented maps in workshops involved interactions and exchanges of ideas between the student teacher presenting his/her own map and his/her peers and lecturer-researcher responding and making critical comments usually in the form of requests for clarifications, recommendations for additions/deletions, or confirmations of presented information. Consequently over the semester, Bobby learnt to interact and respond appropriately to critical comments as he argued the correctness of his maps, provided counter-arguments to points raised by his peers, or sought modifications of maps when justifiable. Through these social negotiations, argument and debate, the student teacher demonstrated growing awareness of the importance of adjusting the level of his mathematical language (manifested as concept labels and linking words) to be consistent with the recommended level of the syllabus' staged outcomes. Furthermore, student teachers voluntarily shared their reflections of their experiences simultaneously encouraging others to do the same. Ensuing discussions therefore, focussed on how their mapping experiences impacted on the way they planned, thought and viewed the development of teaching sequences/learning activities. For example, Bobby discussed initial difficulties as he learnt to concept map problems/activities/units such as the difficulty of identifying appropriate and concise labels for main ideas, clarifying the purpose and focus of maps, and determining the most suitable hierarchies. However, through workshop discourse, Bobby's concerns were eventually clarified. Through the discussion of his reflections, he demonstrated an in-depth engagement and reflective practice with the task of concept mapping which, previously, independently prompted him to schematise the mapping process as "dimensions" to qualitatively clarify the purpose and focus of concept maps and to distinguish between general abstract concept maps (Type 2) such as the teaching-sequence one (Fig. 8.4) and the more contextualised one presented here of the lesson plan (Figs. 8.5, 8.6, 8.7 and 8.8).

Socio-Mathematical Norms

The types of participation norms established in workshops included participation in group/class analysis of key and subsidiary ideas in topics/problems/activities;

the transformation of analysis results into concept maps leading to group/class co-construction of exemplar maps; class critiques of individually constructed maps; and discussions of student reflections and mapping experiences. Finally, established socio-mathematical norms influenced, modulated and directed the dynamics of group/class discussions and critiques in weekly workshops. Undoubtedly, these norms impacted the way Bobby planned and developed his final "best description maps" of a teaching sequence, lessons and assessment as presented here.

Practical Management of the Learning Ecology

The types of practical means by which the lecturer-researcher "orchestrated relations among [the different] elements" (Cobb et al., 2003, p. 9) included selecting appropriate tasks (activities/problems/topics) to introduce concept mapping, providing support to students whilst they were learning for the first time, critiquing their work and setting more tasks to challenge their critical abilities and skills not only of concept mapping but including critical analysis of syllabus outcomes. The lecturer-researcher also facilitated group discussions and critiques during map presentations, and coordinated the sharing of students' reflections as materials for discussion of the impact of concept mapping on their own "thinking about learning" and "thinking about teaching". With workshop presentations and reflection sessions focussing on concept maps, ensuing discourse brainstormed multiple ways in which classroom activities could be supported and facilitated through having their future students present and communicate their mathematical understanding via concept maps. Whilst the actual involvement of school students in concept mapping was not part of the main study, using concept maps by student teachers as pedagogical tools was.

Main Insights and Implications

With the acquired expertise and proficiency in constructing concept maps, the student teacher was empowered to use these tools innovatively (i) to critically analyse syllabus outcomes, and (ii) to design a suitable teaching sequence by hierarchically and visually clarifying prior knowledge and future knowledge and using appropriate mathematical language to effectively communicate staged-appropriate mathematics content. Since completed, practice and final maps encapsulated both the conceptual and epistemological frameworks of a topic, through their construction, the student teacher routinely searched for connections between key and subsidiary concepts, and whilst doing so, he made insightful observations about the qualitative distinction between the nature of maps, depending on their purpose and focus, in terms of a two-dimensional schema, to distinguish between maps that are more abstract as in topic concept maps or those that are more concrete as in the lesson-plan-concept-map. For example, he also distinguished between dimensions of a concept map when used as a meta-cognitive tool to collect his thoughts and ideas about the focus of the map (verb-type) and a final concept map described as his "best-description map" (noun-type). A significant advantage of being proficient in concept mapping is the acquisition of critical skills that can be usefully applied to many situations such as demonstrated through his additional effort to situate the assigned topic within the macro picture of the two-year mathematics curriculum. The student teacher's progressive, pedagogical planning from a macro overview of a 2-year mathematics curriculum to syllabus notes, teaching sequence and subsequently an assessment plan onto the micro view of lesson plans was made explicit for public scrutiny and evaluation by concept mapping the key and subsidiary concepts, and where appropriate illustrative examples and activities.

The insights from the case study imply that concept mapping has the potential to explicate student teachers' understanding of the content of the relevant syllabus in more conceptually-based and interconnected ways for further discussion and clarifications and subsequently assessment of their developing pedagogical competency in communicating and mediating meaning of mathematical concepts and processes.

Acknowledgments This research study was made possible by a research grant from the University of New England. My thanks to Bobby for permission to use his concept maps in the case study reported in this chapter.

References

Afamasaga-Fuata'i, K. (2004a). Concept maps and vee diagrams as tools for learning new mathematics topics. In A. J. Canãs, J. D. Novak, & Gonázales (Eds.), *Concept maps: Theory, methodology, technology*. Proceedings of the First International Conference on Concept Mapping September 14–17, 2004 (pp. 13–20). Spain: Dirección de Publicaciones de la Universidad Pública de Navarra.

Afamasaga-Fuata'i, K. (2004b). An undergraduate's understanding of differential equations through concept maps and vee diagrams. In A. J. Canãs, J. D. Novak & Gonázales (Eds.), *Concept maps: Theory, methodology, technology*. Proceedings of the First International Conference on Concept Mapping September 14–17, 2004 (pp. 21–29). Dirección de Publicaciones de la Universidad Pública de Navarra, Spain.

Afamasaga-Fuata'i, K. (2005). Students' conceptual understanding and critical thinking? A case for concept maps and vee diagrams in mathematics problem solving. In M. Coupland, J., Anderson, & T. Spencer (Eds.), *Making Mathematics Vital*. Proceedings of the Twentieth Biennial Conference of the Australian Association of Mathematics Teachers (AAMT) (pp. 43–52). January 17–21, 2005. University of Technology, Sydney, Australia: AAMT.

Afamasaga-Fuata'i, K. (2006). Developing a more conceptual understanding of matrices and systems of linear equations through concept mapping and vee diagrams. *FOCUS on Learning Problems in Mathematics, 28*(3 and 4), 58–89.

Afamasaga-Fuata'i, K. (2007). Communicating students' understanding of undergraduate mathematics using concept maps. In J. Watson & K. Beswick, (Eds.), *Mathematics: Essential Research, Essential Practice*. Proceedings of the 30th Annual Conference of the Mathematics Education Research Group of Australasia (Vol. 1, pp. 73–82). University of Tasmania, Australia: MERGA.

Ausubel, D. P. (2000). *The acquisition and retention of knowledge: A cognitive view*. Dordrecht; Boston: Kluwer Academic Publishers.

Bobis, J., Mulligan, J., & Lowrie, T. (2004). *Mathematics for children. Challenging children to think mathematically.* Australia: Pearson Prentice Hill.

Cobb, P., Confrey, J., diSessa, A., Lehrer, R., & Schauble, L. (2003). Design experiments in educational research. *Educational Researcher, 32*(1), 9–13.

New South Wales Board of Studies (NSW BOS) (2002). Mathematics K-6 Syllabus 2002.

Novak, J. D. (2004). A science education research program that led to the development of the concept mapping tool and new model for education. In A. J. Canãs, J. D. Novak, & Gonázales (Eds.), *Concept maps: Theory, methodology, technology.* Proceedings of the First International Conference on Concept Mapping September 14–17, 2004 (pp. 457–467). Spain: Dirección de Publicaciones de la Universidad Pública de Navarra.

Chapter 9
Curricular Implications of Concept Mapping in Secondary Mathematics Education

James J. Vagliardo

Recognition of deep-seated conceptual crosslinks in mathematics is often weak or nonexistent among students and faculty who view and study mathematics merely in procedural terms. Too often mathematical course content is presented as an approach to a currently considered problem with the mediation of deeper meaning and the connections to other mathematical ideas left unaddressed. The development of mathematical mindfulness requires that educators substantively address the topics they teach by locating the conceptual essence of fundamental ideas from a cultural-historical context. This important pedagogical work can be enhanced through the skilful use of concept mapping. This chapter provides an in-depth look at how concept mapping can be used in the development of a meaningful secondary mathematics' curriculum that avoids rote learning and favors transcendent cognitive development.

The implications presented emerge from several related uses of concept mapping. The chapter illustrates an approach to mathematics' education that first uses concept maps in conjunction with a direct effort to locate the historically grounded conceptual essence of a significant mathematical concept. Without historical context, mathematics' educators may easily be unaware of the conceptual essence of the concepts they teach. Concept mapping is shown to address this shortcoming. Empirical research is then guided by concept mapping in order to expose the "operating understanding" among students and their teachers revealing specific metonymic inadequacies that exist. By comparative use of concept maps, weak or missing crosslinks are readily identified. Together, these uses of concept mapping inform and guide the design of mathematics' lessons that mediate mathematical understanding in a profound way. Concept mapping is thus shown to provide a useful approach to secondary mathematics' education curricular reform aimed at meaningful learning.

J.J. Vagliardo (✉)
State University of New York, Binghamton, Vestal, NY, USA
e-mail: jjvags@gmail.com

K. Afamasaga-Fuata'i (ed.), *Concept Mapping in Mathematics*,
DOI 10.1007/978-0-387-89194-1_9, © Springer Science+Business Media, LLC 2009

Introduction

Mindful use of an important mathematical concept necessitates substantive knowledge, knowledge that extends well beyond the rote acquisition of standard mathematical procedures. Inversely, mindless use usually involves weak or nonexistent conscious awareness of purpose or meaning involved in activity. Langer (1989) suggests that such mindlessness may be rooted in the development of automatic behavior through repetition and practice. Premature cognitive commitment on the part of a learner, a commitment to an early understanding that lacks the full development that can be achieved through thoughtful contemplation and study of the underlying concepts involved from a historical perspective, may also be the cause. Mindlessness can be induced by organizations with an orientation focused on outcomes, with minor attention given to conceptualization and a focused dependency on rote learning.

Substantive knowledge refers to knowledge that reveals the essence of the concept in question. This notion necessarily avoids the misconception that the cultural historical context which gives rise to an idea, especially a mathematical concept, is of little importance in the development of a deep understanding of that idea. On the contrary, substantive knowledge is grounded in such considerations. An intellectual dedication to a continual search for new and deeper understanding, relating new conceptualizations to current knowledge, may be dependent upon the purity of initial substantive knowledge. Subsequent thought and study is then more likely to locate conceptual crosslinks, and may, over time, lead to the emergence of mental models that reflect the essence of the original idea.

Concept Mapping and Historical Research as a Combined Epistemological Tool

Concept mapping is at the heart of the four investigations presented here: an historical search, a conceptual analysis, clinical research involving mathematics teachers and their students, and the development of a curricular approach to logarithms that addresses the historical and cultural foundation of this important mathematical concept. Each investigation was constructively informed and guided by concept mapping. The logarithm is a concept whose understanding must be mediated by knowledgeable and skillful instruction to be well understood. Results of the clinical component of the study provide evidence of the need for conceptual intervention through improved curriculum design based on concept mapping and historical research as a combined epistemological tool. This is done in the context of the philosophical and theoretical ideas of Lev Vygotsky (1978) and the notions of concept formation and generalization of V. V. Davydov (1990) applied to the development of theoretical scientific thought.

Historical references to the sixteenth and early seventeenth centuries provide the original view of the logarithm revealed by John Napier. In order to better understand the context that gave rise to his work, an extensive reading of the general history of mathematics was completed, starting in ancient Egypt and tracing the development of computational methods and mathematical thought from Greece to India and its subsequent spread to Europe via Arab merchants in the middle ages. The specifics of Napier's thinking cast in the complementary histories of mathematics and philosophy provide the scientific and philosophical foundations for understanding what a logarithm is, what gave rise to the idea, and gives a sense of the important cultural implications of the discovery.

A concept map developed to reveal the discovery of John Napier in its essence provides a surprising consequence with far reaching curricular implications for mathematics educators. The central understanding of the logarithm concept illustrated by this map was compared to those generated from interviews with three faculty members and six of their students and the texts they use as resources. A composite map reflecting all understandings and relationships was also created for purposes of contrasting perceptions. The map further displays traditional instructional content, current curricular focus, and implications for the study of calculus, advanced science, and technology. By means of this composite map, six clinical analysis categories were identified and addressed. Of significance in considering curriculum development are the categories of "Conceptual Representation," "Student Competency and Problem Complexity," and "Application and Importance." Findings in these areas translated into the development of a significant change in pedagogical approach illustrated in a series of introductory logarithm lessons.

Conceptual Analysis From a Cultural Historical Perspective

Scientific ideas are, in the Vygotskian sense, ideas that do not occur spontaneously in the human mind as a result of normal everyday experience but require dedicated theoretical analysis.

> If every object was phenotypically and genotypically equivalent... then everyday experience would fully suffice to replace scientific analysis. ... real scientific analysis differs radically from subjective, introspective analysis, which by its very nature cannot hope to go beyond pure description. The kind of objective analysis we advocate seeks to lay bare the essence rather than the perceived characteristics of psychological phenomena. (Vygotsky, 1978, p. 63)

Scientific ideas spring from and are mediated by a social, cultural, and historical context. John Napier's definition of a logarithm is just such an idea and provides a generalization worthy of historical exploration using the lens of Ausubel's cognitive learning theory as described by Novak and Gowin (1984). I refer to the notions that cognitive structure is hierarchically organized, that this structure is progressively differentiated, and that the process of integrative reconciliation may yield linkage between concepts providing new propositional meaning. The historical record of Napier's work also reveals strong evidence in support of Vygotsky's thoughts on cultural mediation. Napier's discovery is a perfect example of integrative reconciliation. His particular linkage of geometry and arithmetic has impacted the world for nearly four hundred years.

Historical Foundation

In 1619, John Napier wrote, Mirifici Logarithmorum Canonis Constructio, in which he explained his idea of using geometry to improve arithmetical computations. This was the breakthrough that accelerated the discoveries of science and led to the creation of new calculating devices, ultimately leading to modern day electronic calculators and computers.

> Seeing there is nothing, right well-beloved students of mathematics, that is so troublesome to mathematical practice, nor doth more molest and hinder calculators, than the multiplication, division, square and cubical extractions of great numbers, which besides the tedious expense of time are for the most part subject to many slippery errors, I began therefore to consider in my mind by what certain and ready art I might remove those hindrances. – John Napier

The account of the mathematical consideration Napier used in defining a logarithm refers to points moving on two different lines (Fig. 9.1). Consider that point P starts at A and moves along segment AB at a speed proportional to its remaining distance from B. Simultaneously, point Q departs from C and moves along ray CD with a constant speed equal to the starting speed of P. Napier called the distance CQ the logarithm of PB. This idea proved to be an important benchmark in the history of mathematical thought, providing a crosslink between concepts that immediately accelerated the interests of science and economics. The genesis of Napier's discovery is situated in Egyptian, Greek, Hindu, and Arabic thought and reflects the influence of the Renaissance on scientific thinking. From the Egyptian papyrus of Ahmes we see work on the reduction of fractions of the form $2/(2n+1)$ into a sum of fractions with numerators of one, an early form of computational efficiency. As the flow of ideas is transmitted between cultures, the Greeks provide theoretical structure to mathematical thought. Pythagoras establishes the twice split view of mathematics (Fig. 9.2) (Turnbull, 1969) that places arithmetic and geometry on distinctly separate branches. Hindu mathematics, like the Greek, considered arithmetic and geometry as separate categories of mathematics. It is a significant conceptual separation that becomes the focus in the late sixteenth century; ideas actually crosslinked by Napier.

This historical sampling of mathematical thought reveals the setting, motivation, and approach that made the discovery of logarithms possible. The intent of historical reference is to have a sense of what John Napier knew at the time of his discovery.

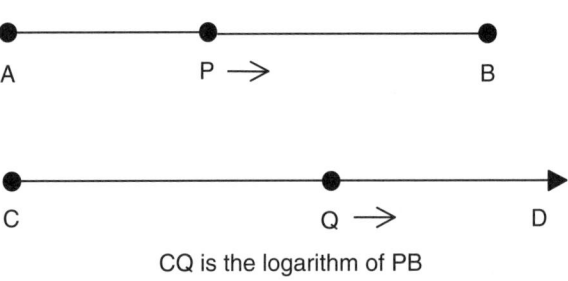

Fig. 9.1 John Napier's development of the logarithm concept

Fig. 9.2 The view of mathematics attributed to Pythagoras

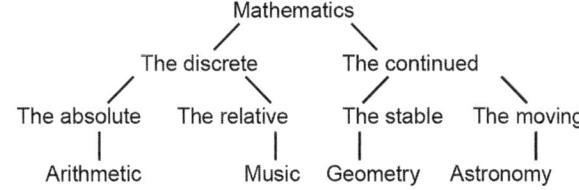

Access to his understandings sharpens the focus of our own as we consider the true nature of what he revealed to the scientific world. The cultural historical approach to the appropriation of knowledge is dependent on such considerations. Significant is the realization that it is only through this historical lens that the true essence of the scientific thought can be understood. In the case of the logarithm, evidence serves to inform us that, motivated by computational efficiency, Napier provided a theoretical link between the worlds of geometry and arithmetic. This novel generalization provided fertile territory for new development, practical and theoretical.

Conceptual Essence of the Logarithm Concept from a Cultural-Historical Perspective

Based on historical reference, a logarithm is a mapping between number sequences with different types of change rates (Fig. 9.3). The arithmetic sequence has a constant rate of change while the rate of change of the geometric sequence either increases or decreases. It is this connection that accounts for the computational power and efficiency the logarithm provides and justifies the importance of this discovery in the historical account of mathematical development. The map situates the conceptual genesis of a new mathematical idea, a realization of a connection between arithmetic and geometry. This relationship is the substantive understanding that requires mediational attention if students are to make sense of their work with logarithms. The mapping across two previously considered disjoint branches of mathematical thought is the core of the logarithm concept, an idea that became of prime interest with the later discovery of exponents and the emergence of calculus.

More Fully Developed Concept Map of a Logarithm

As is often the case, theoretical development leads to new technology. The new technology is applied to practical matters and simultaneously enables related theoretical development. The logarithm gave scientists a new conceptualization that proved immensely valuable in their effort to describe properties of the physical world using the language of mathematics. It should be noted here that the theoretical explanation found in the Mirifici, though completely consistent with calculus, was written prior to the existence of calculus. Points P and Q, moving along at different rates in Fig. 9.1,

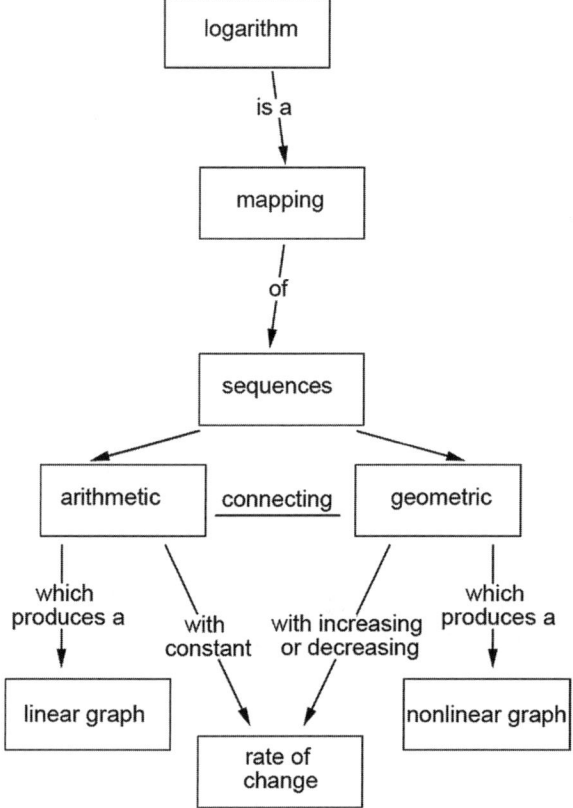

Fig. 9.3 Concept map of logarithm showing the historical conceptual crosslink, genesis of a new mathematical idea

PB decreasing in geometric progression while CQ increases in arithmetic progression, would necessarily fall in the jurisdiction of the mathematics of change, namely calculus. Napier's geometric representation defined logarithms in kinetic terms and foreshadowed their significance in the development of calculus. Therefore it is not surprising that we find logarithms along side other transcendental functions in every modern calculus text. What you do not find in the Mirifici is any mention of logarithms as exponents. Bernoulli and others recognized this connection toward the end of the 17th century. "One of the anomalies in the history of mathematics is that logarithms were discovered before exponents were in use." (Eves, 1969).

In a sense, Fig. 9.3 represents all that was known about logarithms in 1619. Continued mathematical conceptualization provides a concept map of far more extension and depth. The growth of relationships that develop in a scientific discipline rapidly create a complex structure that may mask the purity of the concept at the core.

Figure 9.4 incorporates the cultural historical perspective as the central structure of a more detailed map of the logarithm concept. Generally, concepts arranged on

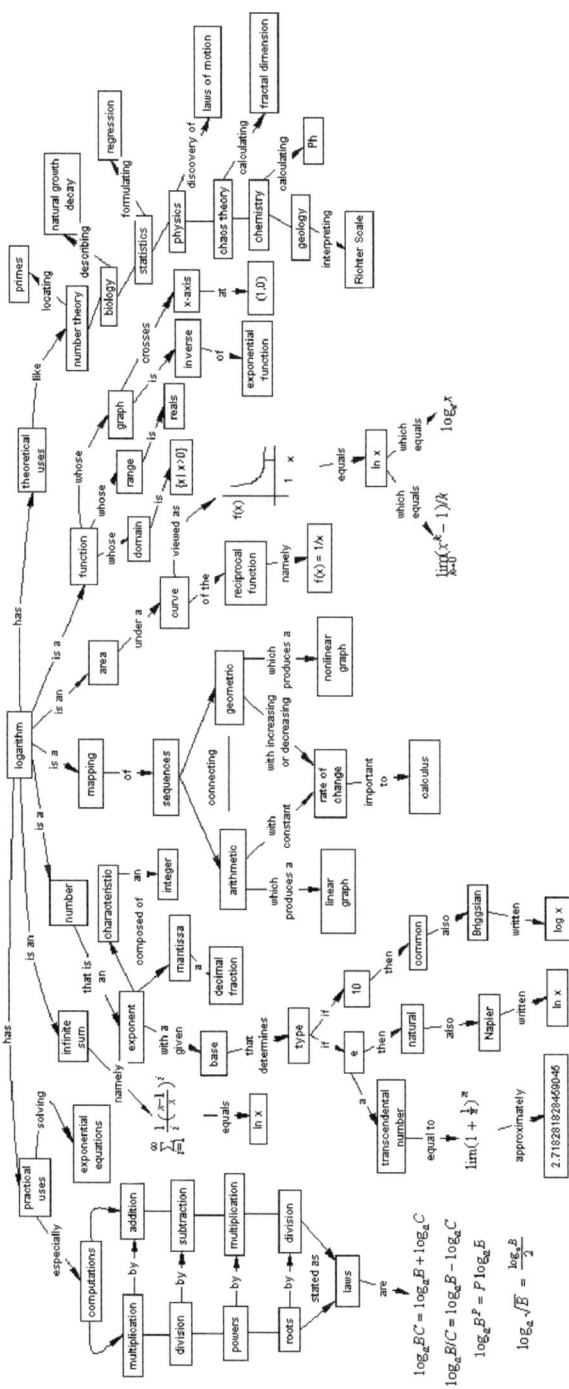

Fig. 9.4 A concept map of logarithm with a cultural historical perspective as its central structure

the left are more strongly associated with the arithmetic notions and those on the right more geometric in nature. Practical uses appear on the left and theoretical connections on the right. It is interesting to note that the exponent emerges with such prominence in the map even though the concept was unknown when the logarithm relationship was first recognized. The same is true for the calculus related content. It too is well represented in the map though it had not yet been developed. This suggests the importance of cultural historical considerations in curriculum development. The concept exists without reference to these later developments.

The map in Fig. 9.4 clearly reveals the extent of schema development that can arise from a simple consideration of points moving on a line. For the mathematics educator, the map complexity can be problematic if the essence of the concept, represented by the central structure, is missing. Instructionally, the concept loses its original meaning. Practical uses become independent procedures involving symbol manipulation. Theoretical connection becomes impossible, for there is no meaningful cross over access without substantive knowledge. Without conscious conceptual understanding of the essence of a scientific idea, mindful use is impossible.

The Problem of Generative Metonymy

Schmittau (2003) has identified the problem of generative metonymy as an impediment to mathematical understanding. Epistemological uncertainty accounts for considerable confusion in mathematical thought and the difficulty of understanding more advanced mathematics. As linguists Lakoff and Johnson (1980, p. 39) remind us, "metonymic concepts allow us to conceptualize one thing by means of its relation to something else." This use of language applied to scientific concepts masks the ideational essence and effects how a learner organizes their thoughts around new ideas. Multiple iterations of metonymy can develop completely inadequate conceptualizations leading to a form of intellectual desiccation.

For students in this study, the ability to reason mathematically and problem solve with logarithms was minimal. As a result of generative metonymy the concept had lost its genetic meaning. Concept mapping applied to this empirical research provides a clear view of student deficits by comparing their composite map with the more fully developed map of the logarithm concept, the one with the cultural historical perspective as its central structure shown in Fig. 9.4. Comparing the teachers' composite view to the map shown in Fig. 9.4 situates each teacher's characterization in various locations. The inherit limitations of developing a full understanding of the idea from each of these vantage points is apparent.

Conceptual Representation – the Teachers' View

Teachers (Fig. 9.5) variously describe a logarithm as: "a number," "a symbol," "a type of notation," "a function," "a different language," and "an exponent." This metonymy makes no mention of the essential characteristic, the mapping between

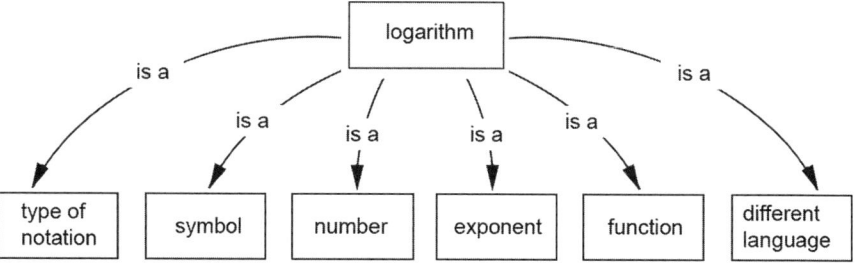

Fig. 9.5 Composite concept map of teachers' characterizations of a logarithm

geometric and arithmetic sequences. The primary understanding of teachers in this study is that a logarithm is an exponent. This is a central theme in their efforts to teach students this concept. The awkward and somewhat nebulous statement that "a logarithm is an exponent that you raise the base to to get a number" is often abbreviated to "a logarithm is an exponent." The statement is highlighted, underlined, and made the key idea in logarithm chapters of each of the twelve textbooks investigated as part of this study. The statement doubly masks the essence of a logarithm and fails to provide the conceptual connection that would make sense of both practical and theoretical applications.

Of related importance are the views of teachers Fred, Maria and Steve. Fred teaches logarithms as a notational convenience. "They [the students] have a hard time with the definition because it's pure notation to them. It's a symbol, $\log_b x$." Maria's students are taught to solve exponential equations using logarithms, a process used as justification for the existence of the logarithmic concept.

"The emphasis should be on what a logarithm is. I start with an equation that is impossible to solve without logs, something like $5^x = 112$." Steve relies on the fact that a logarithm is a function. "I like to think of it as a function that returns a number. The log of x is going to give you back a number. What number? It's going to be the exponent, the log always returns an exponent." These statements represent clear evidence of some teachers' admitted algorithmic focus and the superficiality of understandings they present to students. There is little substance compared to the conceptual essence depicted in Fig. 9.3.

Conceptual Representation – The Students' View

The map in Fig. 9.6 was created from the map in Fig. 9.4 by removing anything not mentioned by students in six hours of interview. In this sense, the map in Fig. 9.6 represents a composite of student understanding for the logarithm concept.

Note that the critical core, the essence of a logarithm, is missing. The laws of logarithms and their use in solving exponential equations remain. Of the scientific applications seen in Fig. 9.4, only pH and natural decay are cited. The student map reveals the limited nature of understanding and accounts for the inadequacy of

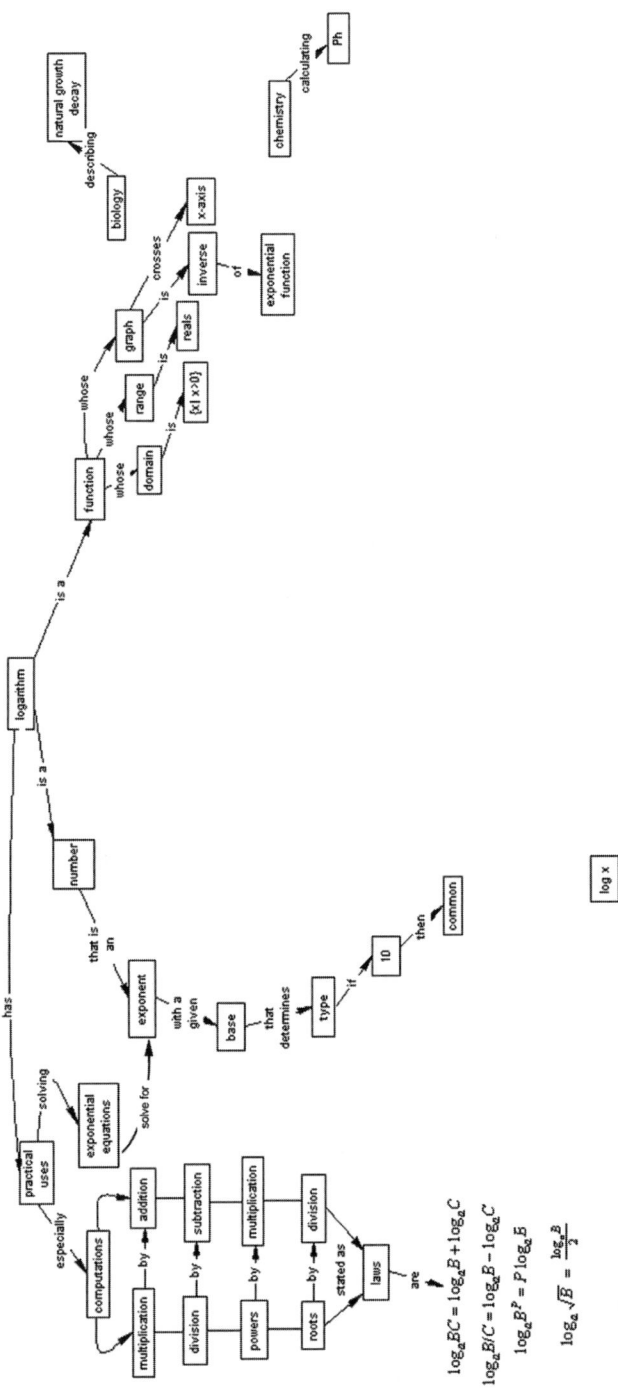

Fig. 9.6 Concept map of logarithm from students' perspective

mathematical reasoning expressed in interviews. How are students to understand the significance of reports of seismographic activity, for example? Will comparisons of 1.1 and 2.2 on the Richter scale be incorrectly interpreted as an earthquake of double intensity, when in fact the increase is tenfold due to the logarithmic nature of the scale? What sense will students make of Mandelbrot's (1977) work on fractals when the formula to determine fractal dimensions depends on logarithms? How will students read with wonder, Bronowsky's (1973) eloquent description of Ludwig Boltzmann's formula $S = K \log W$, entropy is in direct proportion to the logarithm of W, the probability of a given state. It was this formula that settled the theoretical debate over the existence of atoms and made possible current advancements in physics and biology.

The student logarithm map revealed in Fig. 9.6 is the direct result of generative metonymy, mediational inadequacy, and is void of conceptual essence. This lack of substantive knowledge makes mindful student use of logarithms unlikely. Improved conceptual representation can positively impact student problem solving competency and improve their ability to apply their knowledge to related scientific work. This curricular work in mathematics represents the positive contribution to be made by using concept mapping in conjunction with cultural historical research.

Curriculum Proposal

A review of twelve high school math texts, books that assumed first contact with the logarithm concept and calculus texts, which presupposed earlier work, served to confirm much of the clinical findings of this study. General observations of the texts revealed weak or nonexistent emphasis on the conceptual essence of the mathematics presented. Absent was any apparent depth of research in support of the historical context that gave rise to the mathematical ideas discussed. Text content variety seemed to inhibit methodological consistency at the expense of a coherent model of mathematical thought. Subject relevance was inadequately addressed.

Comparing specific content found in these texts with the findings of this logarithm study would lead one to conclude that teachers do not vary pedagogically from the approach found in their text books. Chapters involving logarithms were either presented jointly with discussions of exponents or were immediately preceded by such a chapter. Introduction to the logarithmic function nearly always involved describing the inverse of the exponential function. Content incorporated 3 or 4 laws of logs followed by an algorithmic problem set. The newer the publication the more likelihood of calculator based exercises. The algebraic exercise of focus was the exponential equation and topic relevance addressed primarily via four applications: population growth, radioactive decay, the Richter Scale, and inflation.

One book reviewed was very different. Mathematics: A Human Endeavor by Harold R. Jacobs' contained a radically different approach to presenting logarithms to students. This author clearly understood the historical foundation of the concept and with this insight develops student facility with arithmetic and geometric sequences prior to addressing logarithms. Instead of "Exponents and Logarithms,"

Jacobs' chapter is titled "Large Numbers and Logarithms," well chosen words that at once present to students the concept and the context that gave rise to the idea, defining its initial scientific consequence. Number and relation, the essence of mathematical thought, are captured in one title. Given the heretofore-stated conceptual essence, historical foundation, and importance in the study of mathematics and considering the understandings that exist among mathematics educators and students, Jacobs' chapter on logarithms is excellent.

Perhaps, however, it is possible to present the concept to students in a manner very similar to the geometric rendering of the definition offered by Napier. As Schmittau (1993) has stated, "pedagogical mediation must facilitate the appropriation of the scientific concept through a mode of presentation that reflects the objective content of the concept in its essential interrelationships" (p. 34). By doing so with logarithms, it may be possible to close the gap of graphical understanding that is evident among students and simultaneously reveal the computational consequence that enabled the scientific community. The four lessons that follow are presented toward that end. The first lesson addresses the essence and origin of the concept. The second lesson presents the link between the conceptualization of a logarithm as a relation between moving points and the graphical representation of that relation. The third lesson shows the impact of Napier's discovery on computational efficiency and accuracy. The final lesson cites two sources that develop the use of logarithmic scales for use in mathematical reasoning. Lessons presented are each grounded in an understanding of arithmetic and geometric sequences. The computational significance of the mapping between these sequences reveals the essence and power of logarithms.

Lesson 1: "Introducing the Logarithm"

Figure 9.7 displays points moving along two lines, points which left their starting positions at the same time.

The diagram is similar to Napier's original, differing only in that the points on line 1 are accelerating (Napier's decelerated) and that the points have been carefully constructed so that the constant movement of a point along line 2 is accompanied by a point on line 1 that is doubling its distance from starting position. Students do not need to know this since it is the intent of the lesson to work with the concept

Fig. 9.7 Points moving along two lines

of logarithm in the abstract. The doubling metric will allow students to later ascend from this abstraction to the concrete where any acceleration rate will be easy to understand since they will be simple instantiations of the correct abstraction. The base 2 logarithms used will also allow an easy bridge to the computational lesson that follows. The lesson should proceed with students considering the differences they recognize in points along line 1 versus those along line 2. It should become apparent that the distances are growing rapidly along line 1 and steadily along line 2. Connections to previous student understanding of arithmetic and geometric sequences, can be accented. The discussion should lead to defining the logarithm of the distance a point has moved along line 1, as the distance that the corresponding point has moved along line 2. That is to say, the logarithm of distance PQ is distance P'Q' in Fig. 9.1. This is not quite the same definition Napier gave us, but similarly captures the essence of a logarithm and may more clearly introduce students to the concept in a substantive and mindful way. For those students who ask "So what?" the argument can be made at this stage that the points along line 1 will soon accelerate out of sight but their corresponding points will still be visible along line 2 enabling us to keep track of where they are along line 1 as a result of this logarithmic relationship. Lesson 4 addresses the "So what" question more directly.

Lesson 2: "The Logarithmic Graph"

To gain a better understanding of the nature of a mathematical relationship it is useful to view the relationship as a graph, in this case a graph of the related values between geometric and arithmetic sequences. By simply rotating line 2 counterclockwise 90 degrees in Fig. 9.1 and joining the starting point to that of line 2 the logarithmic relationship is revealed graphically. By means of this lesson activity students can generate a partial graph of the log function and discuss the related growth rates along the axes. Discussion can explore hypotheses about what happens to the graph between plotted points and later, when metrics are incorporated, what this means about other logarithmic values. Metrification will also allow discussion about bases of logs, the

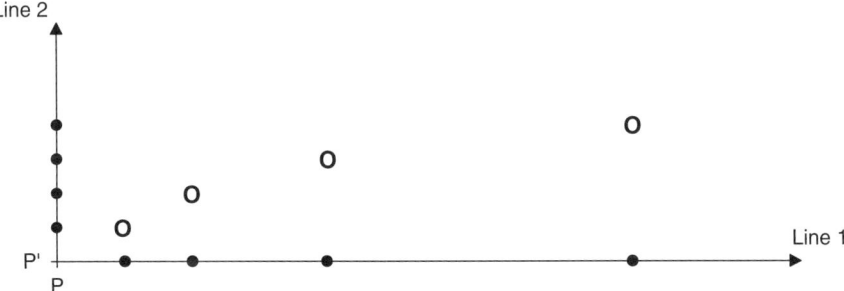

Fig. 9.8 Point plot of the logarithmic relationship

Table 9.1 Distances from the starting points in Lesson 1

Distance from the start on line 1	Corresponding distance on line 2
2	1
4	2
8	3
16	4

relationship of logs to exponents, laws that guide the computational use and the other common curricular content teachers generally include (Fig. 9.8).

The important point is that the fundamental concept be revealed and modeled before the metrification so that the essence of a logarithm is not instructionally compromised. Ascendancy to the concrete should fine tune the well established conceptual understanding, an understanding that requires a historical reference on the part of the teacher. Real world application of the concept can then proceed mindfully.

Lesson 3: "Logarithms, So What?"

The set of points along lines 1 and 2 given in lesson 1 have been given reasonable distances in Table 9.1, reasonable since the points along line 1 were intentionally set up visually by doubling the distance from the starting point.

Students will be reminded that according to the definition, the values in the second column are the logarithms of those in the first column. Recalling the properties of arithmetic and geometric sequences will allow students to expand this list of values in both directions. Logarithms of the larger numbers are to be used to discuss the historical advantage that developed as a result of using logs for computation. The list of values less than 2 will allow students to complete the graph of the logarithmic function yielding the x-intercept and the negative portion of the graph. The asymptotic nature of the graph can also be addressed. Just as it was easier to work with the steadily moving points along line 2, once you understand the geometric/arithmetic relationship it is easier to compute using the column of logs than to manipulate the very large distances that appear in the first column of the table.

For the student who utters "Logarithms, so what?" Table 9.2 has the answer.

Inherent in logarithmic relationship is the ability to multiply a number in the millions by a number in the billions by simply adding 20 and 30 and reading the answer from the table. Ease of computation is what Napier provided the scientific community in 1619. Even with a modern calculator in hand, it may still be faster and more accurate to use this table than to type in all those digits. The computation in question is recorded below. Compared to the standard multiplication algorithm, this calculation would require more than 150 operations to complete, with the chance for error very likely. With a previously established conceptual understanding of sequences students can develop the laws that govern the computational use of logarithms as a concrete consequence. Notational expressions of these laws should be simple to grasp for the conceptual foundation of the relationship has been well established.

Table 9.2 Using logarithmic values to ease calculations

Number	Logarithm	Calculation
2	1	$N = 1048{,}576 \times 1{,}073{,}741{,}824$
4	2	$\log N = \log 1048576 + \log 1{,}073{,}741{,}824$
8	3	$\log N = 20 + 30$
16	4	$\log N = 50$
32	5	therefore, from the table
64	6	$N = 1{,}125{,}899{,}906{,}842{,}620$
(values intentionally missing)		
1,048,576	20	
1,073,741,824	30	
(values intentionally missing)		
1,125,899,906,842,624	50	

Lesson 4: "Logarithmic Scales and Logarithmic Graph Paper"

In addition to filling out the tables of values so that the log function can be more completely graphed, students should be introduced to logarithmic scales and logarithmic graph paper. The idea is the same. Using the values 0.0, 0.3, 0.48, 0.6, 0.7, 0.78, 0.85, 0.9, 0.95, 1.0 as the base 10 logarithms of 1 through 10, students are to produce a number line in which the positions of the numbers are proportional to their logs. The scale to be produced is shown in Fig. 9.9. Students can clearly see the relationship between the arithmetic and geometric sequences of values that comprise the logarithm concept. Computational and analytical exercises can follow this activity meaningfully.

These scales allow us to investigate the nature of a mathematical relationship that may exhibit an accelerating rate of change. The teachers who were part of this study often noted the facility logs provide when solving exponential equations. Logarithmic scales provide a similar tool, graphically. Jacobs provides an excellent lesson using the graph produced in lesson 2 above to create a pair of logarithmic scales in order to produce a slide rule with C and D scales. Students then use this slide rule to perform calculations.

These same logarithmic scales can be used to produce semi log and log log graph paper. This graph paper can be in turn be used to investigate the dimension of fractals, brightness of stars, and a host of other scientific applications. The physics department of the University of Guelph, Ontario, Canada has a well developed lesson on their web site showing that "if a graph of log y vs. log x for a set of data

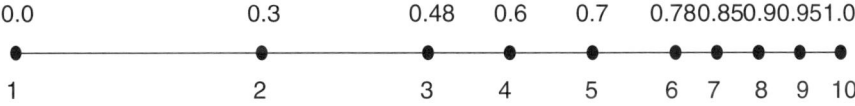

Fig. 9.9 A logarithmic scale

is a straight line then the data does indeed follow the relation $y = a \, x^b$." This is but a sample of the activities that can be used to demonstrate scientific use of this powerful mathematical concept. The table below (Table 9.3) is offered to teachers who are looking for other scientific applications to share with their students.

Table 9.3 Teachers' list of logarithm related science

*Seismographic studies and the Richter scale ($R = \log(i/i_0)$)
*Cancer research and the growth rate of cancer cells
*Radioactive half-life and nuclear decay
*Population growth
*Loudness of sound
*Oceanographic studies of sunlight intensity at a given depth
*Brightness of stars
*Fixing the age of moon rocks
*Radiocarbon dating
*Compound interest
*Position of a piano key and the pitch it produces ($n = 12 \log_2 (p / 440)$)
*Photographers f-stop setting ($n = \log_2 (1/p)$)
*Maximum velocity of a rocket given the ratio of its mass with and without fuel
   ($v = -0.0098 \, t + c \ln R$)
*Expiration time of a natural resource (coal, crude oil, etc.)
*Calculation of pH
*Kepler's third law of planetary motion
*Gauss' formula for the number of primes less than A, for very large A. (A/log A)
*Area between a hyperbola and its asymptote. (say $y = 1/x{-}1$)
*Development of security key cryptosystems
*Mercator series
*Fractal dimensions, zoom factor, and Chaos theory
*Stirling's formula for approximating factorials, for large n
*Euler's constant (0.5772156649...)
*Power series $x{-}x^2/2 + x^3/3{-}x^4/4 + -$
*Statistical inference and correlation coefficients
*The Boltzmann constant

Curriculum Design Questions to Use in Conjunction with Concept Mapping

Schmittau's (personal conversation, 2001) five guiding questions aid in the design of a curriculum and instructional sequence that would better enable student substantive understanding and mindful facility with a mathematical concept like logarithm.

Question 1: Essence . . . . What is it?
Question 2: Origin . . . . How did it come about?
Question 3: Methodology . . . . . . How should we look at it?
Question 4: Models . . . . Are there models that reveal it?
Question 5: Relevance . . . . What other connections are there?

Thoughtful consideration of these questions when planning a unit of mathematics centered on any important concept can provide profound instructional impact. The usefulness of concept mapping to guide and inform such curricular decisions increases the likelihood of evolving a quality instructional design.

The essence of a logarithm, as has been said many times in this work, is the marriage of arithmetic and geometry. A natural place for students to begin, therefore, is a study of the nature of arithmetic and geometric sequences. Such a unit should precede work with logs. The origin of the idea has been shown to be the geometric problem posed by Napier in the Mirifici. Selectively replicating that problem, with a slight modification for simplicity as presented in lesson 1, is intended to overcome the conceptual disconnect so apparent in student and teacher interviews presented here. This means the introduction of the concept can focus on the essence of the original idea. The relationship, as evidenced by the concept map presented in Fig. 9.3, would suggest looking at the idea computationally, algebraically, and geometrically. The model that best captures the nature of the relationship is the graph of the function as well as the use of logarithmic scales. Syntactic representational competence should follow naturally, allowing students to make sense of written mathematics and to express themselves correctly in written form. Since notion of exponents developed historically after logarithms, the lessons provided here are not justified on the basis of exponents, an approach that seems unnecessary. Perhaps exponents should be presented to students after work with logarithms. Concept relevance is evidenced by the long list of scientific applications and the role the idea played historically in the development of a wide range of calculating machines. These curricular suggestions attempt to capture the conceptual essence, historical relevance and cultural implications of this significant mathematical idea. The goal of promoting substantive knowledge and student mindfulness may be well served with this instructional approach.

Conclusions and Implications

The focus guiding this study precipitated an extensive historical search for the origin of the logarithm concept, an in-depth clinical study involving students and teachers, and curricular considerations based on these findings. It is hoped that study results serve to inform mathematics educators specifically about teaching logarithms and more generally about using concept maps to create mathematics curriculum that makes sense. Teaching on the basis of a curriculum constructed in this way avoids rote learning, encourages meaningful understanding of concepts fully developed and historically based, and enhances recognition of conceptual crosslinks in other mathematical domains.

The curricular implications of concept mapping in secondary mathematics education are clear. Substantive knowledge and mindful use of a scientific concept is dependent upon a clear understanding of the cultural and historical source and development of the idea. Vygotskian notions on cognitive development direct mathematics educators to identify the essence of concepts to be taught, reflecting with clarity the cultural historical context that produces new mathematical thought. Concept mapping, when combined with historical research, serves as an important epistemological tool that can render to consciousness the conceptual essence of a mathematical idea. In conjunction with empirical data, concept maps can also be used to expose the operating understanding of important mathematical concepts held by students and their teachers. These direct uses of concept mapping in turn provide the means for educators to identify substantive focus for curriculum design and provide pedagogical direction toward the mindful use of learned mathematics on the part of students. Meaningful instruction in mathematics avoids the temptation of a rote learning approach, recognizing that focus on functional efficiency often leads to cognitive deficiency. Complex problem solving depends on the thoughtful application of meaningful mathematical ideas. In this regard concept mapping can instruct and guide mathematics educators toward a pedagogy of significant cognitive consequence.

References

Bronowsky, J. (1973). The ascent of man. Boston, MA: Little, Brown.

Davydov, V. (1990). Types of generalization in instruction: Logical and psychological problems in the structuring of school curricula. *Soviet studies in mathematics education* (Vol. 2). Reston, VA: National Council of Teachers of Mathematics.

Eves, H. (1969). *An introduction to the history of mathematics* (3rd ed.). New York: Holt, Rinehart and Winston.

Lakoff, G., & Johnson, M. (1980). *Metaphors we live by.* Chicago, IL: University of Chicago Press.

Langer, E. (1989). *Mindfulness.* Reading, MA: Addison-Wesley.

Mandelbrot, B. (1977). *The fractal geometry of nature.* New York: W. H. Freeman.

Novak, J. D., & Gowin, D. (1984). *Learning how to learn.* Cambridge, UK: Cambridge University Press.

Schmittau, J. (1993). Vygotskian scientific concepts: Implications for mathematics education. *Focus on Learning Problems in Mathematics, 15*(2 and 3), 29–39.

Schmittau, J. (2003). Cultural-historical theory and mathematics education. In A. Kozulin & others (Eds.), *Vygotsky's educational theory in cultural context* (pp. 225–245). Cambridge, UK: Cambridge University Press.

Turnbull, H. W. (1969). *The great mathematicians.* New York: New York University Press.

Vygotsky, L. (1978). In M. Cole, V. John-Steiner, S. Scribner, & E. Souberman (Eds.), *Mind in society: The development of higher psychological processes.* Cambridge, MA: Harvard University Press.

Chapter 10
Using Concept Maps and Gowin's Vee to Understand Mathematical Models of Physical Phenomena

Maria S. Ramírez De Mantilla, Mario Aspée, Irma Sanabria, and Neyra Tellez

The fundamental presence of mathematics in the development and comprehension of physics and its applications in physics learning has many forms. Beyond being an indispensable component in the definition of metric concepts that express physical magnitudes, it allows the construction of mathematical models to represent multiple physical phenomena. Mathematical models have several applications in physics. First year physics students often ignore the important role mathematics and particularly mathematical models play in the learning of physics. Students limit themselves to the use of a menu of equations often misunderstood as a set of "cook-book" procedures applied to solve physics problems, without a real understanding of the reason for using a particular function or model to solve a problem. This chapter reflects on the outcome of a research project undertaken at "Universidad Nacional del Táchira" (UNET), Venezuela, which focuses on the different ways teachers and students could use concept mapping and Gowin's Vee for the mathematical modelling of physical phenomena. We have designed a strategy for the teaching and learning of mathematical models most used in first year physics courses. This strategy uses concept maps to improve understanding of basic conceptual structures involved in the mathematical modelling process of physical phenomena, and Gowins' Vee as a tool that facilitates the process of building a student's own knowledge of a mathematical model for a particular experiment.

Introduction

First year physics students at the Universidad Nacional del Táchira (UNET), Venezuela, face difficulties in science learning. They experience difficulties understanding a whole body of information and in building their own knowledge about complex conceptual structures. They also find it difficult to link concepts and to handle adequate representational techniques either to show or sum up complex information. To help solve these problems, we have found concept maps a powerful tool that

M.S.R. De Mantilla (✉)
Universidad, Nacional Experimental del Táchira, San Cristóbal, Estado Táchira, Venezuela
e-mail: marimant@unet.edu.ve

K. Afamasaga-Fuata'i (ed.), *Concept Mapping in Mathematics*,
DOI 10.1007/978-0-387-89194-1_10, © Springer Science+Business Media, LLC 2009

facilitates students' comprehension of physics at the university level. Also, students face difficulties in the physics laboratory in making connections between lectures and laboratory and in understanding how to plan and guide science fair research. In this sense, we have used Gowin's Vee to help students plan and carry out some experiments in the physics lab.

On the other hand, we are conscious that learning mathematics is different from learning physics, especially for first year university students. The structure of the "scientific method" in mathematics is distinctly represented by the activity, ability and use of deduction. In physics, this feature is also found, but it is not the only activity present, as this natural science also requires the application of induction, inherent in the scientific method to build up knowledge about the real world. The language used to express physical knowledge about the world has universal elements of mathematical language used by all scientists, as well as idiomatic language, which is specific to each country. The symbolic, syntactic and semantics terms of the language of mathematics are acquired by someone after those terms are acquired as part of a specific idiomatic language. This implies that, it is more difficult for students to comprehend this mathematical language than learning to talk and express themselves in their native language.

It is a commonly known that first year university students of natural science have not finished incorporating to their cognitive structure the fundamental elements of the language of mathematics evident when they are asked to use them within specific contexts in their studies of physics. Furthermore, they must complete the course showing an appropriate knowledge and understanding of the language of physics. This learning difficulty in mathematics causes a delay in the learning of physics.

In our experience, students show little understanding of the important role mathematics and mathematical models play in the learning of physics especially when building mathematical models to explain physical phenomena in kinematics and dynamics of particles. Instead, students limit themselves to the use of equations and theorems, often misunderstood as a set of "cook-book" procedures used to solve physics problems, without a real understanding of their significance or the reason for using a particular function or model to solve a problem or explain a physical phenomenon.

In order to overcome this undesirable situation, our research team at the Universidad Nacional Experimental del Tachira, (UNET), Venezuela, worked with first year engineering students, using a strategy for the teaching and learning of basic mathematical models essential to the comprehension, explanation and prediction of real-world phenomena most commonly studied in first year physics courses. These mathematical models are based on essential concepts of function theory. The strategy we propose uses concept maps to improve understanding of concepts and basic conceptual structures involved in the mathematical modeling process and Gowin's Vee as a tool that facilitates the process of building a student's own knowledge of a mathematical model for a particular experiment.

This chapter is organized to explain, from our perspective, what concept maps and Gowin's Vee are, the strategy we used in the introductory physics laboratory course, the phases involved in the implementation of the strategy, and, finally to provide some results from each phase. The following concept map (in Fig. 10.1) summarizes the structure of this chapter.

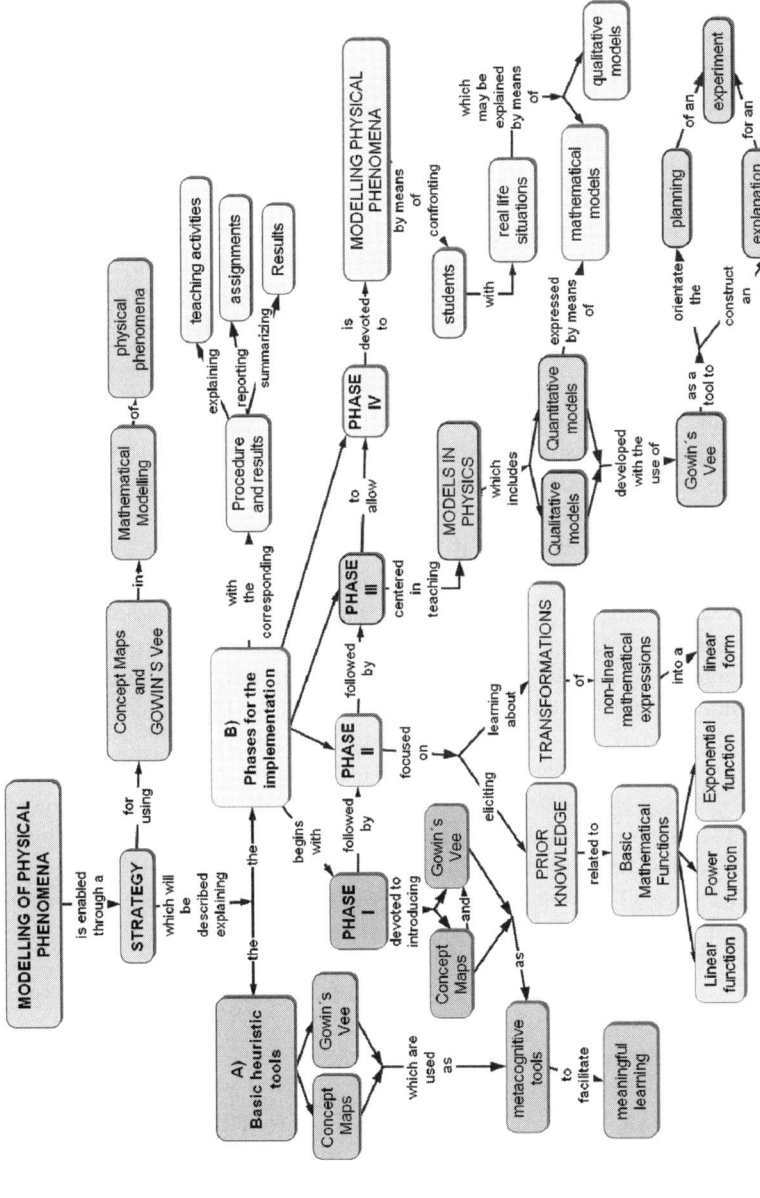

Fig. 10.1 Concept map of the strategy for the teaching & learning process of *modelling of physical phenomena*

Theoretical background: Concept Maps and Gowin's Vee

Research undertaken over two decades with our students, has convinced us that a student's knowledge is a result of an autopoietic phenomenon taking place within an individual's cognitive structure. Cognitive abilities are without any doubt, the protagonists or active agents in this personal knowledge self-construction process.

The most important role in this process of personal knowledge construction is played by metacognition. This is understood as the capacity an individual mind has to be aware of (a) possessing a certain set of cognitive abilities, (b) the way in which the individual builds his or her own knowledge about the world and (c) self development potential. Metacognition is then the capacity of thinking about how we think, about how we know, and about how we learn. Along with the importance assigned to metacognition, we have considered a set of six cognitive abilities that, in our experience constitutes a minimum set needed for the learning process of basic physics. We name it "the basic group of cognitive abilities" comprising the ability to memorize, comprehend, apply, analyze, synthesize and evaluate.

The learning of physics may be considered as the elaboration of a discourse developed entirely in a rational plane. In consequence, metacognition as well as the basic group of cognitive abilities, which we have taken into consideration, are defined in the most objective way possible, trying to avoid any unnecessary emotional ingredient that may colour any cognitive ability.

On the other hand, we have found in concept maps, a heuristic tool designed many years ago (Novak & Gowin, 1988), a very powerful tool used by physics teachers to express concepts and meaningful relationships between those concepts while presenting the different topics we teach to the students in a clear way and with a minimum of complexity. But what seems to be most important is that concept maps constitute an effective aid to promote meaningful thinking, a tool to promote the self-construction of knowledge on the student's behalf.

In a similar way, our attempts to introduce the heuristic tool known as Gowin's knowledge Vee or Gowin's Vee have convinced us of the advantages of using it as either a teaching tool or as a student learning tool. It helps to promote the self-construction of knowledge related to mathematical process, which allows the comprehension of physical phenomena and the development of experiments inherent to physics. These two heuristic tools are explained next.

What is a Concept Map?

The first difficulty a student faces when attempting to comprehend a text is to understand what it is all about. That is, to grasp the global sense of the communication, understand its elements and the relationships between them. Imagine a physics student, seeking for information about *frame of reference*, finds the information in the following two different ways, see Fig. 10.2. Probably it is easier for many students to grasp the whole sense of the concept *frame of reference* when faced with a

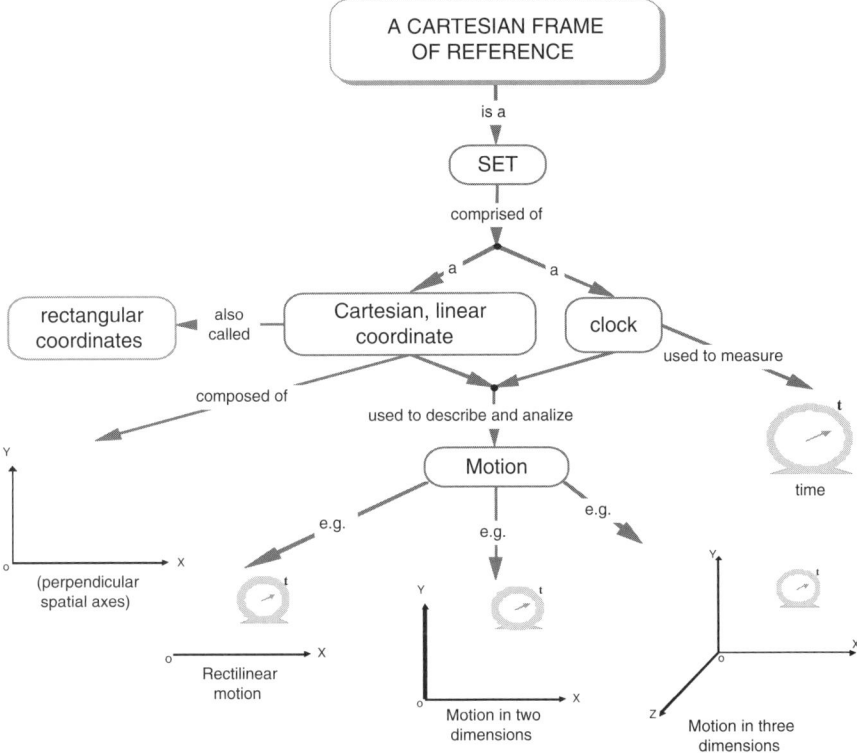

Fig. 10.2 Cartesian frame of reference

graph like the one illustrated above. This is due to the powerful visual effect that a graph has to facilitate understanding of a concept or a conceptual structure.

This graph is essentially a *concept map*. It is a map-like illustration that shows *meaningful relationships* between concepts (events, objects). Observe that this is a knowledge representation about a particular *main idea* (in this case, *frame of reference*), in the form of a graph comprising boxes connected with labeled lines. Words or phrases that denote *concepts (events, objects)* are placed inside the boxes, and *relationships* between different concepts are specified on each line. *Propositions* (node – link – node triads) are a unique feature of concept maps, compared to other graphs. Propositions consist of two or more concept labels connected by a linking relationship that forms a *semantic unit* (Novak & Gowin, 1988). And the reader, if not familiar with concept maps, may know some of the concepts mentioned before as separate entities but have no clear sense of the relationships between some of them. Obviously, this makes it difficult to understand the whole conceptual structure *concept map*. Let us then discuss some of the concepts involved in this definition: *object, event, concept, proposition* and *meaningful relationships*, to be able to understand what a concept map is.

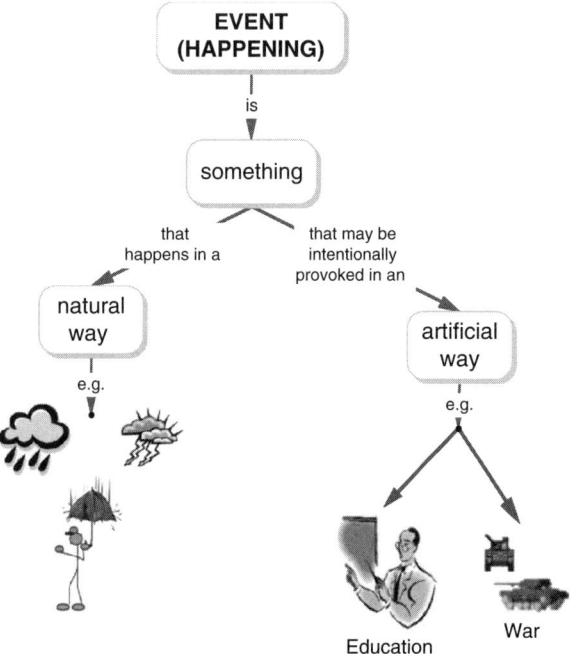

Fig. 10.3 Event

According to Novak, the construction of new knowledge begins with the understanding of the terms *event* and *object* (Novak & Gowin, 1988). *Event* may be represented (see Fig. 10.3) in this manner. And also, according to Novak's definition, *object* may be explained as provided in Fig. 10.4.

Now that we understand what *objects* and *events* are, we can define *concepts* as *perceived regularities in events or objects, or records of events or objects, designated by a label* (Novak & Gowin, 1988). Also, concepts are mental representations of objects or events with the following characteristics: (a) correspondence between the concepts and what it represents, (b) absence of ambiguity and (c) optimal use of the language involved. In concept mapping, *concepts* are usually written inside cells or boxes although the actual design of cells is arbitrary. The important thing is that the graph highlights concepts visually in a clear and distinctive way.

But *concepts* are not isolated in a concept map; they are connected by labeled lines or arrows called *links,* which consist of words, phrases or verbs that explain meaningful relationships between concepts by words or signs/symbols. Arrows, if used, designate the directionality of the relationship. Otherwise, the concepts must be arranged in a hierarchical way, from the most abstract and inclusive concepts on the top of the graph to the most concrete and specific and it is assumed that the direction of the relationship is downward. This facilitates the reading of concepts and the links among them as whole sentences.

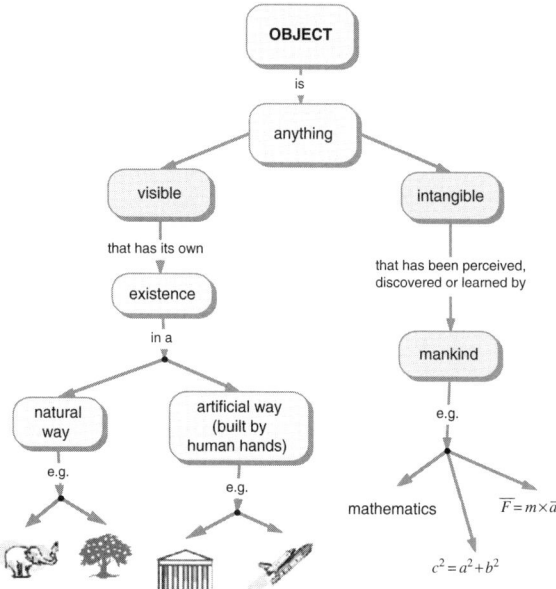

Fig. 10.4 Object

Relationships among concepts are diverse. Some examples are presented in Fig. 10.5. Cañas, Safayeni, and Derbentseva (2004) classify them as static or dynamics relationships. Static relationships between concepts help to define, describe and organize knowledge for a given domain. Classifications and hierarchies are usually captured in relationships that indicate belongingness, composition, and categorization.

These relationships comprise: *inclusion* (a concept is part of another one), common *membership* (two concepts are part of another) and *intersection* (a concept is the meaning generated by crossing other two concepts). This intersection may be

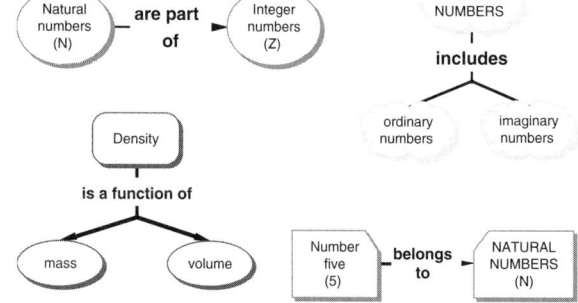

Fig. 10.5 Different types of relationships between concepts

probabilistic (e.g. *Polygons may be regular*), or based on similarity between two concepts. Dynamic relationships between two concepts show how changes of one concept cause change of the other concept in a proposition (e.g. *Concept A leads to concept B*). Dynamic relationships are those based on causality (e.g. *Volume is an inverse function of the density for a given mass*), or those based on correlation/probability (e.g. *Concept maps may help students to achieve meaningful learning*). Scientific knowledge is based on both static and dynamic relationships between concepts. The teacher who is learning about concept maps for the first time must know that there are different types of relationships, and should take them into consideration when he or she attempts concept mapping. Finally, a *link* may be *simple* (showing the connection between two concepts) or a *crosslink* (showing the relationships between ideas in different branches of the map).

Let us now talk about *propositions*. The basic unit of representation in concept maps is a *proposition* defined as two concepts plus a relationship, which is stated with a label on the link between concepts. Two or more concept connected by a linking relationship forms a *meaningful statement* also called a *semantic unit*. Also it can be said that propositions are units of meaning constructed in the cognitive structure. Each proposition is a sentence that has a unique standardized *truth value* (true or false) that is the basis for arguing whether the graph makes true or false assertions. Consider the following sentences: (*a*) "*Complex numbers include natural numbers*" and (*b*) "*Mars has a unique moon*". These sentences are *propositions* with a truth value (propositions *a* is true and *b* is false). In comparison, consider the following sentence: "*Mathematics problems are easy to solve*". This sentence is not a proposition as there is no way to know whether it is true or false. Some mathematics problems may be easy to solve ... but not all. Moreover, what is easy for one person may be difficult for someone else.

The scientific knowledge, that is to be reflected in concept maps, deals with the truth, that is, knowledge widely accepted as true by a given scientific community. That is the reason why Novak insists on constructing maps with real meaningful propositions. Many conceptual structures in mathematics and physics could be understood in an easier way if they were introduced by means of concept maps. An example of a concept map about polygons is shown in Fig. 10.6.

Novak (Novak & Gowin, 1988) argues that a map is always built as an answer to a *focus question* (e.g. What is a polygon?). The *main idea* (the polygon) is the supraordinate concept, a *subsumer*, and the most general and more inclusive concept. In order to expand on this concept (i.e., concept *polygons*), subordinate (less important) concepts related to it begin to emerge. Describing appropriate relationships of the main concept with subordinate concepts, make possible the construction of propositions. Similarly, collateral structures begin to appear from the connections among propositions. As a summary for this section, Fig. 10.7 represents an answer to the focus question: *What is a concept map?*

The Proceedings of the International Congresses on Concept Mapping, CMC, held in Pamplona (2004) and Costa Rica (2006) are valuable sources of information about concept maps. There is also a special issue of *Focus on Learning Problems in Mathematics* devoted entirely to concept maps with an article by the authors

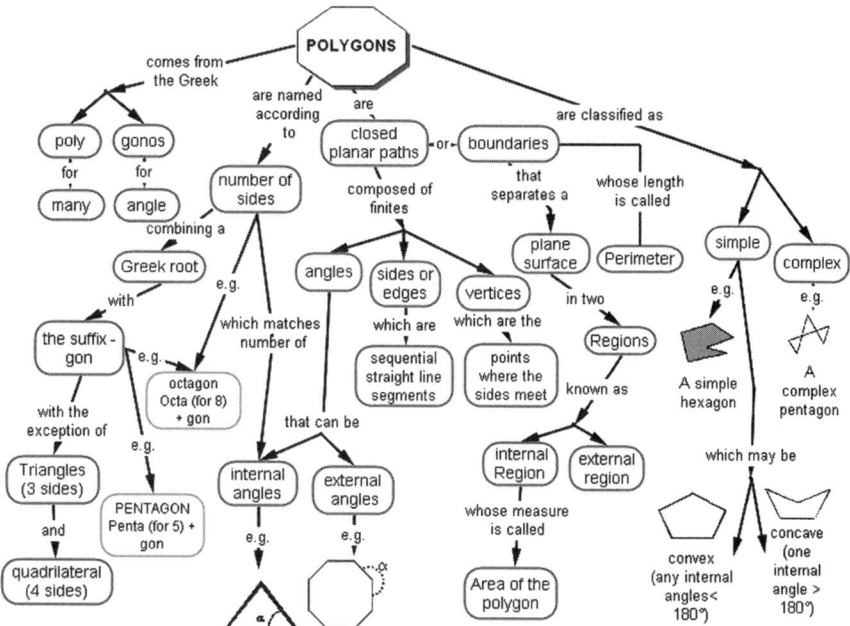

Fig. 10.6 Concept map "polygons"

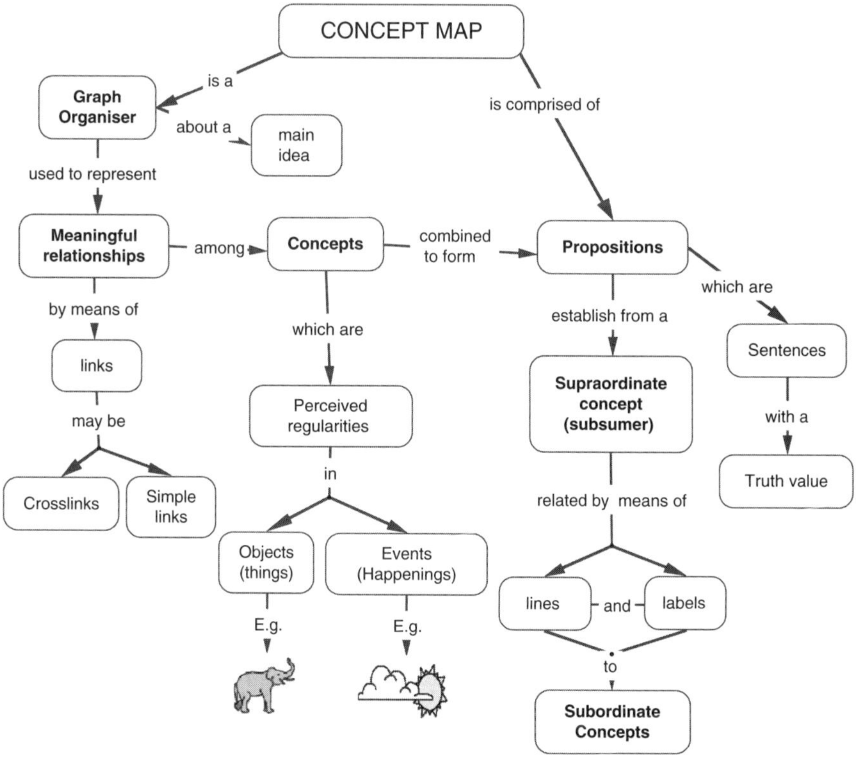

CONCEPT MAP

Graph Organiser — is a — CONCEPT MAP — about a — main idea

CONCEPT MAP — is comprised of — Propositions

Graph Organiser — used to represent — Meaningful relationships

Meaningful relationships — among — Concepts — combined to form — Propositions

Meaningful relationships — by means of — links

links — may be — Crosslinks / Simple links

Concepts — which are — Perceived regularities

Perceived regularities — in — Objects (things) / Events (Happenings)

Objects (things) — E.g.

Events (Happenings) — E.g.

Propositions — establish from a — Supraordinate concept (subsumer)

Propositions — which are — Sentences

Sentences — with a — Truth value

Supraordinate concept (subsumer) — related by means of — lines — and — labels

lines and labels — to — Subordinate Concepts

Fig. 10.7 Concept map of a "concept map"

(Ramírez, Aspée, & Sanabria, 2006) to answer these questions: *What are concept maps? How are they constructed? What is the theory that supports concept maps? What are they used for?* And finally you will find an explanation of how concept maps were used with large groups of university students to facilitate the teaching-learning process (see also Ramírez de M. & Sanabria, 2004).

Gowin's Vee

This is a heuristic tool known as Gowin's knowledge Vee or Gowin's Vee, designed initially by Novak and Gowin (1988) to be used in science laboratories to help teacher and students to clarify the nature of the work to be done in the lab and the main goals and objectives of a given experiment. It is a V shape diagram that explains researcher's decisions about two domains: conceptual and methodological. This diagram shows the conceptual aspects to be considered in order to carry on an experiment or research, and the methodological procedures the researcher follows in order to solve the problem orientated by means of a focus question. Figure 10.8 is a concept map designed to explain the knowledge construction process using a Gowin's Vee.

Gowin shares with Novak basic concepts such as objects and events, and insists on the need for a focus question that will orientate a research or the development of an experiment in a science lab. This diagram helps students to focus discussions on answering a main question, or research question that is placed in the middle of the V-shaped diagram.

The Vee diagram separates the conceptual side, which is written on the left, from the methodological side, which will be filled on the right. It is a visual aid that represents the relations between the research question and the objects or events to be investigated. This diagram also shows the interrelationships between the conceptual framework and the methodological procedures that will be used in the research process. The diagram focuses on the research question and moves downward through the vertex where the student will describe the specific events or object being studied.

Table 10.1 describes the main concepts involved in a Gowin's Vee Diagram as Gowin defines them (Novak & Gowin, 1988). Students' use Gowin's Vee to understand what they are going to do in the lab; what question they must answer; what machines, objects or lab equipment students could work with; and finally, which events are going to be investigated. This will help them to organise themselves in order to plan, carry out the experiment, collect raw data, transform and graph data, and finally analyze the results obtained. The construction of the conceptual side begins with a student studying the regularities about the events or object being investigated, which will lead him or her to concepts, hypothesis models and theories. On the other side, students will be making decisions about the methodological procedure they will follow in order to study a particular phenomenon. Gowin's Vee is a useful device to establish relationships between the conceptual and practical sides of a laboratory activity and it also helps to structure the discussions that precede and follow the practical activity.

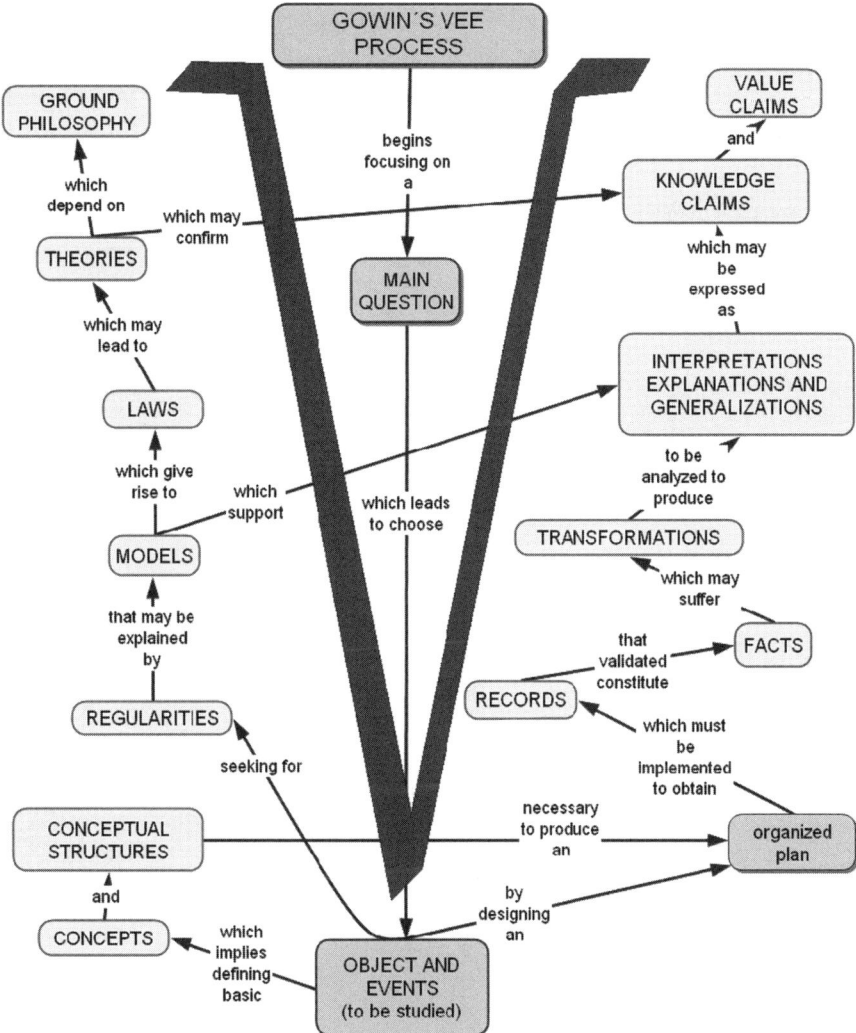

Fig. 10.8 Concept map of Gowin's Vee

Phases of the Implementation of the Strategy

This section explains the strategy used for the teaching and learning process for the topic "*Mathematical Modeling of Physical Phenomena*". This was a constructivist teaching/learning approach designed to improve first year physics students' confidence in, and attitude to, modeling of physical phenomena. This strategy was based on the use of the heuristic tools concept maps and Gowin's Vee to help students in their self-construction of knowledge. The program encouraged students' use of concept mapping in order to explore basic mathematical function concepts, as well as the concept of *models* in physics during the first part of the laboratory

Table 10.1 Gowin`s Vee some definitions

Focus question	Initiate activities between the two domains and are embedded in or generated by theory; FQ's focus attention on events and objects
Events & objects	Phenomena of interest apprehend through concepts and record-marking: occurrences, objects
Methodological domain	
Value claims	The worth, either in field or out of field, of the claims produced in an enquiry
Knowledge claims	New generalizations in answer to the telling questions, produced in the context of inquiry according to appropriate and explicit criteria of excellence.
Interpretations, explanations & generalizations	Product of methodology and prior knowledge used for warrant of claims
Results	Representation of the data in tables charts and graphs.
Transformations	Ordered facts governed by theory of measurements and classification
Facts	The judgment, based on trust in method, that records of events are valid
Records	Records of event or objects
Conceptual domain	
Concepts	Signs or symbols signifying regularities in events and shared socially
Conceptual Structures	Subsets of theory directly used in the inquiry
Constructs	Ideas which support reliable theory, but without direct referents in events or objects
Principles	Conceptual rules governing the linking of patterns in events; propositional in form; derived from prior knowledge claims
Theories	Logically related sets of concepts permitting patterns of reasoning leading to explanations
Philosophies	(E.g. Human understanding by Toulmin)
World views	(E.g. nature is orderly and knowable.)

workshops. Later on, students were encouraged to analyze and carry on experiments using Gowin's Vee to model physical phenomena. The strategy was implemented throughout four successive phases outlined at the beginning of this chapter (see Fig. 10.1).

Phase I: Learning About Concept Maps and Gowin's Vee

As a result of a diagnostic survey carried out by the research team with physics students, it was evident that they needed training in the construction and use of concept maps and Gowin's Vee. In order to do so, written materials for Gowin's Vee were prepared and these were supplemented with materials from the text "*Concept maps as a heuristic tool to facilitate learning*" (Ramirez de M., 2005).

Phase II: Eliciting Basic Conceptual Knowledge of Mathematical Functions

Though the students have completed two mathematics courses; it was not uncommon for them to have difficulties stating, graphing and using mathematical functions necessary for modeling physical phenomena. Therefore, it was necessary to clarify students' prior knowledge of the linear, power and exponential functions.

Phase III: Acquisition of the Concept "model" in Physics

In this phase, students were taught the meaning of *model*, in a way that this concept is understood within the physical sciences. This was followed by the teaching and learning for the concepts *qualitative and quantitative models* and *mathematical model.*

Phase IV: Students Training in the Application of the Concept "Model" to Physical Phenomena

In this phase, emphasis was put on the creation of mathematical models for specific physical phenomena. The main goal was to involve students in varied learning experiences to allow for the emergence of autopoiesis of procedural knowledge for the construction and handling of mathematical models. This is the most important phase in the implementation of the overall strategy, as it is the stage in which student can reach sufficiency in the use of mathematics to build up satisfactory models for the world of physics.

The Evaluation of the Strategy

Based on an interpretation of the Complementary Principle for dialectic approaches used in education, we adopted two perspectives to evaluate our strategy. Firstly, a quantitative research approach was used to analyze students' written work to count the number of right answers in the questions that have been stipulated as unknowns to be solved. Secondly, a qualitative analysis of interviews conducted with students allowed the assessment of the degree of comprehension, about the understanding of potential uses of both heuristic tools. These two perspectives were directed towards the same final objective, which was the evaluation of the strategy used. The application of two distinctly different analytical approaches has a parallel in the physical sciences. It is evident, for instance, with the corpuscular and wave models for the nature of light, in which the Complementary Principle allows the acceptance of both models as contributors to the knowledge of this physical entity (light) without imposing the need that one of them prevails over the other. This was the case with the two dialectic perspectives adopted to evaluate the strategy, with each providing different information. It was possible to obtain evidence for the number of successful learning cases using the quantitative approach, and also to determine a general consensus about the comprehension of models and ability to apply them, using the qualitative approach.

Data was collected from a trial study undertaken with one hundred students of the course *Introductory Physics Laboratory* in the first semester of 2003. These labs are one-credit courses where students meet for two hours every week for a 16-week semester.

The Strategy – Trial and Results

This section focuses on the strategy used, a description of the teaching activities and a brief summary of some of the results obtained in each phase.

Phase I: Learning About Concept Maps and Gowin's Vee.

Teaching activities – Students were provided with a text about concept maps designed for a constructivist learning of a topic (Ramírez de M, 2005). They were also trained in the use of *Inspiration^{TM} 8.0* and *CmapTools*, both of which are concept mapping software programs that help students organize their ideas in a visual manner. In addition, students were provided with appropriate written text materials to learn about Gowin's Vee. An eight-hour workshop was conducted to introduce students in the design of concept maps and Gowin's Vee for specific purposes within the context of real situations and examples from simple everyday experiments.

Assignment – Students were asked to produce concept maps for a single topic related to physics concepts already known by them. Similarly they were asked to construct a Gowin's Vee diagram for a simple everyday experiment, which they had been involved with.

Discussion and results – In this phase students acquired the necessary skills to build concept maps and Vee diagrams (the results indicated that 84% of the students managed to do the assigned maps successfully). The other heuristic tool, Gowin's Vee, seemed to be more difficult for them to understand and to use in order to follow up and report a simple experiment (62% of the students completed them correctly).

Phase II: Eliciting Basic Conceptual Knowledge of Mathematical Functions

In this phase, the focus was on two activities: (i) comprehension of the three basic mathematical functions and (ii) transformations of power and exponential function in order to graph them on semi-logarithmic and logarithmic paper.

Basic Mathematical Functions

Teaching activities to review mathematical notions about basic mathematical functions – Students were shown a group of mathematical functions, which should be known by them, as they were part of the topics taught at the introductory university mathematics courses prior to studying physics. This group of functions comprised the linear function $y = mx + b$, the power function $y = b. \, x^m$ and the exponential function $y = b. \, e^{m.x}$. These functions were taught in a single presentation in a traditional way, without using concept maps or any other heuristic tool.

Assignment – Students were asked to express their understanding of the three functions mentioned above, by using the concept mapping technique, previously taught to them.

Discussion and Results – The concept maps produced by students were compared with "expert maps" previously prepared. These maps were grouped into two categories:

(a) *Satisfactory*: Those maps which clearly show the main idea, subordinate concepts and relationships among them for a given mathematical function (65%).
(b) *Insufficient* (35%): In this category, the distinction was made between maps exhibiting conceptual mistakes in mathematical prior knowledge (12%), and maps showing an incorrect application of the concept map heuristic tool (23%).

In the examples that follow, the reader needs to take into account that this research was conducted in Spanish. Concept maps have been translated into English to illustrate students' work. Samples of students' concept maps classified as *satisfactory* are presented in Figs. 10.9, 10.10 and 10.11.

Transformations of Basic Mathematical Functions

Teaching activities: Transforming Exponential Equations and Power Equations into Linear Equations – The teacher drew attention to transformations of power and exponential functions which allow the drawing of linear graphs.
His explanation is provided below:

First: The function $y = b.\ e^{m.x}$ can be transformed into the equivalent function $Ln\ y = Ln\ b + m.\ x$ which is linear in the variables $Ln\ y$, and x. In this work, $ln\ x = log_e\ x$ denotes a natural logarithm.
This last function shows a linear tendency when a graph of $Ln\ y$ and x is plotted on a millimetre paper sheet and also shows a straight line in a graph y vs x on a semi logarithmic paper.
Second: The function $y = b.\ x^m$ can be transformed into the equivalent function.
$Ln\ y = Ln\ b + m.\ Ln\ x$, which is linear in the variables $Ln\ y$, and $Ln\ x$.
This last function shows a linear tendency, when plotted in millimetre paper sheets, using the variables $Ln\ y$ and $Ln\ x$ and also shows a linear trace when we make a graph in logarithmic paper of the variables y and x.

Assignment – Students were asked to explain how it is possible to deduce from a set of *y, x* values on logarithmic and semi-logarithmic paper, which one of the functions (power or exponential) was compatible with the assigned set of values. They were again asked to provide explanations by means of a concept map.

Discussion and Results – From the results, 58% of the students provided satisfactory maps. Some students (20%) confused the log and semi-logarithmic papers and arrived at wrong conclusions. The other students made maps, which did not show a clear conceptual structure about transformations of non-linear functions. From previous studies, it has been observed that it is not easy for students to express in a map what they have learnt in class from the teacher's presentation. Figure 10.12 shows a map classified as "satisfactory" produced by a student.

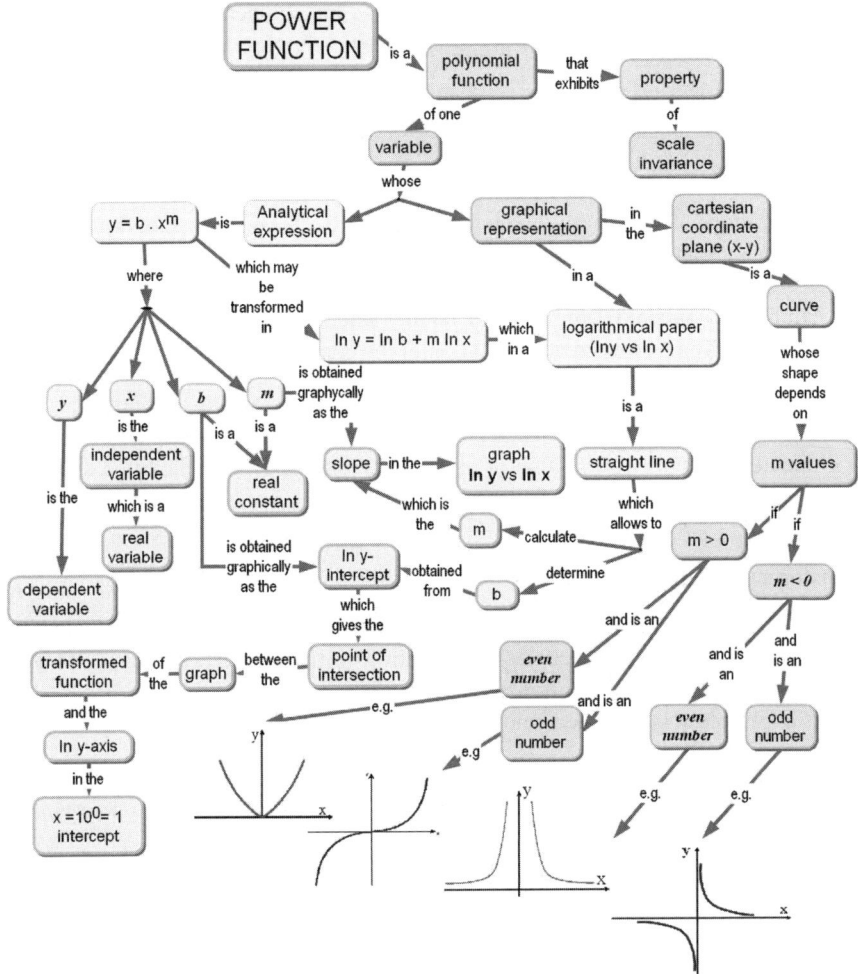

Fig. 10.9 Concept map of the power function made by a student

Phase III: Acquisition of the Concept "Model" in Physics

This phase focused on three aspects: (1) the concept of *model* in physics science, (2) qualitative models and (3) the concept of *mathematical model* for a physical phenomenon.

The Concept of *Model* in Physics Science

Teaching activities – Based on the epistemological conceptions of model, presented by Hertz (1900), Feynman (1970), and Bunge (1997), a short text was designed with standard definitions of model as well as qualitative and quantitative models, as they are understood in physics.

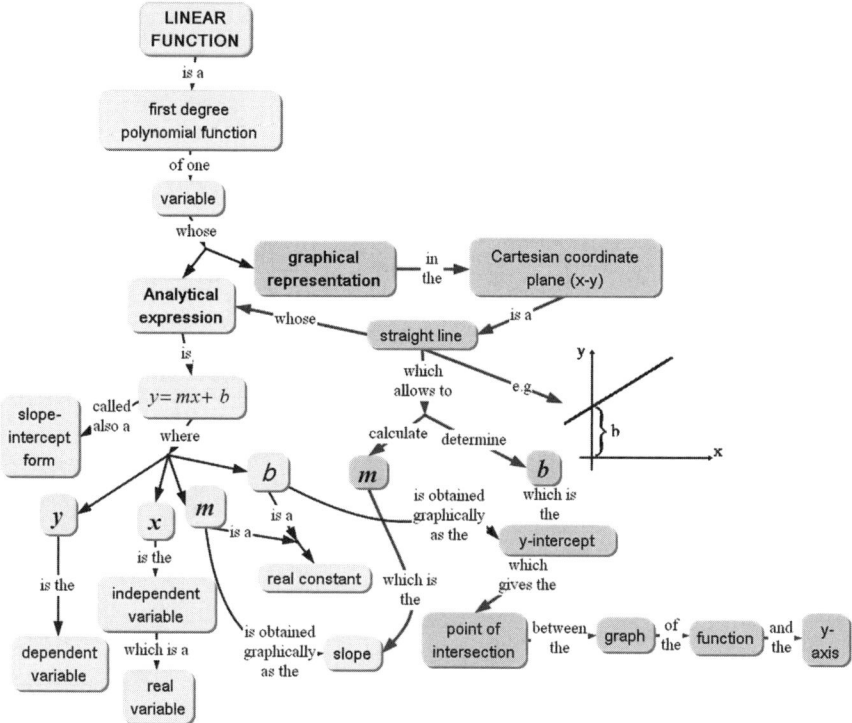

Fig. 10.10 Concept map of the linear function designed by a student

Assignment – Students were asked to work in small groups, read the text and generate a concept map for the concept "*Model in physical science*".

Discussion and Results – Most groups, corresponding to 80% of the students, managed to produce maps demonstrating acceptable understanding of the concept "*Model*" This is not surprising and supports previous findings which indicate that the easiest way for beginners to design a concept map is from the information given in a written text. In comparison, beginners find it more difficult to make a map from the contents given in a presentation or a lesson. A lower percentage (58%) was obtained when students were asked to build concept maps in Phase II about mathematical functions following their class notes and what the teacher said in the lab class. Two examples of satisfactory concept maps for this topic are shown in Figs. 10.13 and 10.14.

Qualitative Models and Gowin's Vee for Constructing an Explanation

Teaching activities – Students were presented with a phenomenon, which could be explained with a qualitative model. It is related to a physical system called the black box tunnel. In physics, a black box is a system whose internal structure is unknown, or need not be considered for a particular purpose.

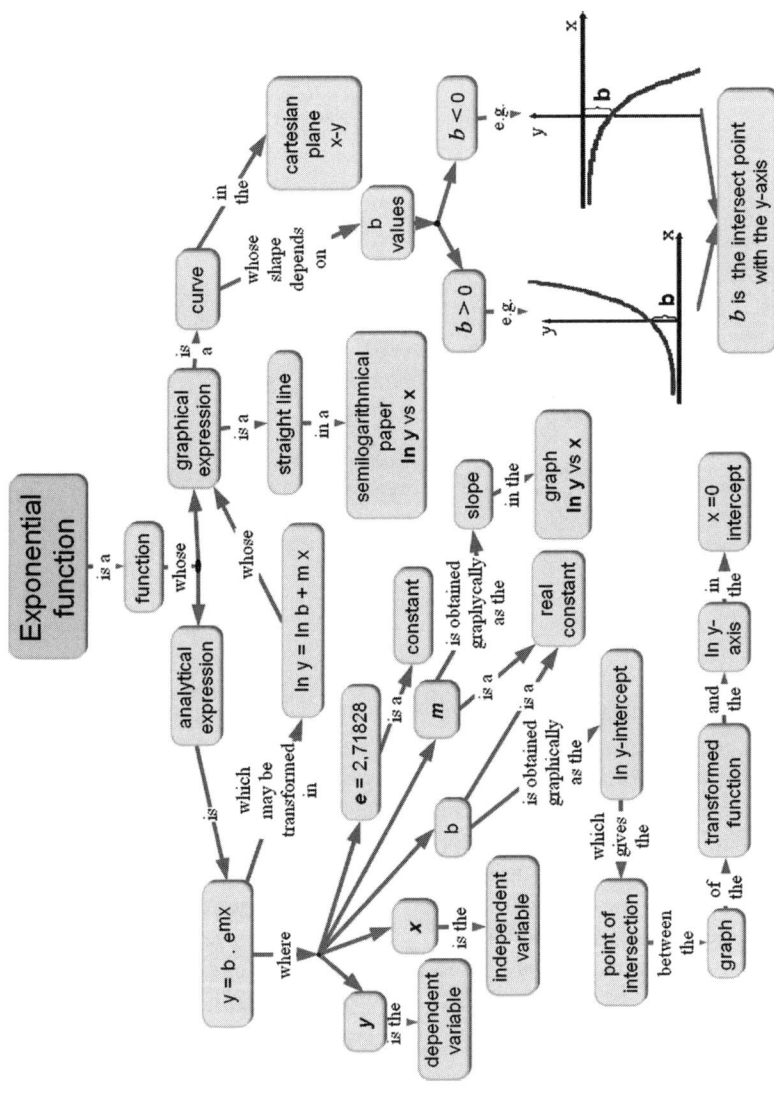

Fig. 10.11 Concept map of the exponential function made by a student

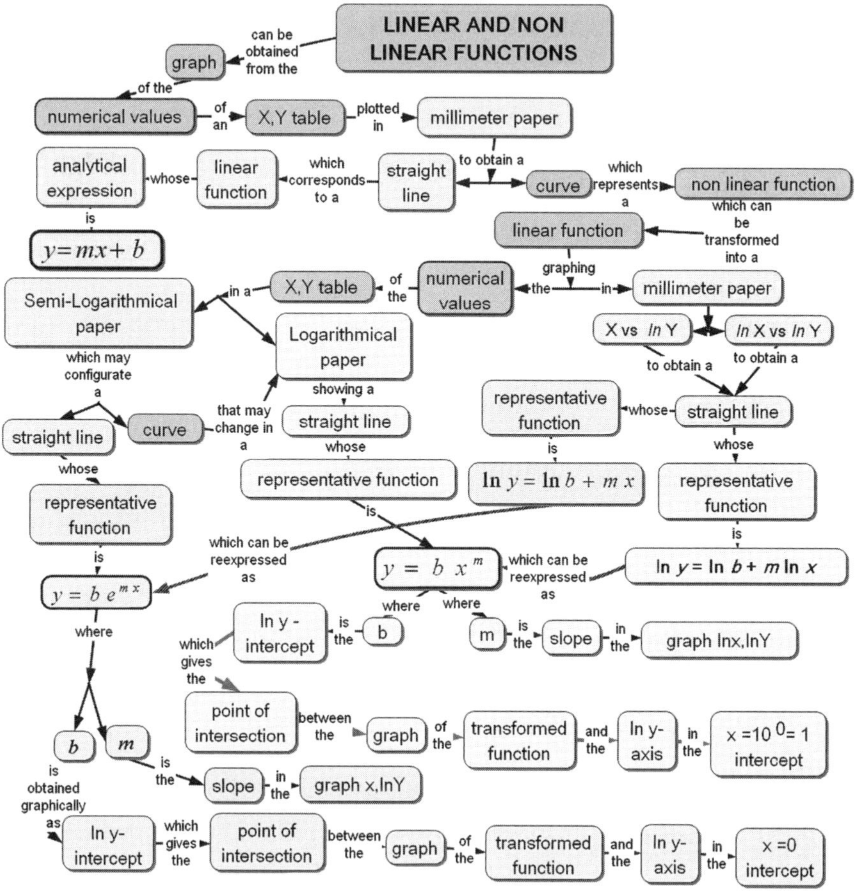

LINEAR AND NON LINEAR FUNCTIONS

can be obtained from the → graph

of the

numerical values —of an→ X,Y table —plotted in→ millimeter paper

analytical expression —whose→ linear function —which corresponds to a→ straight line —to obtain a→ curve —represents a→ non linear function

which can be transformed into a

is

$$y = mx + b$$

linear function

graphing

numerical values —the→ —in→ millimeter paper

Semi-Logarithmical paper —in a→ X,Y table —of the→

X vs *ln* Y ←→ *ln* X vs *ln* Y

Logarithmical paper —to obtain a→ to obtain a

which may configurate a

showing a

representative function —whose→ straight line

straight line — curve —change in a→ straight line

that may

is

whose

whose

representative function

$$\ln y = \ln b + m x$$

representative function

is

$$y = b\ e^{m x}$$

representative function —is→

which can be reexpressed as

$$y = b\ x^{\ m}$$

which can be reexpressed as

$$\ln y = \ln b + m \ln x$$

where

where

ln y - intercept —is the→ b

m —is the→ slope —in the→ graph lnx,lnY

where

point of intersection —between the→ graph —of the→ transformed function —and the→ ln y-axis —in the→ $x = 10^{\ 0} = 1$ intercept

b m —is the→ slope —in the→ graph x,lnY

is obtained graphically

is obtained graphically as

ln y-intercept —which gives the→ point of intersection —between the→ graph —of the→ transformed function —and the→ ln y-axis —in the→ $x = 0$ intercept

Fig. 10.12 Map of linear and non-linear functions made by a student

The following procedure was followed:

(a) Students were given a written text with the following description.

A   B

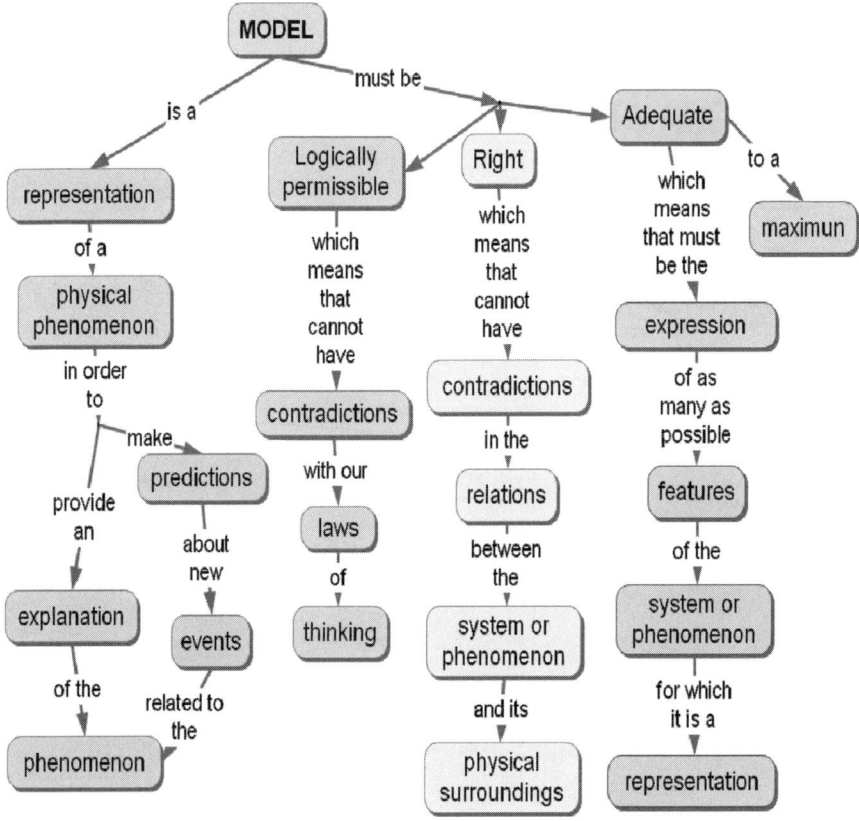

Fig. 10.13 Concept map about *model*

There is a tunnel with open ends in both sides. Some experiments made with small toys cars allow us to state the following facts:

- If a car enters the tunnel in point B at a low speed, the car comes back to point B.
- If the same car enters the tunnel in point B with a higher speed, then it goes across the tunnel and emerges in point A.
- If the same car enters the tunnel in point A moving slowly, the car returns to point A.
- If the same car enters the tunnel in point A with a higher speed, then it does not goes across the tunnel or return to point A, rather on disappears inside the tunnel.

Create a qualitative model that could explain this phenomenon and might be used to make predictions about the way the tunnel works.

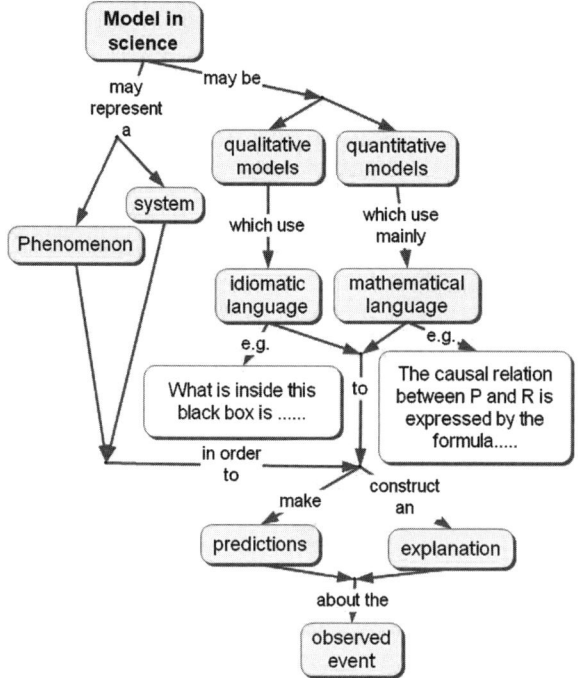

Fig. 10.14 Concept map about *model* in science

(b) Students were given a written guide explaining how they could proceed in order to make a qualitative model.

> In order to build a qualitative model of a phenomenon or physical systems you must first observe it. Then you must imagine what it looks like and find out if there is something similar that you already know. This is to find out if there is something we can remember with an analogous behaviour, which has been observed previously. This may help us to make a representation of the phenomenon or physical system. We cannot forget that whatever we produce as an answer must comply with the general characteristics of a model.

(c) An adaptation of Gowin's Vee was presented to help students with their reasoning about the phenomenon under consideration (see Fig. 10.15).

Assignment – Working in small groups, students were asked to build a qualitative model that could explain the structure and functions of the tunnel and could be used to make predictions about the way the tunnel works. It was suggested that they should follow the steps and answer the questions mentioned in Gowin's Vee described above.

Discussion and Results – Fifteen groups out of the twenty-five groups (of four students each) managed to produce acceptable tentative models for the phenomenon observed. Six groups had an idea of what was going on, but could not use the

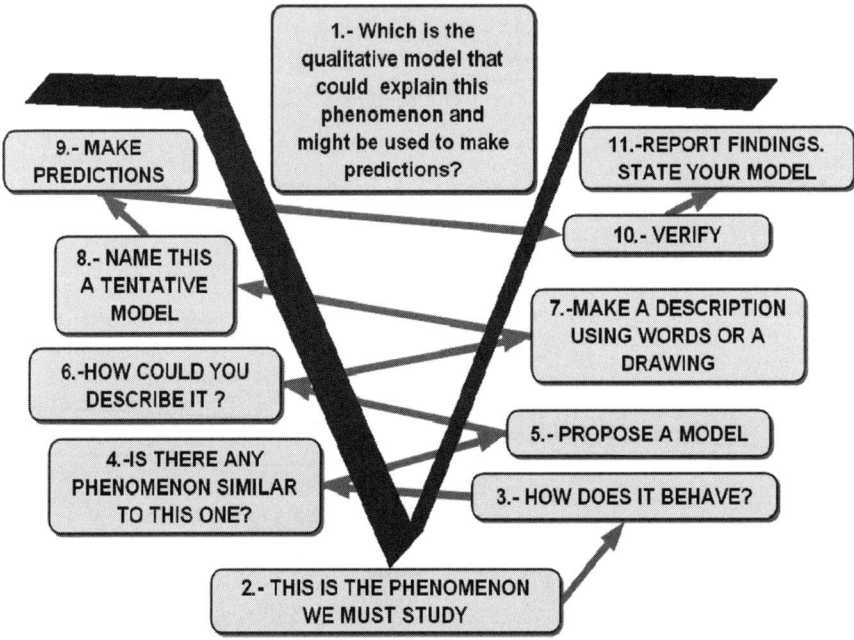

Fig. 10.15 Gowin's Vee for the construction of qualitative models

Gowin's Vee provided (Fig. 10.15) as a tool to orientate the process of reasoning. They simply produced a written report explaining what they thought was going on. Two groups produced wrong qualitative models that did not account for the events explained.

The other students could not manage to generate adequate qualitative models corresponding to the phenomenon under consideration. It became clear from the interviews conducted with those students than they could not connect the theory about models (which they successfully reflected on in their concept maps) with the interpretation of a real phenomenon and the need to build a qualitative model to explain it. Figure 10.16 illustrates Gowin's Vee produced by a group of students for the first part of the experiment and the corresponding explanations to account for the phenomenon observed.

Figure 10.17 illustrates the Vee diagram the same students made for the second part of the experiment (when the car enters the tunnel in point A).

Quantitative Models

Teaching activities – A teacher worked alongside students to develop an experiment about a pendulum that swings through a small angle (in the range where the function $\sin \theta$ can be approximated as θ) in order to answer the focus question: *What are the*

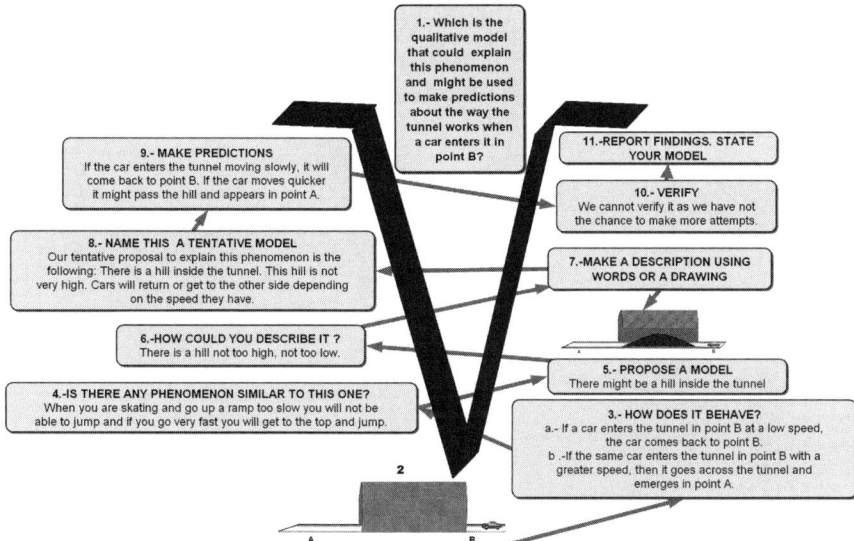

Fig. 10.16 Qualitative model for a car entering the tunnel in point B

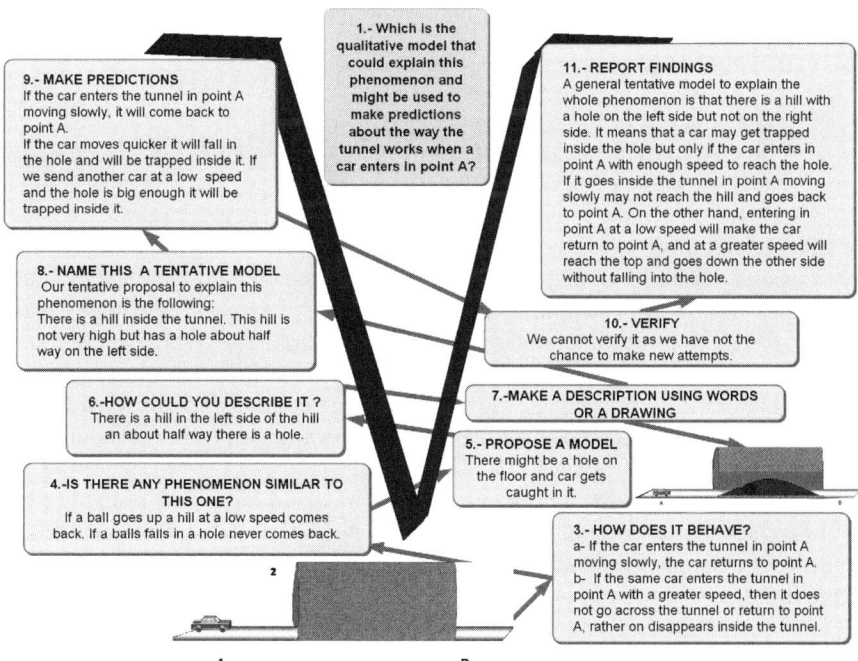

Fig. 10.17 Qualitative model for the car entering in point A and general explanation

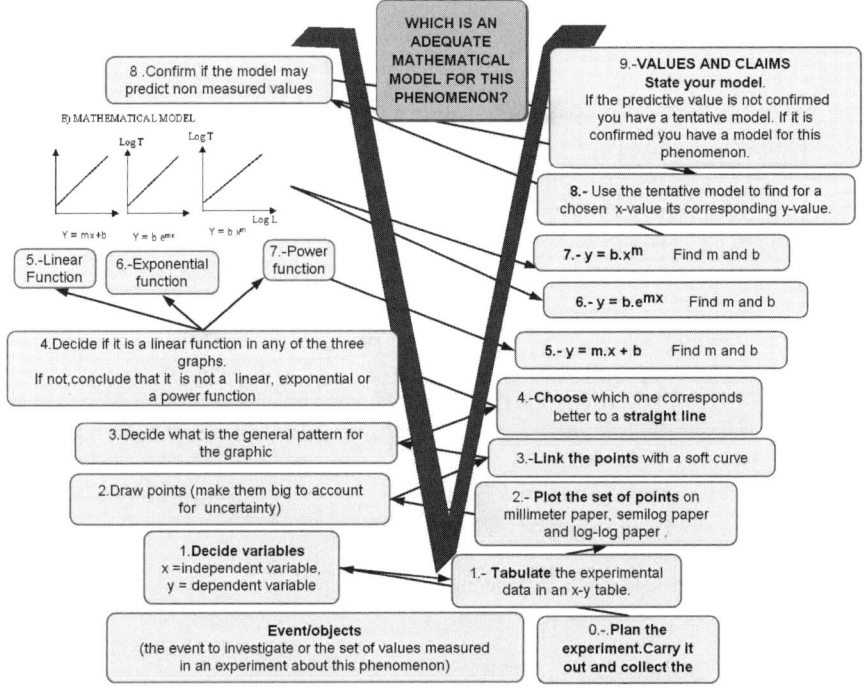

Fig. 10.18 Adaptation of Gowin's Vee for quantitative models

relationships between the mass of bob, the length of pendulum, the displacement of the pendulum (amplitude) and the period of a pendulum?.

The teacher and students discussed what they could do in order to study the different variables involved. They considered various options in studying, for instance, mass against length of pendulum, until students found out that they had to study separately the effect of changing a variable such as displacement (amplitude), mass of bob, or length of pendulum on period of a pendulum. The teacher showed and explained the adaptation of Gowin's Vee we use (see Fig. 10.18) and put an emphasis on the importance of using Gowin's Vee guiding questions for each Vee element in order to plan, carry out the experiment, collect raw data, transform it and graph it and analyse the results obtained.

Assignment – Once the information was recorded and organized, students were asked to plot the set of points and were asked to find the appropriate mathematical model to explain the different relationships between the variables and finally express the corresponding quantitative models. The teacher helped them to complete Gowin's Vee diagram up to the value claims and conclusions (Sanabria & Ramírez, 2006).

Discussion and Results – Seventy-four out of a hundred (74%) students managed to produce the adequate mathematical function (power function) for the set of values T against length of the string (L). The others incorrectly generated an exponential model from their analysis of the data graphed on a semi-logarithmic paper as they were convinced that an exponential function was an appropriate model. It is worth mentioning that of this group, sixty students managed to define adequately the meaning of the variables to give an appropriate quantitative model to explain the phenomenon observed.

The others did not understand why it was necessary to change the y-value for period (T) and the x-value for length of the string (L), showing that they had misunderstood the physical meaning of the variables involved in the experiment as well as the need to modify the mathematical power function variables to express it as a mathematical model to account for the phenomenon observed. That is, students thought that expressing the results of their experiment as the mathematical function $y = 1.98\ x^{0.5}$ is the same as answering the mathematical model to express the relation between T and L is $T = 1.98\ L^{0.5}$ which is not. It was necessary for the students to distinctively express the meaning of the x-value and the y-value. Figure 10.19 shows the Vee diagram made by a student for this experiment.

It needs to be restated that all work was done in Spanish and there may have been some changes in the real meaning of some expressions when translating some vees of students' work. In general terms, it was found that students' use of Gowin's Vee

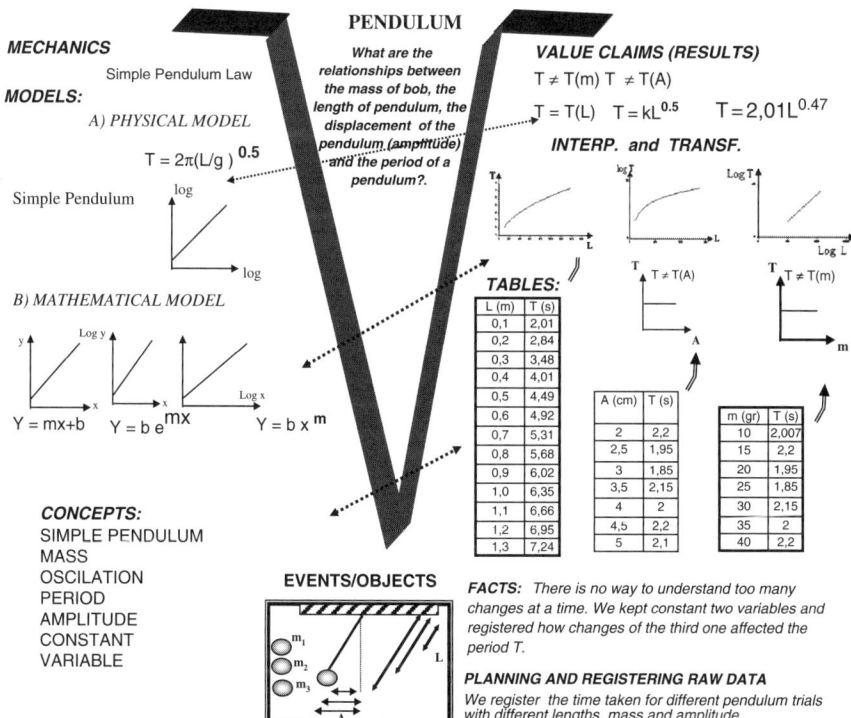

Fig. 10.19 Translation of the Vee diagram for the experiment of a pendulum made by a student

to help them construct a *quantitative model* proved to be more effective than the use of the same tool to make *qualitative models*. Further research is needed to find a more effective way of using Gowin's Vee as an aid to develop qualitative models.

Phase IV: Modelling Physical Phenomena

This phase was devoted to the development of real life experiments that may be explained by means of quantitative and qualitative models using the lab activities organized for the whole semester.

Teaching activities – In this phase the teacher presented to the students every phenomenon and set up the restrictions, if necessary. Then it was made clear, which machines, objects or lab equipment students could work with, and formulated the appropriate question for the students to answer, letting them decide what to do in order to answer each question. In every lab, the teacher emphasized that final work and results should be given using concept maps and Gowin's Vee diagrams.

Assignment – Students were asked to carry out the different experiments, using concept maps, when necessary, and Gowin's Vee diagrams. They were asked to create *qualitative models*, which account for (i) the internal structure and functioning of a car seat belt and, (ii) a toy friction car. They were not allowed to examine either of these physical systems or to pull the pieces apart.

Also, as an integral part of the studies, every week they were asked to develop *quantitative models* for the different experiments we studied in this physics laboratory. Figure 10.19 shows the Gowin's Vee diagram produced by a student for the simple pendulum experiment.

Discussion and Results – By the end of the course, students showed signs of an acceptable conceptual understanding of modeling physical phenomena through concept mapping and Gowin's Vee diagrams. Final grades for this course showed that 81 students passed the course compared with 19 that failed it. This result was an improvement over previous courses where the passing rate for all students was 65%, lower than the 81% obtained as a result of the application of this strategy. Our findings confirm those reported by Afamasaga-Fuata'i (2006), which stated that verifying and justifying mathematical solutions were greatly facilitated through the combined usage of concept maps and Vee diagrams.

Conclusions

In the last two decades, this research team has tried different teaching and learning strategies to introduce physics concepts to our students. Statistical analysis of the strategies used in previous years for the topic *Mathematical Modeling of Physical Phenomena* gave a mean passing rate of 65% for students taking regular courses. On the other hand, a qualitative analysis of results allowed us to propose that the main problems faced by students failing the physics course were mainly dueto:

- Lack of motivation to study this topic.
- Insufficient prior knowledge of the linear, power, exponential and functions.
- Students' difficulties in explaining the process followed in order to generate their models.
- Students' difficulties in communicating the results of their experiments.

The strategy described in this chapter was designed to overcome these difficulties, based on the use of the heuristic tools concept map and Gowin's Vee. This strategy uses concept maps to improve understanding of concepts and basic conceptual structures involved in the mathematical modeling process of physical phenomena, and Gowin's Vee as an adapted tool that facilitates the process of building student's own knowledge of a mathematical model for a particular experiment.

The results after the application of the strategy showed 81% of students passed the course. A qualitative analysis of process and results allows us to propose that improvement in overall performance along the lab course may be due to:

- An increase in student's motivation to develop the experiments with the aid of the heuristic tools concept maps and Gowin's Vee.
- Consciousness of the need to improve knowledge about mathematical functions and the plotting of curves in order to find adequate models to explain physical phenomena.
- An improvement in students' ability to communicate results and to interpret their findings while studying physical phenomena.

Beyond the rather broad measures of "passing" or "success", evidence from a variety of assessments suggests that the quality of learning was high relative to that for students in previous courses. Students finished with significantly higher levels of confidence in their abilities to understand, plan, carry out and analyze an experiment. There were also many benefits reported by students including ownership of knowledge, the development of skills to build concept maps and the use of Vee diagrams to plan and develop ideas and basic experiments in order to model physical phenomena.

Even though the results demonstrated the benefits obtained from applying this strategy, it is worth noting that in order to obtain satisfactory results it is necessary to take into consideration two facts:

1. A considerable amount of time must be devoted to train teachers in the application of the strategy and to train students in the use of concept maps and Gowin's Vee; and
2. Initially, there will be an increased workload for the teacher in order to develop an instructional sequence for the learning experiences.

This research group has continued to use and improve the strategy described in this chapter throughout successive trials with satisfactory results. By the end of each semester students get more familiarity with the use of Gowin's Vee and concept maps to help them understand physical phenomena and construct adequate mathematical models to account for them.

This strategy reinforces the notion that in order to explain physical phenomena, the process of proposing appropriate mathematical models, verifying and justifying them is greatly facilitated through the combined usage of concept maps and Vee diagrams. Finally, the ability to model physical phenomena is facilitated through the development of this strategy that serves to promote the process of "thinking about thinking", or more precisely metacognition. This strategy can be adapted for other purposes and in other educational contexts.

Novak's and Gowin's heuristic tools have been applied for more than twenty years and, like a good wine, have done nothing but improve with time. Concept maps and Gowin's Vee are readily accessible tools which can be used to support teaching and learning in science. Thus, it is our hope that teachers begin building their own knowledge base about these heuristic tools and trial effective ways of using them with students. That is the challenge.

References

Afamasaga-Fuata'i, K. (2006). Developing a more conceptual understanding of matrices and systems of linear equations through concept mapping and Vee diagrams. *Focus on Learning Problems in Mathematics*, Summer, *28*(3 & 4), 58–89.

Bunge, M. (1997) *Teoría y Realidad*. Barcelona: Ariel.

Cañas, A. J., Safayeni, F., & Derbentseva, N. (2004). Concept maps: A theoretical note on concepts and the need for cyclic concept maps. Retrieved December 10, 2004, from University of West Florida, Institute for Human and Machine Cognition Web site: http://cmap.ihmc.us/Publications/ResearchPapers/Cyclic%20Concept%20Maps.pdf

Feynman, R. (1970). *Lectures on physics*. Massachussets: Adison-Wesley.

First International Congress on Concept Mapping (CMC) (2004) *Proceedings of the CMC*. Retrieved January 09, 2005, from University of West Florida, Institute for Human and Machine Cognition Web site: http://cmc.ihmc.us/CMC2004Programa.html

Hertz, H. (1900) *The principles of mechanics presented in a new form*. USA: Dover.

Novak, J. D., & Gowin, D. B. (1988). *Aprendiendo a aprender*. Barcelona: Martínez Roca.

Ramírez de M. M. (2005). El Mapa Conceptual como Herramienta Heurística para facilitar el Aprendizaje. San Cristóbal: Unet Fondo Editorial.

Ramírez de M. M., Aspeé, M., & Sanabria, I. (2006). Concept maps: An essential tool for teaching and learning to learn science. *Focus on Learning Problems in Mathematics*, Summer, *28* (3 & 4), 32–57.

Ramírez de M. M., & Sanabria I. (2004). *El* Mapa Conceptual como Elemento Fundamental en el Proceso de Enseñanza-Aprendizaje de la Física a Nivel Universitario. Retrieved December 03, 2004, from University of West Florida, Institute for Human and Machine Cognition Web site: http://cmc.ihmc.us/papers/cmc2004-086.pdf

Sanabria, I., & Ramírez de M. M. (2006). Una estrategia instruccional para el laboratorio de física I usando la "V de Gowin". Revista Mexicana de Física S, 52(3), 22–25.

Second International Congress on Concept Mapping (CMC) (2006). *Proceedings of the CMC*. Retrieved February 02, 2008, from University of West Florida, Institute for Human and Machine Cognition Web site: http://cmc.ihmc.us/CMC2004Programa.html

Chapter 11
Applying Concept Mapping to Algebra I

William Caldwell

Concept mapping has great potential for increasing meaningful learning in mathematics at all levels. The fundamental quality that supports this kind of learning is the necessity for concept mappers both to clarify for themselves, and to relate simply and explicitly the connections that exist among the various concepts being mapped. This requires that the concepts first be connected to the current knowledge of the person doing the mapping. This meaningful learning takes place for the teacher as well as the student. The teacher also can use the student maps to determine to what extent the student grasps the ideas and their relationships to one another. Thus concept mapping plays a role in teaching, learning, and also in assessment.

Introduction

Concept mapping can be a useful tool in measuring student progress in middle grades mathematics courses, as well as being a valuable planning and teaching tool for teachers of these courses. To explore these ideas, we will focus on Algebra I, a specific middle school/high school course that typically is taught in the eighth or ninth grade, and represents a significant hurdle for students because of the abstract concepts it includes.

A *concept map* is, mathematically, a directed graph in which the vertices (nodes) represent individual concepts, and the edges are labeled with descriptive *connecting phrases* and are assigned directions to describe the relationship between the two concepts in adjacent nodes. For example, the nodes "Rational Numbers" and "Real Numbers" in a map might be connected as follows:

W. Caldwell (✉)
University of North Florida, Jacksonville, FL, USA
e-mail: wcaldwel@unf.edu

K. Afamasaga-Fuata'i (ed.), *Concept Mapping in Mathematics*,
DOI 10.1007/978-0-387-89194-1_11, © Springer Science+Business Media, LLC 2009

The connecting phrase is the verb "are," and the direction of the arrows indicates the direction in which the map should be read. Notice that reversing the arrows in this example yields an incorrect relationship between the concepts.

A concept map is developed to answer a particular question (often called a *focus question*) through describing relationships that exist among a selection of concepts pertinent to the question. These will be mathematical concepts in our maps.

We first must agree on the Concept Mapping principles we will adhere to. The most important of these principles, but one that can be difficult to follow, is the protocol that *every basic sequence (i.e., node – connecting phrase – node) in a concept map should comprise a correct mathematical sentence when read in the direction of the connecting links.* To clarify this, consider first the map

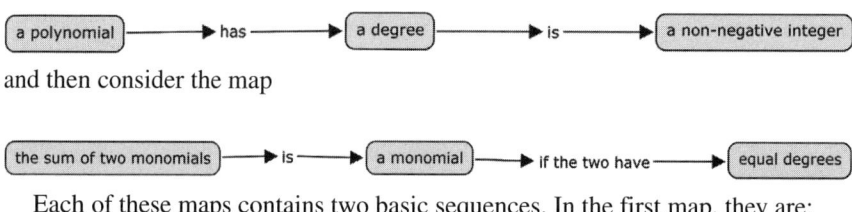

and then consider the map

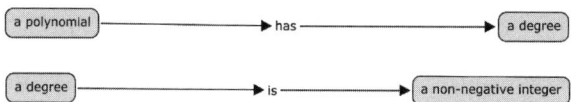

Each of these maps contains two basic sequences. In the first map, they are:

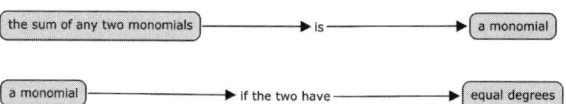

These two basic sequences each yield satisfactory mathematical sentences when read in order.

On the other hand, the two basic sequences in the second example are:

Neither of these two basic sequences yields a correct mathematical statement when read in the directed order. The first yields a sentence, but is mathematically incorrect; and the second is a phrase that is not a complete sentence. Thus, the map does not follow our protocol, even though when the full original map is read in sequence, it does give a correct mathematical sentence.

Another important principle we will follow is that we will consider a concept map to be "Good" when it is *Correct, Clear,* and *Complete.* The following conditions should be satisfied.

Correct: Is every basic sequence in the map a correct mathematical sentence?
Clear: Are the relationships among the concepts described clearly? Do they give precise and relevant information that helps to form an answer to the focus question?

Complete: Do the specific concepts used in the map and the relationships described among them provide an answer to the focus question that is suitable for those to whom the map is directed?

Developing an Algebra I Course Through Concept Mapping

The typical Algebra I course usually is focused on five or six major mathematical concepts, and relies upon some specific prerequisite knowledge. The picture of Algebra I we will discuss with concept mapping will be structured upon the following five basic concepts:

Equations and their solutions
Inequalities and their solutions
Exponents
Polynomials
Graphs

Each of these can be described through identified sets of sub-concepts, and the concepts related through concept maps. Note that we will discuss developing these concepts at the Algebra I level – although all five concepts appear at various levels throughout the mathematics curricula in high schools and colleges, and thus could be discussed at many levels of expertise.

Course Prerequisites

The prerequisite knowledge required for our course will be fundamental ideas about:

Constants and variables
Real numbers
The number line
The Cartesian plane

Concept maps of prerequisites showing the essential sub-concepts and relationships among them also will be necessary for us to describe the course. The prerequisites we select as fundamental are:

A beginner's understanding of the terms "constant" and "variable."
Familiarity with the basic arithmetic of the real numbers, including the integers and rational numbers; and, some knowledge of irrational numbers.
Experience with the number line and the Cartesian plane.

To use concept mapping as a resource for planning a course and developing individual lessons, the teacher will follow the process above – first identify the major concepts that will form the framework for the course, and then give a general picture of the basic prerequisite knowledge required for the student to be successful in dealing with those concepts at the level at which the course will be delivered. For each of the major framework concepts decided upon, a list of related sub-concepts that are essential to understanding the major concept will be developed. For example, suppose the teacher is developing the major concept "polynomials." The specific sub-topics identified will be influenced by the level of the course being planned; the identified sub-concepts for "polynomials" in Algebra I might be:

Monomials (for use in the definition of "polynomial.")
Operations on polynomials
Degree of a polynomial
Factoring polynomials (quadratics and a few special cases)
Roots of polynomials
Graphs of polynomials (linear and quadratic ones)

The teacher then will write a narrative statement of ideas and techniques surrounding each sub-concept, and describe how they will be included in the development of the major concept. In the list above, some of the surrounding ideas have been given in parenthetical comments following the sub-concepts; they will be expanded upon in the statement. Here is an example of such a narrative statement involving the first three of the sub-concepts in the list above.

Narrative on the Development of Polynomials

Monomials will be introduced in terms of the *constant coefficient*, the *variable*, and the *exponent*. The idea of "degree" will be introduced, and the processes of multiplying any two monomials and of adding any two polynomials of equal degrees will be presented. Students will learn to evaluate monomials for specific assigned values of the variable. Polynomials will be defined as being either a monomial, or as the algebraic sum of two or more monomials. Then, the addition of polynomials will be defined in terms of the operations on monomials that comprise them. (A property like the distributive law for addition over multiplication of real numbers will be used in defining multiplication of polynomials.) The degree of a polynomial will be defined, and the degree of the sum or product of two polynomials will be related to the degrees of the original two.

This paragraph describes not only the content the teacher expects to include in the development of the three sub-concepts, it also suggests the order in which they will be presented, and relates some of the connections among those concepts. In this statement, there are specific concepts that must be introduced to the student, viz., monomial, constant coefficient, variable, exponent, and degree. There also are processes that must be explained – adding monomials of equal degree, multiplying any two monomials, forming polynomials as the algebraic sum of monomials, and evaluating monomials and polynomials. For each of these concepts and processes, a definition or a process explanation will be thought out by the teacher, and this information then will be translated into process or concept maps. Then, the maps will be linked to one another as appropriate to present a comprehensive picture of how monomials will be treated in the course. Here is an example of how the teacher might proceed.

Working With Monomials

Definitions *Monomial in x. An expression of the form cx^n, where c is the constant coefficient (represents a particular but perhaps unspecified real number), x is the variable (a "place holder" that may be assigned a specific real number value when "evaluating" the monomial for that particular value of x), and n is the exponent (a non-negative integer). Specific examples of monomials in x are $3x^5$, and $(-4)x^{17}$. In the general form of a monomial, it is not necessary to give the constant the label "c," – any letter (although usually one selected from the beginning part of the alphabet) could be used. Thus bx^{14} is also a monomial in x. The label of the variable can be changed, also. For example, $7y^8$ is a monomial in the variable y. (It is common practice to use letters from the latter part of the alphabet to denote variables.)*

Evaluating a monomial in x. The monomial cx^n is thought of as c times the nth power of x. To evaluate this at x = 4, we calculate $c \times 4^n$. Therefore, if c = 6 and n = 3, we are calculating 6×4^3, so we get 384. We can evaluate cx^n for any real number replacing x.

Constant monomial. A monomial of the form cx^0. The special case where c = 0 is called the monomial 0.

Degree of a monomial. The value of the exponent for a monomial different from the monomial 0. For example, $7x^{23}$ has degree 23. The monomial 0 does not have a degree.

At this point, the teacher could develop a concept map describing the relationships among the concepts comprising the definitions above. Such a map is given in Fig. 11.1.

Once this map has been determined, the teacher will clarify the role that monomials will play in the development of polynomials. First, the (binary) algebraic operations of addition and multiplication of monomials must be defined. Since multiplication of two monomials always results in a monomial. This process is addressed first, and then the meanings of the quotient, sum, and difference of two monomials are addressed. The quotient of two monomials results in either a monomial or in the reciprocal of a monomial, which is a type of rational expression; and the sum or difference of two monomials is either a monomial, or is a polynomial with two terms. Once the results of all four of these processes are determined, a composite process map is developed. Figure 11.2 represents such a map. Notice that in this map we have cases where the result of performing an operation on two monomials does not result in a monomial.

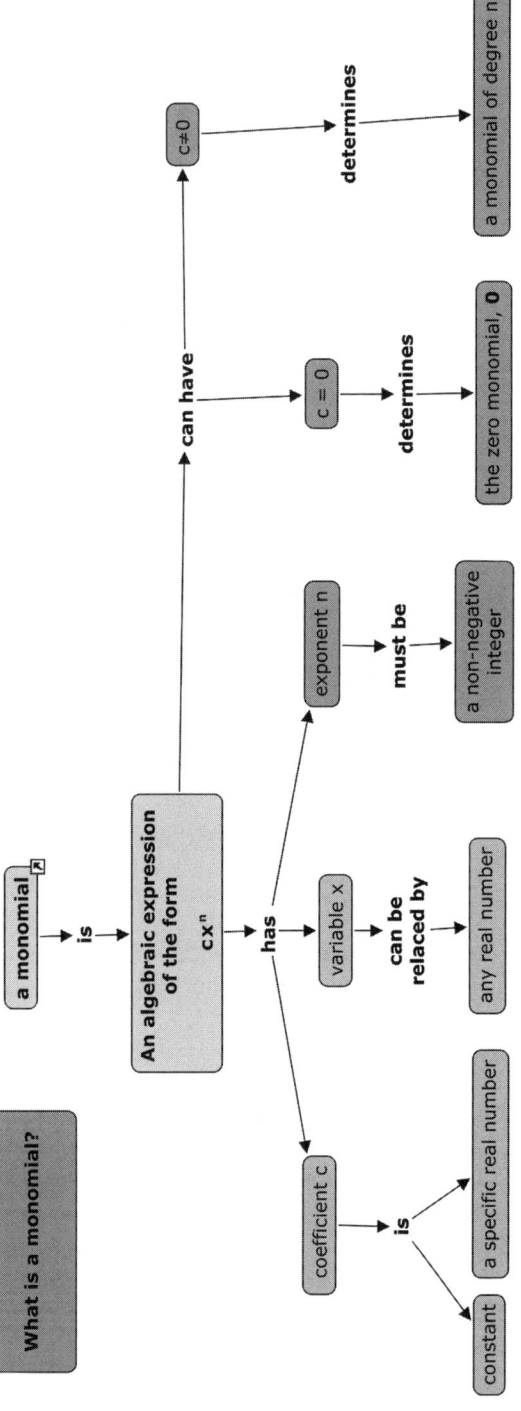

Fig. 11.1 Definition and characteristics of a monomial

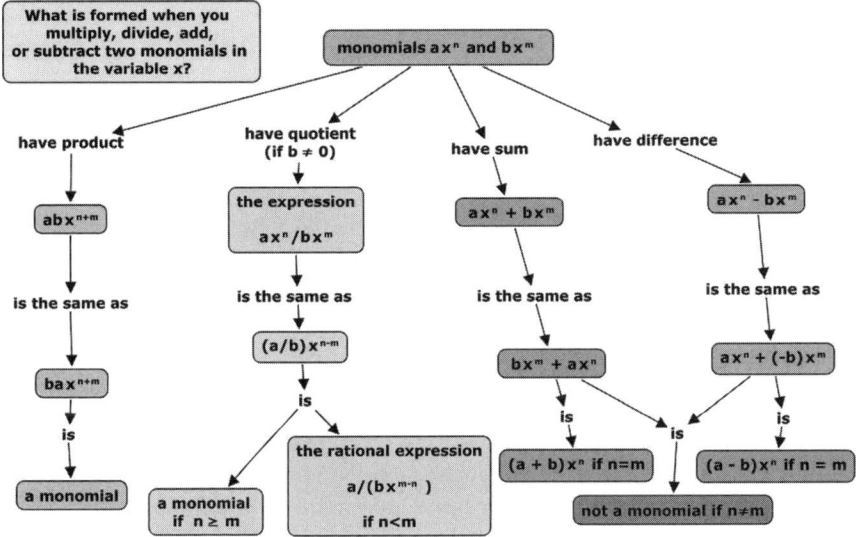

Fig. 11.2 Algebraic operations on monomials

This places us at the point of defining polynomials and continuing the development of this basic concept in our Algebra I course.

Polynomials are Developed From Monomials

Definitions *Polynomial in the variable x. A monomial, or the (algebraic) sum of two or more monomials in x. The monomials that comprise a polynomial are called the monomial terms of the polynomial. Examples of polynomials are $7x^{14}$, and $4 + 5x^7 + 3x^{10} + 2x^8$. The first of these has one term, the second has four.*

When defining the addition and multiplication of polynomials, it is useful to agree on a standard form in which polynomials will be written.

Standard form for a polynomial in x. A polynomial is in standard form if its individual terms have distinct degrees, and are listed in either increasing or decreasing order of degrees. The second example above is not in standard form. Its two possible standard forms are $3x^{10} + 2x^8 + 5x^7 + 4$, and $4 + 5x^7 + 2x^8 + 3x^{10}$. The polynomial $5x^{10} + 3x^{10}$ also is not in standard form. Its standard form is $8x^{10}$, as can be seen from the "sum" portion of the map in Fig. 11.2. Clearly, any polynomial has a standard form.

A concept map describing the definition of polynomial in terms of monomials is given in Fig. 11.3.

The word "added" when applied to polynomials must be discussed by the teacher, since before they take algebra, students think of addition as being related only to numbers, not to symbols.

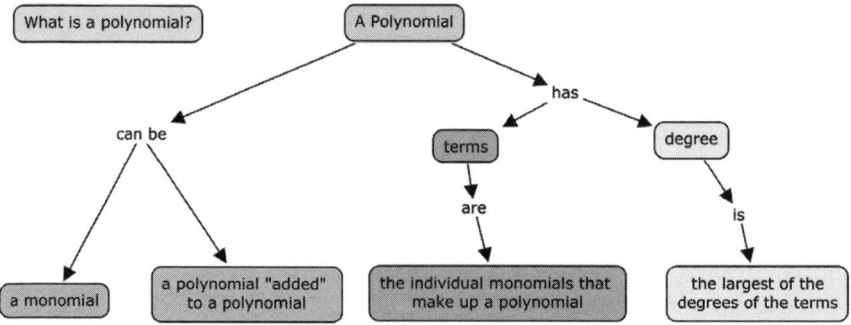

Fig. 11.3 Definition of a polynomial

The map provided in Fig. 11.2 discussed addition of monomials; this idea can be generalized to addition of polynomials since polynomials are defined in terms of monomials. The map in Fig. 11.3 may serve as a guide to the teacher in preparing a discussion of the definition of polynomial, and to a lesser degree a tool for students to organize the concept of polynomial in their minds.

To continue the development of the polynomial component of the course, we next address what is meant to say that two polynomials are the same. By this, we mean that they have identical monomial terms. For example, the polynomials

$$10x^{12} + 5x^8 + 3x^4 - 2x + 6 \text{ and } 4x^{12} + 5x^8 + 3x^4 - 2x + 6$$

are not the same, because, for example, the second one does not have the term $10x^{12}$, whereas the first one does.

On the other hand, the polynomials

$$3x^4 + 5x^8 + 10x^{12} - 2x + 6 \text{ and } 6 + 10x^{12} + 5x^8 + 3x^4 - 2x$$

are the same because they do in fact have identical terms – the polynomials would match exactly if written in order of increasing degrees of monomial terms.

Definition *Two polynomials are identical if (and only if) they have exactly the same set of (monomial) terms.*

Operations on Polynomials

Notice that the concept map in Fig. 11.3 shows that the sum of two monomials is a polynomial, and the one in Fig. 11.2 shows that the algebraic addition and multiplication of two monomials both are commutative operations. That addition and multiplication of polynomials also are associative operations is merely stated as being true in Algebra I, so we will assume that commutativity and associativity hold when we perform addition and multiplication operations on all polynomials.

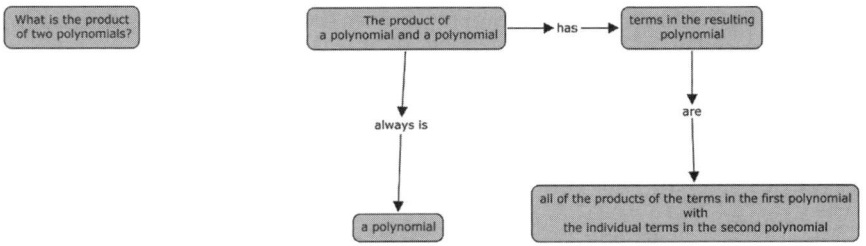

Fig. 11.4 The product of polynomials

First we address addition. From Fig. 11.3 we can conclude that *any* finite sum of monomials is a polynomial, and that when we add any two polynomials we get a polynomial. It is easy then to rewrite any sum of polynomials into a standard-form polynomial using the properties of addition of monomials given in the map in Fig. 11.2, and assuming that the associative and commutative properties for addition are satisfied. Multiplication of polynomials is described in Fig. 11.4. Note that this definition usually will not give a polynomial in standard form, so that the simplification process – changing the resulting polynomial into standard form – will be discussed again by the teacher when this is presented.

Degree of a Polynomial

The remaining idea we must address from our narrative on polynomials is the *degree* of the sum and product of two polynomials.

The degree of a polynomial is defined in the concept map in Fig. 11.3, and the degree of a monomial is given in Fig. 11.1. If we examine the map in Fig. 11.2, we can develop the degree of the sum and the product of two polynomials from information given there. We see from these that the degree of the polynomial that is the sum of two non-zero monomials is the larger of the degrees of the two. This tells us that the degree of the sum of *any* two polynomials will be the larger of the degrees of the two polynomials being added. Also from Fig. 11.2, the degree of the product of two non-zero monomials is the sum of the degrees of the two monomials. From the map in Fig. 11.4, the term in the product of two non-zero polynomials that will have the largest degree will be the product of the term in the first polynomial having largest degree among the terms of that first polynomial, with the term in the second polynomial having the largest degree among that polynomial's terms. Thus, the degree of the product of any two *non-zero* polynomials will be the sum of the degrees of the two. We see from Fig. 11.2 that the product of the zero polynomial any other polynomial is always the zero polynomial, so such a property of degrees (the degree of the product is the sum of the degrees of the two factors) would not hold if one of the two polynomials were the polynomial **0**. That is why we do not assign a degree to the zero polynomial.

The polynomials investigated in Algebra I are to a great extent linear and quadratic polynomials, with some brief coverage of special cases of polynomials of higher degree. In the course, linear polynomials (polynomials of degree less than or equal to 1) are investigated individually (for graphs, slopes, and intercepts), in pairs (for finding intersection points), and sometimes in larger groupings (as in linear programming). Quadratic polynomials (polynomials of degree 2) are covered deeply, as well, with some graphing presented, and a heavy emphasis is given to solving quadratic equations.

Evaluating Polynomials and Solving Polynomial Equations

In the monomial cx^n, we think of the components c, x, and n as being numbers, as was described in Fig. 11.1. We also think of the monomial as being a mathematical product of the coefficient c and the nth power of some unspecified number x. Thus, when we replace the variable x with a specific real number, we can calculate the resulting product. This is called *evaluating the monomial* at the specified number. For example, if we evaluate the monomial $4x^3$ at $x = 5$, we get 4×5^3, or 500. Polynomials are evaluated at a specified number by evaluating each of the monomial terms at that number value, and then adding the results. This can be put into a concept map as in Fig. 11.5.

Polynomial equations arise in the process of using algebra to provide mathematical models of real world problems. Finding an answer to the real world problem through such a mathematical model will require applying algebraic techniques to the equations, in order to find what is called a *solution* to the polynomial equation.

Definition *A solution to a polynomial equation in Algebra I is a real number such that when each of the two polynomials in the equation is evaluated at that number, the resulting two values are equal. (Complex number solutions to polynomial equations usually are not addressed in Algebra I.)*

Particular attention is given to polynomial equations of the form $p(x) = 0$, where $p(x)$ is any polynomial, and 0 is the zero polynomial $0x^0$. Finding solutions to polynomial equations of this form is called *solving the polynomial $p(x)$*, and the solutions found are referred to as *solutions to the polynomial $p(x)$, or roots of the polynomial $p(x)$.*

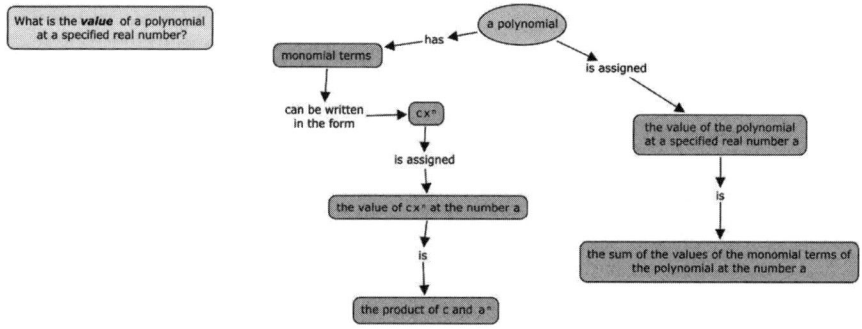

Fig. 11.5 Evaluating a polynomial at a fixed real number

Quadratic Polynomials

In Algebra I, much attention is given to quadratic polynomials and their solutions. As we develop this segment of the course, again a list of concepts to be covered is proposed, and a brief narrative written to describe how the class presentation of the topic will unfold.

Suppose that the following concept list is decided upon:

Quadratic Polynomials
Definition
Factors
Roots
Discriminant
Solving techniques
Quadratic Formula

Once the teacher has prepared a narrative describing how these ideas will be presented, the teacher then would develop a concept map relating the topics reflecting the manner in which the teacher views the connections among them. Such a map will be useful in developing the day-to-day order of presentation of the ideas, and will assist in the development of individual lesson plans. A possible such map is given in Fig. 11.6.

Factoring polynomials plays a major role in Algebra I, and some teachers will want a more general coverage of factoring in addition to that given in Fig. 11.6. Figure 11.7 gives a possible such map.

The teacher planning the course would develop the narrative and concept maps for the remaining topic in the original list – graphs – and would determine the best order in which to present the material, setting up examples, necessary definitions, and connecting concept maps as the course unfolds. In this way, a concept map describing the entire course could evolve. A sample master map for a complete Algebra I course is given in Fig. 11.8. It is complex; however if it is examined relative to the major topics identified and how each of those is connected to other ideas treated in the course, it becomes easier to read.

The value in such a map is that it provides a template for all teachers in a particular school to use as they teach the course. Indeed, the faculty could develop such a set of "expert" concept maps to serve as a model for the course their school would offer.

We have looked at the development of only a very small portion of the master map – a part in the segment on *polynomials and rational functions*. In the process, we have developed a number of maps of concepts related to the three nodes within that map segment.

To complete the Algebra I course development using a concept mapping approach will require the careful analysis of each of the remaining topical segments, and the development of the necessary concept maps related to the nodes within those segments. That process will require the teacher to think carefully through each of the

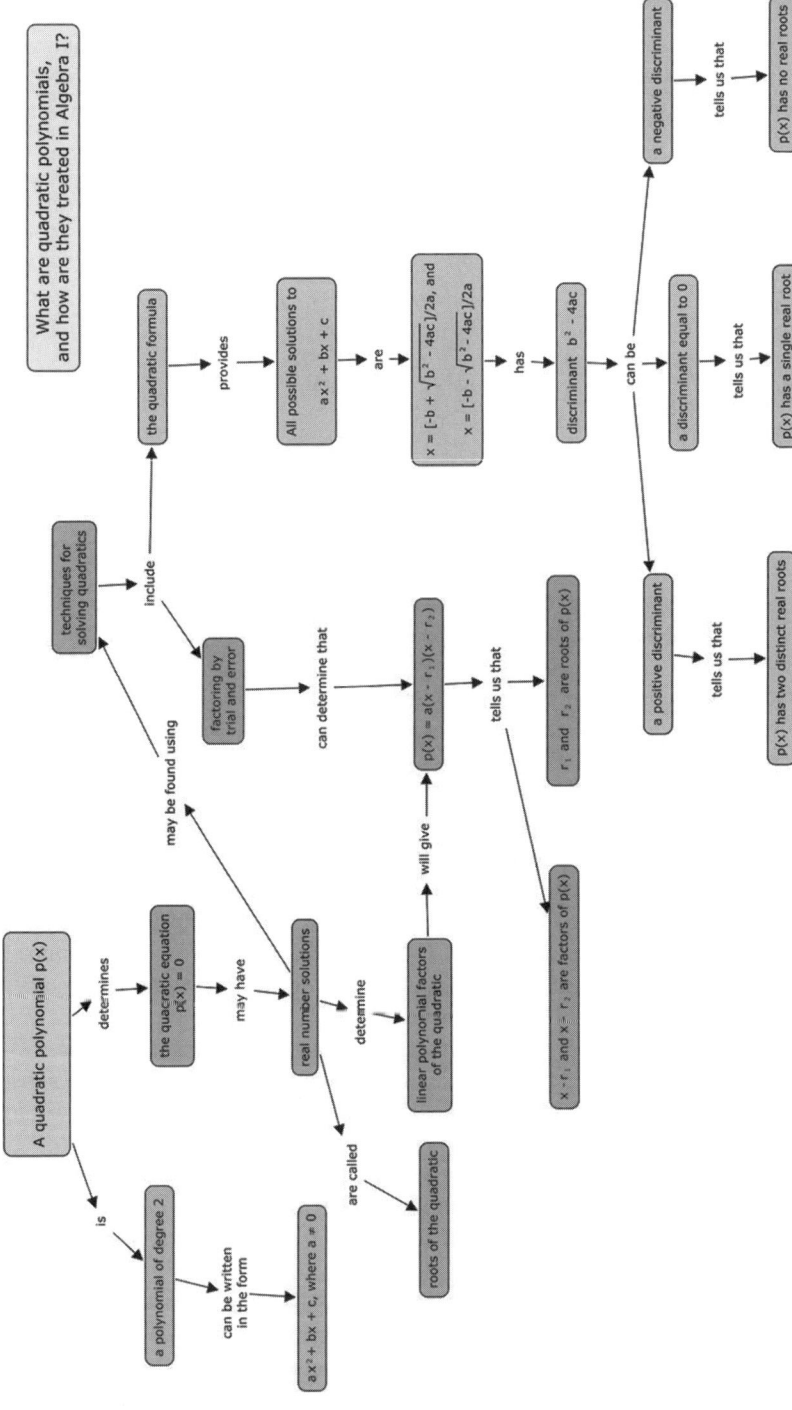

Fig. 11.6 Quadratic polynomials and their solutions

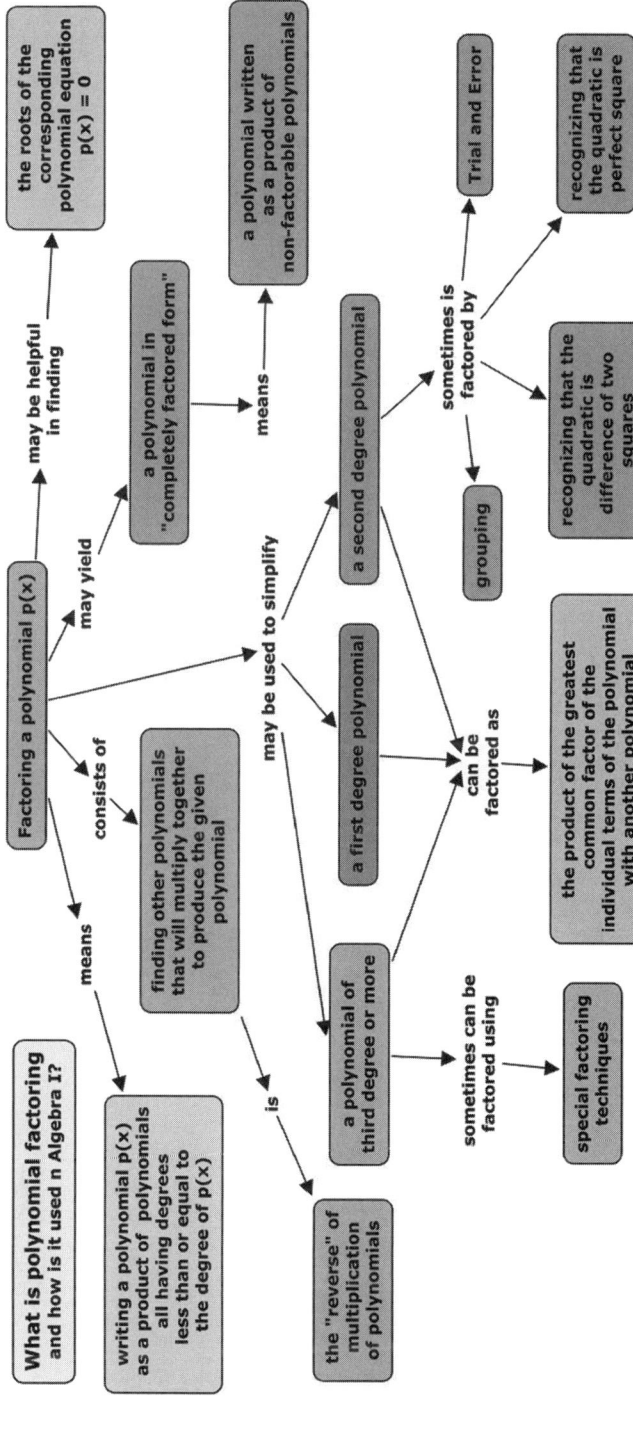

Fig. 11.7 Definition of factoring a polynomial

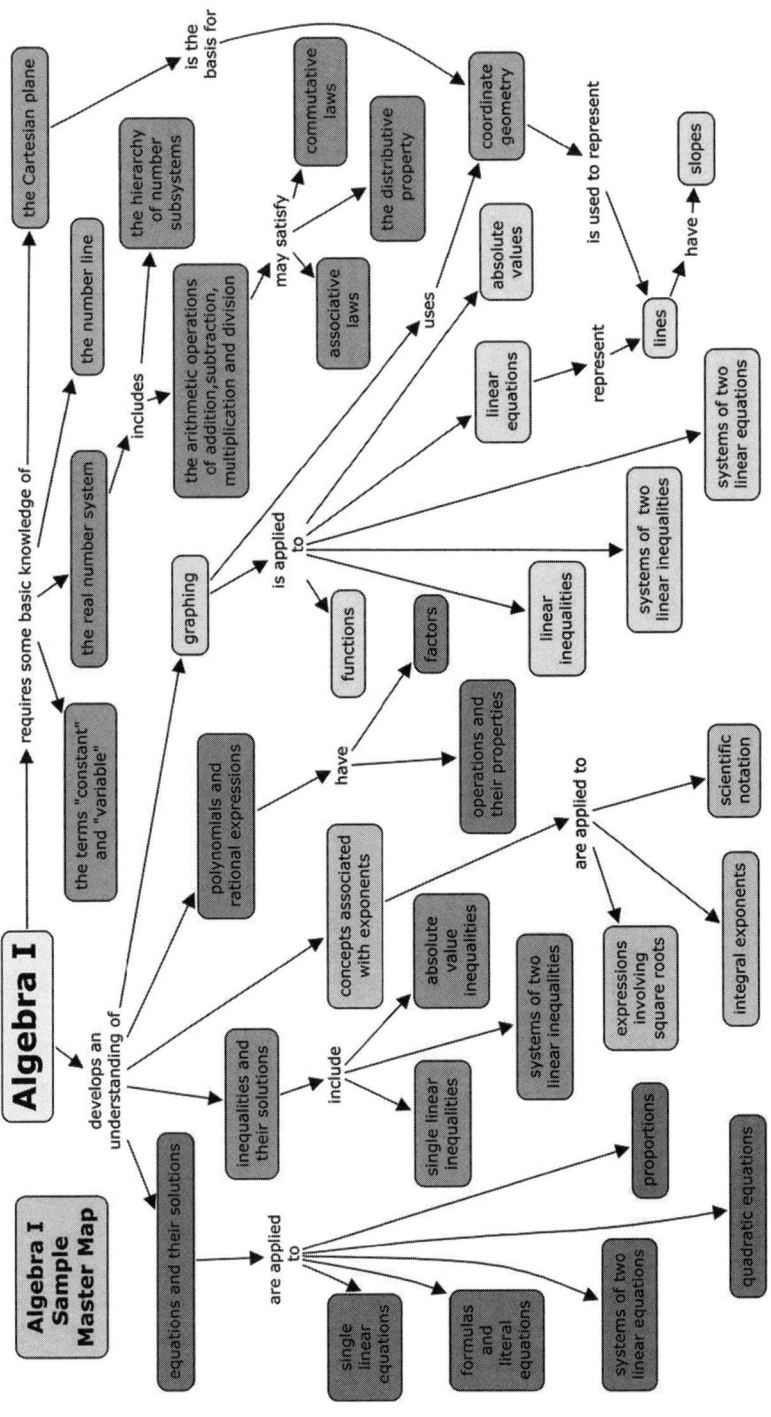

Fig. 11.8 A possible master concept map for Algebra I

individual concepts and their meanings, and then to explore the many connections that exist among the concepts and sub-concepts identified. The maps produced may differ from teacher to teacher, but all who go through the process will gain a greater understanding of the topics in the course, and will recognize any existing areas in the course where their own understanding needs strengthening.

In addition to being a useful tool in course planning, concept mapping can be quite valuable in organizing lessons. This value can come into play when a teacher is using a textbook in which the order of presentation in the text is not comfortable for them. One teacher dealt with this situation by reviewing the textual material carefully, identifying the concepts included in that material, determining the primary concept(s) being addressed therein, and then using the main concept they could identify as the root concept in a concept map to describe the entire unit. Specifically, the table of contents for the unit in the text presented the topics in the following order:

Area of a Circle
Circumference of a Circle
Circle Vocabulary
Area of a Triangle
Substitution
Distributive Property
Order of Operations

A review of the unit, though, led the teacher to decide that the goal of the unit was to have the students learn how to deal with complex numerical expressions. The map in Fig. 11.9 was produced, and then used to decide how the material would be addressed in daily lesson plans of the teacher. As can be seen, the order in which topics were covered using the concept map emphasized the relationships the teacher visualizes as the optimal way of connecting the concepts, rather than the textbook author's presentation.

The three properties that we use to evaluate concept maps - that they are clear, correct, and complete - allow us to use concept mapping in student assessment. For example, suppose that we have been discussing the idea of function in Algebra I. By giving the students a list of topics related to functions and asking them to describe the connections among them in a concept map, we can get a useful picture of how the students grasp the main concepts the teacher is presenting.

Figure 11.10 is a concept map that gives a comprehensive picture of the idea of *function* in Algebra I, and could serve as a guide to the teacher in forming the classroom discussion of the concepts surrounding functions, as well as being a source for ideas on how to have the students use concept maps to show their understanding of the concepts involved.

For example, the teacher might give the students a concept list such as:

Function
Domain
Range

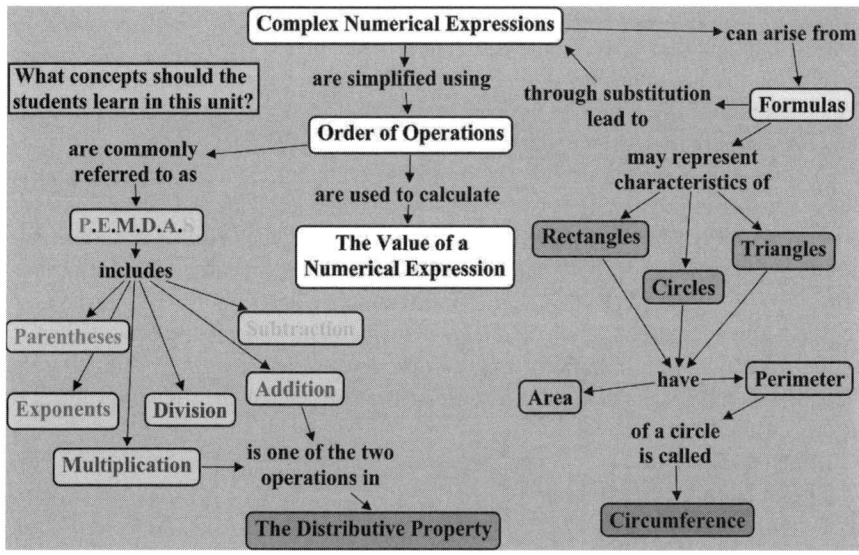

Fig. 11.9 Parsing of a mathematics text unit based upon its concept organization

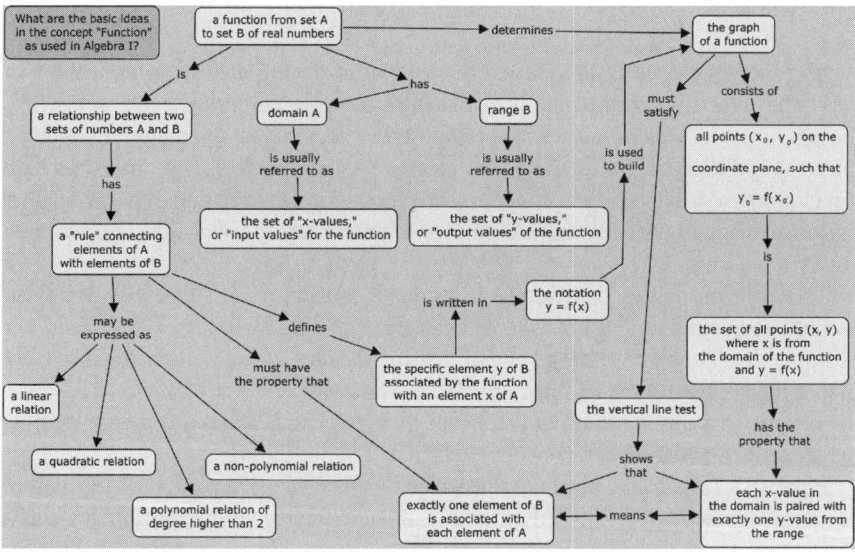

Fig. 11.10 Function and its underlying concepts

The teacher then could ask the students to produce a concept map describing the relationships and connections they see among these concepts. Using the clear-correct-complete assessment model, the teacher would evaluate the maps and determine the level of understanding each student has of the connections among the

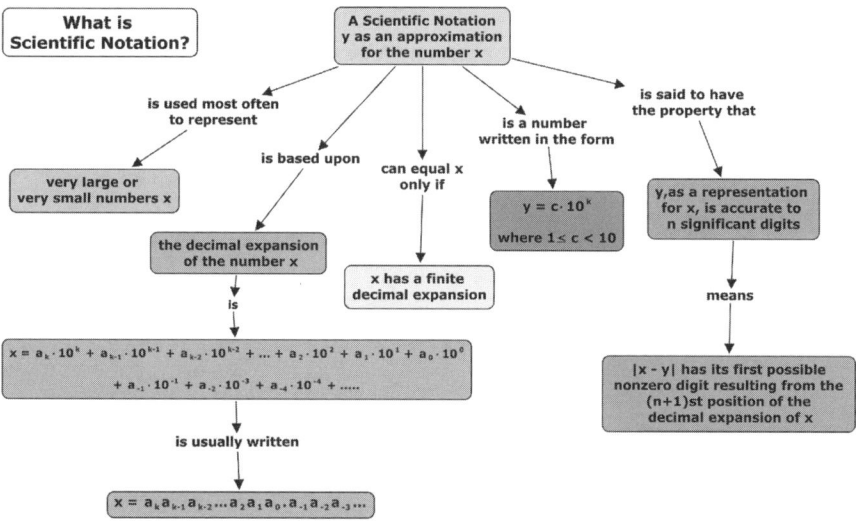

Fig. 11.11 A formal concept map definition of scientific notation

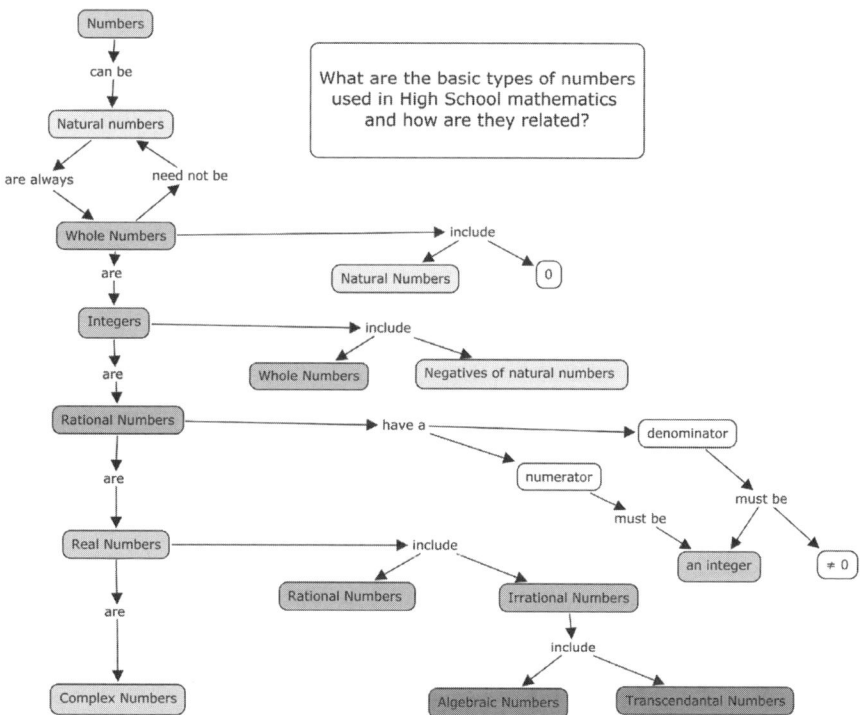

Fig. 11.12 A concept map of relationships among sets of numbers encountered in Algebra I

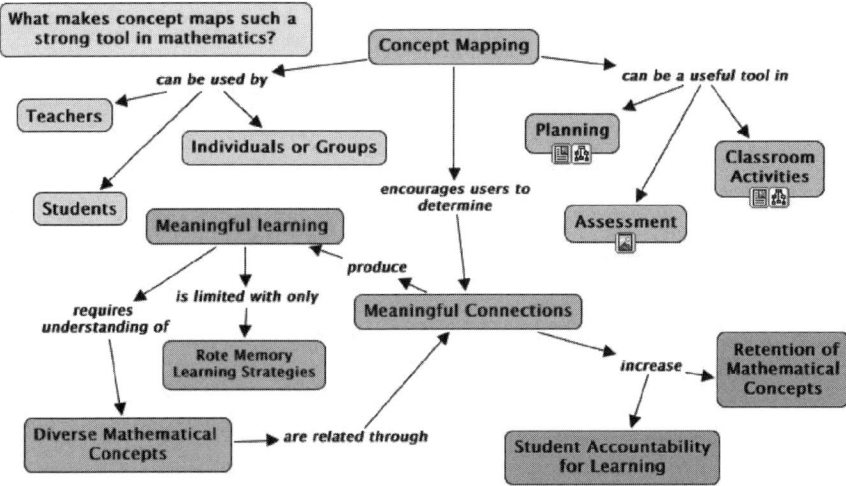

Fig. 11.13 The value of using concept maps in mathematics courses

ideas. This will allow the teacher to determine what kind of assistance is needed to help the student fill critical gaps in knowledge and understanding.

Concept mapping techniques and structures also can be used to express definitions, as in Fig. 11.11, and to clarify relationships among sets, as in Fig. 11.12.

Most important, however, is that teachers have the opportunity to introduce this important tool to their planning, and present this visual learning technique to all of their students. Indeed, the concept map in Fig. 11.13 expresses the value of the concept mapping approach to mathematics teaching in general.

Part IV
University Mathematics Teaching and Learning

Chapter 12
Enhancing Undergraduate Mathematics Learning Using Concept Maps and Vee Diagrams

Karoline Afamasaga-Fuata'i

Data from a group of six students who participated in a study, which investigated the impact using concept maps and Vee diagrams (maps/diagrams) has on learning and understanding new advanced mathematics topics, is presented. Constructing comprehensive topic concept maps, and Vee diagrams of problems as ongoing exercises throughout the semester was a requirement of the study. Students quickly found that learning about the new tools and a new topic was demanding. However, they also found that the concurrent use of the two tools in learning a new topic contributed substantively in highlighting the close correspondence between the conceptual structure of the topic and its methods. Having students display their developing understanding and knowledge on maps/diagrams greatly facilitated discussions, critiques, dialogues and communications during seminar presentations and one-on-one consultations. Maps/diagrams were qualitatively assessed three times during the study period. It was found that there was noticeable growth in students' in-depth understanding of topics as indicated by increased valid propositions and structural complexity of concept maps, and multiplicity of methods and suitability of guiding theoretical principles on Vee diagrams. The chapter discusses the results and provides implications for teaching mathematics particularly for promoting meaningful learning and effective problem solving in mathematics classrooms.

Introduction

The vision of classrooms filled with students actively engaged in working and communicating mathematically is one that is encapsulated by various *Teaching and Curriculum Standards* (NCTM, 2007; AAMT, 2007; NSW BOS, 2002). Research shows that students' perceptions of what is important to learn in mathematics is influenced by routine classroom practices and assessment programs, that is, students' mathematical learning reflect the way they have been, and are, taught mathematics (Thompson, 1984; Knuth & Peressini, 2001; Schell, 2001). In addition,

---

K. Afamasaga-Fuata'i (✉)
University of New England, Armidale, Australia
e-mail: kafamasa@une.edu.au

K. Afamasaga-Fuata'i (ed.), *Concept Mapping in Mathematics*,
DOI 10.1007/978-0-387-89194-1_12, © Springer Science+Business Media, LLC 2009

pedagogical decisions teachers make about teaching and assessment are influenced by their mathematical beliefs (Ernest, 1999; Pfannkuch, 2001). Typically, an authoritative perspective views mathematics as a body of knowledge to be taught by transmission and learnt by simply receiving the information. In contrast, cognitive and social perspectives view mathematics learning and understanding "as the result of interacting and synthesizing one's thoughts with those of others" (Schell, 2001, p. 2) suggesting mathematics knowledge is a social construction that is validated over time, by a community of mathematicians. Hence making sense is both an individual and consensual social process (Ball, 1993). Ideally, classroom practices should equip students with the appropriate language and skills to enable the construction of the mathematics that is taught, and critical analysis and justification of the constructions in terms of the structure of mathematics (Richards, 1991). Lesh (2000) argues that, "mathematics is not simply about doing what you are told . . . . . . mathematics is about making sense of patterns, and regularities in complex systems . . . it involves interpreting situations mathematically." (p. 193) while Balacheff (1990, p. 2) points out that "students need to learn mathematics as social knowledge; they are not free to choose the meanings they construct. These meanings must not only be efficient in solving problems, but they must also be coherent with those socially recognized. This condition is necessary for the future participation of students as adults in social activities."

Existing problems with mathematics learning in Samoa (as in most other countries) are perceived as related to students' perceptions of mathematics, ability to communicate mathematically, and critical problem solving (Afamasaga-Fuata'i, Meyer, Falo, & Sufia, 2007). Firstly, in Samoa the narrow view most undergraduate students have, reflects their school mathematics experiences, found to be mostly rote learning, a problem consistently raised by national examiners. Even the top 10-percent of Year 13 (equivalent to Year 12 in Australia) students consistently struggle with applications of basic principles to solve inequations/equations and/or graph functions (Afamasaga-Fuata'i, 2005a, 2002; Afamasaga-Fuata'i, Meyer, & Falo, 2007). Secondly, students justify methods in terms of sequential steps instead of the conceptual structure of mathematics. Thirdly, students may be proficient in solving familiar problems, however, the lack of critical analysis and application becomes evident when they are given novel problems (Afamasaga-Fuata'i, 2005b). Such approaches are symptomatic of authoritative classroom practices in which students typically do not question, challenge or influence the teaching of mathematics (Knuth & Peressini, 2001, p. 10). The examination-driven teaching of secondary mathematics in Samoa naturally inculcates a narrow view of mathematics (Afamasaga-Fuata'i, 2005a, 2002), As a result, problem solving skills students acquire over the many years of secondary schooling may not necessarily be situated "within a wider understanding of overall concepts" and would probably not be "long-lasting" (Barton, 2001).

As one way of addressing the demonstrated problems about students' mathematical performance, a series of studies was conducted to explore how their mathematical understanding could be improved beyond the algorithmic and procedural proficiency typically characterising their mathematical experiences after years of schooling. Since students also experience difficulties when trying to communicate and argue mathematically (Richards, 1991; Schoenfeld, 1996), transfer, and apply what they know to solve novel problems (Afamasaga-Fuata'i, 2007, 2006, 2005a, 2002), it became increasingly important, for Samoan mathematics education, that research is conducted to examine students' conceptions and evolving understanding of the mathematics that is embedded in familiar/simple problems in anticipation of solving more challenging and novel problems. More importantly, by having students identify the relevant conceptual bases of a problem and its solution would reveal the extent and depth of students' integrated or fragmented understanding of the relevant mathematics.

The data reported here was collected from a group of six undergraduate students who used concept maps and Vee diagrams (maps/diagrams) to communicate their growing understanding of a new topic they had not encountered recently in their mathematics courses. Consequently, the challenge was for them to learn about a new topic through independent research and then construct maps/diagrams to illustrate their developing understanding. New socio-cultural practices in the classroom setting (i.e., socio-mathematical norms) were established, to define the expectations for all students participating in the study. For example, classroom interactions and presentations of an individual's work in a social setting were in accordance with the principles of the social constructivist and socio-linguistic perspectives, which view the process of learning as being influenced and modulated by the nature of interactions and linguistic discourse undertaken in a social setting (Ball, 1993; Schoenfeld, 1996; Ernest, 1999; Richards, 1991; Knuth & Peressini, 2001).

This chapter's focus questions are: (1) In what ways did the activities of concept mapping and Vee diagramming influence students' mathematics learning? (2) What roles did concept maps play in learning about the structure and nature of mathematics learning? (3) What roles did Vee diagrams play in facilitating the problem solving process and generation of multiple methods?

The main assumptions of the study included (1) mathematics has a conceptual structure that guide its methods of solutions (Schoenfeld, 1991); (2) knowledge is a human construction; (3) meta-cognitive skills are required for controlling and monitoring the problem solving process; and (4) educational and classroom practices tend to sanction a particular view of mathematics learning. The theoretical framework of the study is examined next followed by the methodology and analysis of the data collected.

Theoretical Framework

The difference, between an authoritative perspective of mathematics learning and Ausubel's cognitive theory of meaningful learning, socio-linguistic and social constructivist perspectives, is the extent to which classroom discourse and social interactions are supported (Wood, 1999). That is, students learn mathematics in meaningful ways, by developing their understanding through the construction of their own patterns of meanings and through participation in social interactions and critiques (Novak & Cañas, 2006); Novak, 2002). In contrast, rote learning tends to accumulate isolated propositions rather than developing integrated, interconnected hierarchical frameworks of concepts (Novak & Cañas, 2006; Ausubel, 2000; Novak, 2004).

The theoretical principles of meaningful learning, describe the process in which the learner deliberately chooses to relate new information to existing knowledge by assimilation through progressive differentiation (i.e., a reorganization of existing knowledge under more inclusive and broadly explanatory principles) and/or integrative reconciliation (i.e., integration of new and old knowledge into existing knowledge structures through a degree of synthesis) (Ausubel, 2000; Novak, 2004, 1998, 1985). It is argued that, making connections between old and new knowledge, may be cognitively facilitated by organising and constructing, and visually communicated through, maps/diagrams.

A hierarchical concept map is a graph consisting of interconnecting concepts (*nodes*), which correspond to important concepts in a domain and arranged hierarchically, *connecting lines* indicate a relationship between the connected nodes, and *linking words* describe the meaning of the interconnections (explanation). A *proposition* is the statement formed by reading the triad(s) "node–linking-words→node" (Novak & Cañas, 2006). For example, the triad "Functions – *may be described using*→equations" forms the proposition, "*Functions* may be described using *equations*." Having students identify main concepts and organize them into a concept hierarchy of interconnecting nodes with propositional links can indicate the existing state of students' cognitive structures or patterns of meanings.

Gowin's epistemological Vee was developed as a strategy to assist in the understanding of meaningful relationships between events (phenomenon), processes or objects. It visually illustrates the interplay between what is known and what needs to be known or understood. With the Vee-structure situated in the event/object to be analysed, the two sides represent on the left, the thinking, conceptual aspects underpinning the methodological aspects displayed on the right. All elements interact with one another in the process of constructing new knowledge or value claims, or in seeking understanding of these for any set events and questions. A completed Vee diagram represents a record of an event or object that was investigated or analysed to answer particular focus questions (Novak & Gowin, 1984).

Gowin's Vee was subsequently adapted to guide the process of solving a mathematics problem (Fig. 12.1). The Vee diagram was used, in this study, as a tool to not only assess students' proficiency in solving a problem but as well the depth and extent of the conceptual bases of this proficiency by having students identify the mathematical principles and concepts underpinning listed methods and procedures. Hence, the structure of the Vee (Fig. 12.1) with its various sections and guiding questions provide a systematic guide for students' thinking and reasoning from the problem statement (*Event/Object*) to identify the given information (*Records*) and questions to be answered (*Focus Questions*).

The reasoning process may continue onto the identification of relevant *Concepts* and *Principles* that potentially suggest, or can guide the development of appropriate

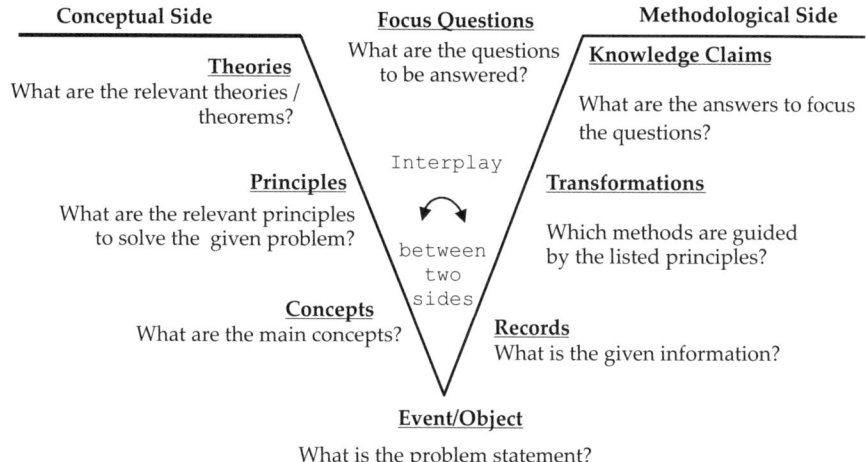

Fig. 12.1 The mathematics problem solving Vee diagram (Afamasaga-Fuata'i, 1998)

methods and procedures (*Transformations*) to generate an answer (*Knowledge Claims*) to the focus questions. Alternatively, if methods are easily obtainable (i.e., for familiar problems), then students are challenged to identify the relevant principles and concepts underpinning the methods. The arrow indicates that there is a continuous interplay between the two sides, as students reason through the various sections of the Vee. This (continuous interplay) is a necessary process to ensure that the conceptual and theoretical underpinnings are abstracted and displayed on the left while the given information, interpretations and subsequent transformations to find answers are displayed on the right side.

The social constructivist perspective views learning as the construction of knowledge to make sense of our experiences whilst socially interacting with others, and that together students and teachers can learn from each other's strategies. Also, the socio-cultural theories view classroom practices as means of reinforcing certain views of what it means to learn and succeed in mathematics. Collectively, learning mathematics involves both individual and social processes. According to Schoenfeld (1991) and Ernest (1999), when an effort is made to *change* classroom practices into activities that involve: *questioning, analyzing, conjecturing, refuting, proving, extending,* and *generalizing* as students solve problems, those rituals and practices can actually shape the behaviour and understanding of students *by making it more natural for them to think and reason mathematically.* Overall, both the meaningful learning and social constructivist approaches support the meta-cognitive development of students' understanding and the active construction of mathematical thought whilst publicly presenting, for example, maps/diagrams, within a social setting.

Methodology

The study was conducted, with the class meeting three times a week for one-hour sessions over 14 weeks, as an exploratory teaching experiment (Steffe & D'Ambrosio, 1996) using the meta-cognitive tools of concept maps and Vee diagrams (Novak, 1985, 2004) with students presenting their work for group and one-on-one social critiques. After completing practice sessions, in constructing maps/diagrams and presenting in a social setting for critiques, students selected their new topics as content for the application of the meta-cognitive tools. The six students (Student 1–6) chose the topics *Laplace's transform, trigonometric approximations, least squares polynomial approximations, multivariable functions and their derivatives, partial differential equations*, and *numerical methods of solving first order differential equations.*

The newly established socio-mathematical norms required students to undertake independent research of their new topic; be prepared to justify their constructions publicly in-class; address concerns raised by, and negotiate meanings with, their peers and researcher during critiques; and to continuously improve their maps/diagrams for subsequent presentations. Students were expected to demonstrate and communicate their understanding of the new topic clearly and succinctly so that the critics (peers and researcher) could make sense of it.

Students engaged in the cyclic process of presenting→revising→critiquing→ presenting for at least three iterations over the semester. The data collected consisted of progressive concept maps (4 versions) and Vee diagrams of four problems (at least 2 versions each).

Data Analysis

Concept maps and Vee diagrams are analysed qualitatively. For concepts maps, the analysis was in terms of the structural complexity, and validity of concept labels and propositions. Vee diagrams, on the other hand, the analysis was in terms of the extent of the appropriateness and relevance of the Vee entries (i.e., overall criteria), and the mutual correspondence between the listed conceptual information and displayed methodological information (i.e., specific criteria) in the context of the problem.

Concept Maps

The qualitative approach adopted with the map analysis is a modification of the Novakian scheme (Novak & Gowin, 1984) and by using only counts of occurrences of each criterion. Collectively, the criteria assess the concept maps in terms of the *structural complexity of the network* of concepts, *nature of the contents* (entries) of concept boxes (nodes) and *valid propositions.*

The *structural criteria* indicate the extent of integrative cross-links between concepts, progressive differentiation between levels, and the number of average hierarchical levels per sub-branch, multiple branching nodes, sub-branches and main branches. Progressive differentiation at a node is indicated by multiple outgoing links from the node whilst integrative reconciliation is represented by cross-links, which connect concepts across concept hierarchies or (sub-)branches.

The *contents criteria* of a node indicate the nature of students' perceptions of the mathematical knowledge in terms of mathematical concepts as distinct from illustrative examples, procedural steps, or descriptive linking words. Illustrative examples are those used to illustrate concepts. Inappropriate nodal entries are procedural entries, redundant concepts and linking-word-type. Redundant entries indicate students' tendency to learn information as isolated from each other instead of identifying potential integrative cross-links with the first occurrence of the concept or consider a re-organization of the concept hierarchy. Linking-word-type indicates students' difficulties to distinguish between a "mathematical concept" and descriptive phrases.

Valid propositions are defined as mathematically correct statements or propositions constructed from at least two valid, interconnected nodes with appropriate linking words that describe the nature of the (string of) inter-relationship(s). A proposition is invalid if linking words are missing, incorrect nodes have inappropriate entries.

Vee Diagrams

Vee diagrams are qualitatively analysed to determine whether or not the conceptual and methodological sides mutually support each other. That is, do the listed principles support the given solution? Are the listed principles the most relevant for the displayed solution? Is the knowledge claim supported by the listed principles and transformations? As Gowin (1981) points out: "The structure of knowledge may be characterized (in any field) by its telling questions, key concepts and conceptual systems; by its reliable methods and techniques of work. . ." (pp. 87–88). Hence, students' understanding of the production of mathematical knowledge claims would be demonstrated by the entries on the Vee.

Data Collected and Analysis

The data collected is presented beginning with the concept map data and then followed by the Vee diagram data before a discussion of the main themes that emerged from both sets of data.

Concept Map Data

Data for the six students' progressive concept maps (first and final maps) are shown in Tables 12.1 and 12.2 below. The six cases are presented next.

Student 1's Topic – Laplace's Transform (LT)

From his research, Art selected a few concepts for his first map to provide a definition for LT, and to illustrate how they are used in solving initial value problems. His first concept map had 17 nodes of which 14 were valid with 3 inappropriate ones due to procedural, redundant and link-word-type entries. Only 35% of the propositions were valid with only one integrative crosslink, see Table 12.1. The high proportion of invalid propositions was due to missing or inappropriate linking words. At the first group critique, critical comments focused on the need to reconsider the hierarchical order of concepts, missing relevant concepts and inappropriate concept labels. Comments from subsequent critiques over the semester pinpointed areas of confusion, which guided Art to sections of his map that needed re-organization and re-structuring to enhance its intended meaning. By the end of semester, Art's final concept map showed an increase in valid nodes (from 14 to 24) with significantly more valid propositions (from 35 to 78%), more sub-branches (from 6 to 10), higher average hierarchical levels per sub-branch (from 4 to 6), an additional main branch (from 3 to 4) and an increased number of multiple branching nodes (from 4 to 8) (Table 12.2).

Overall, Art's final concept map had become more integrated and complex as his understanding expanded and became more enriched as a result of critiques, revisions and individual research. For example, he wrote: "with concept maps, its uses that I have experienced from the semester is that they broaden my understanding of my

Table 12.1 Concept maps' contents and valid propositions criteria

Referente Student Map	Art 1 1	Art 4	Ada 2 1	Ada 4	Lou 3 1	Lou 4	Asi 4 1	Asi 4	Afa 5 1	Afa 4	Les 6 1	Les 4
Concepts	14	24	8	19	13	43	12	51	13	43	36	84
Examples	0	0	2	0	0	0	1	0	2	2	2	4
Definitional	0	2	2	6	0	0	3	0	0	0	0	0
Inappropriate	3	0	2	2	0	0	1	0	0	1	2	17
Total	17	26	14	27	13	43	17	51	15	46	40	105
Concepts	82%	92%	57%	70%	100%	100%	71%	100%	87%	93%	90%	80%
Examples	0%	0%	14%	0%	0%	0%	6%	0%	13%	4%	5%	4%
Definitional	0%	8%	14%	22%	0%	0%	18%	0%	0%	0%	0%	0%
Inappropriate	18%	0%	14%	7%	0%	0%	6%	0%	0%	2%	5%	16%
Valid prop	6	18	6	17	12	41	6	42	14	47	40	87
Invalid prop	11	5	8	14	0	1	10	17	2	2	5	28
% Valid	35%	78%	43%	55%	100%	98%	38%	71%	88%	96%	89%	76%

Table 12.2 Concept maps' structural criteria

Referent	Art	Art	Ada	Ada	Lou	Lou	Asi	Asi	Afa	Afa	Les	Les
Student	1		2		3		4		5		6	
Map	1	4	1	4	1	4	1	4	1	4	1	4
Sub-branches	6	10	4	8	3	15	6	20	4	19	14	32
Hierarchical levels	4	6	4	7	8	8	4	7	4	9	10	15
Main branches	3	4	2	3	1	6	3	4	3	5	4	5
Integrative Cross-links	1	1	0	0	4	6	0	4	0	4	7	18
Multiple Branching nodes at:												
Level 1	2	2				2	2	2	2			
Level 2	2	3,2	2			3		3,2				
Level 3	2,2	2,2	2			2,2	2,3	2,6		4	2	2
Level 4		2	2	5	2	2		3,2,3	2	2,4,2		
Level 5		2		2		2,2,2,3	2	3		2	2	2
Level 6		2			2	2,2				2	2,2	2
Level 7		2				2,3		2,2,2		2	2,2	2
Level 8						2		3	2	3,2		
Level 9										2		2,3
Level 10				2						2		2,2
Level 11										2	2	2
Level 12										2,2	2	3,2,3,2
Level 13										2		2
Level 14										2		2,2
Total # multiple Branching nodes	4	8	3	3	2	14	4	13	3	15	10	16

chosen topic ... (constructing concept maps) allows the writer to easily understand his own topic through substantial and more comprehensive links and to simply make changes from comments in class presentations."

Student 2's Topic – Trigonometric Approximations

Ada found his topic hard, but after reading a few textbooks, he chose to approach his topic using his background knowledge of Taylor's polynomial. For example, he chose to demonstrate the concept of approximations of values of a compound trigonometric function by successively approaching the point. Thus, the first map was mainly procedural but with time and critiques, his final map evolved into a more conceptual one with the demonstration of method of application relegated to a Vee diagram.

For example, Ada wrote in his report: "I was forced to look for key concepts involved in Taylor's polynomial and how they are interrelated to other branches of mathematics. I sought how the terms in the series functioned and what relationship they had to practical applications like speed, acceleration and distance, forming the ability to use this tool in other situations. ... Overall, it was a difficult but helpful experience in which I have a deeper understanding of Taylor's polynomial but as yet many unanswered questions." Ada's final concept map had relatively more valid concepts (from 8 to 19). Unfortunately, 6 nodes had definitional phrases, which require further analysis to form more succinct conceptual entries. There was also an increase in valid propositions (from 6 to 17) but the inordinately high invalid propositions (from 8 to 14) are due to missing linking words, and inappropriate-nodes (definitional phrases).

Student 3's Topic – Least Squares Polynomial Approximations (LSPA)

Lou's first concept map consisted of one main branch with only two multiple branching nodes, and four integrative cross-links, see Table 12.2 for data and the map in Fig. 12.2. From the first group critique, she realized that her map did not provide sufficient concepts to explain the main ideas relevant to her topic particularly the concept of *errors* in spite of having included the concept of *squared differences*. Hence, with more readings, and research, she added in concepts of *errors, five-point-least-square-polynomial, smoothing formula, data smoothing* and *nth degree* to name a few, for her first revisions.

However, as she wrote in her report: "Despite the clustered and plentiful information given in my map, the main concept of errors is lost. This is because there was less emphasis on understanding the topic. Rather, a collection of various concepts seemed more important at the time. Hence, an improved map would require meaningful concepts, mathematical formula, neater presentation, and simple examples. I learnt here that the basic idea behind the topic is that there is an error and everything falls around the minimising of this error."

With more critiques, further research for additional concepts and subsequent revisions, Lou realized that the concepts of *Least-square polynomial P(x), Function F(x)* and *Error = Y(x) – P(x)* have to be positioned appropriately and the case for *continuous data* required further clarification.

By her third revision, Lou noted that her revised map "showed a clear hierarchy of linking concepts . . . hence it was easier to follow what the map is trying to tell us. However, there is still work to be done on clarification, organization and available information." She also learnt that "organization plays a huge role in making the map comprehensive."

With more critiques and revisions, Lou's final concept turned out to be a "a much more effective one in terms of understanding the concepts related to the topic (LSPA). So, the idea of errors was clear, its application and determination was also specified, and the table for clarification of Newton's formula, was also a great improvement. "

In summing up her experiences in the study, she wrote: "I have now seen an evolvement from a very basic map to a more complicated one. The surprising fact discovered is that the basic map (i.e., first map) was more confusing than the resulting one (i.e., final map)." This is quite a revealing statement about the value of her final map as a more meaningful, comprehensive and informative piece of work. Part of Lou's final map is shown in the right map in Fig. 12.3 for comparison to her first attempt.

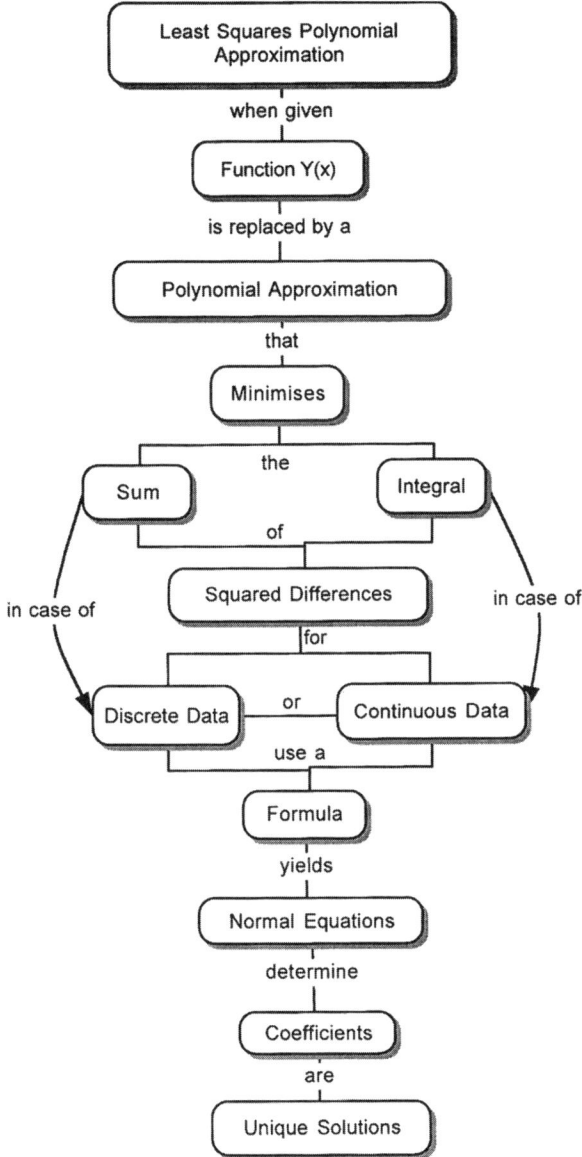

Fig. 12.2 Lou's first concept map

Student 4's Topic – Multivariable Functions

Asi's first concept map had 12 valid nodes with 4 invalid nodes due to a definitional phrase and inappropriate entries. The invalid propositions (10 out of 16) were due to missing linking words or inappropriate nodes (see Tables 12.1 and 12.2). In spite of Asi's efforts, the group found her first concept map presentation confusing due

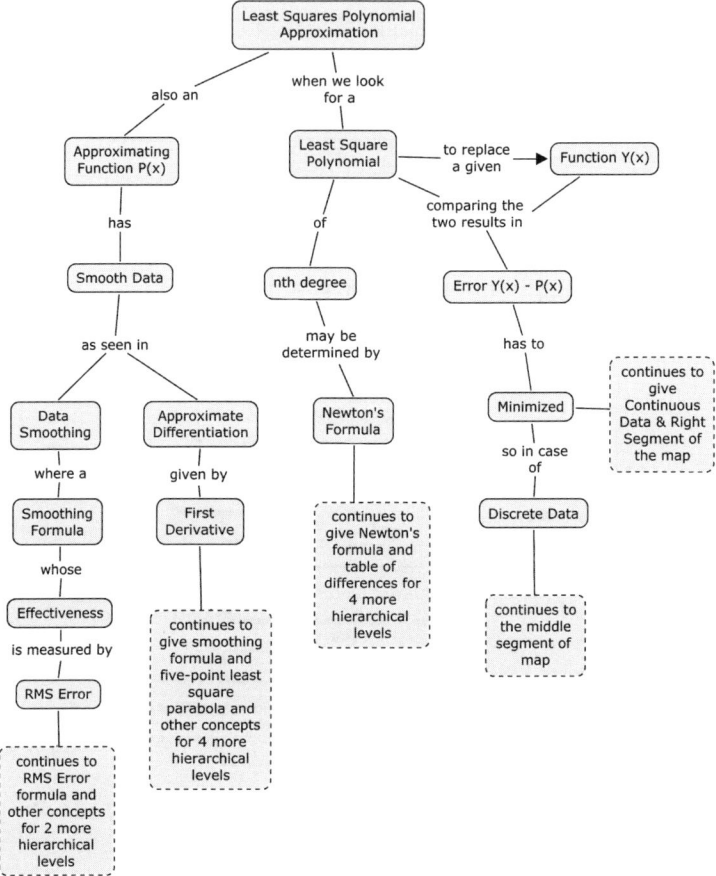

Fig. 12.3 Partial view of Lou's final concept map

to vague and inappropriate linking words. Asi then revised and reorganized her concept hierarchy to make the map more meaningful. Subsequent one-on-one and group critiques over the semester eventually resulted in a final map which was more differentiated with increased multiple branching nodes (from 4 to 13), and sub-branches (from 6 to 20) with a higher average hierarchical levels per sub-branch (from 4 to 7). In response to critical comments, Asi reorganized the concept hierarchy, revised linking words to make them more descriptive of interconnections, created more sub-branches, and provided meaningful integrative cross-links to improve the clarity and organization of information. Overall, the final map had significantly more valid nodes (from 12 to 51) and valid propositions (from 6 to 42). Asi wrote in her final report: "To me, using concept maps has given me a chance to learn more of my research topic."

Student 5's topic – Numerical methods

Afa's first concept map had 15 valid nodes of which 2 were examples, 3 multiple branching nodes, and 4 sub-branches with average hierarchical levels of 4 per

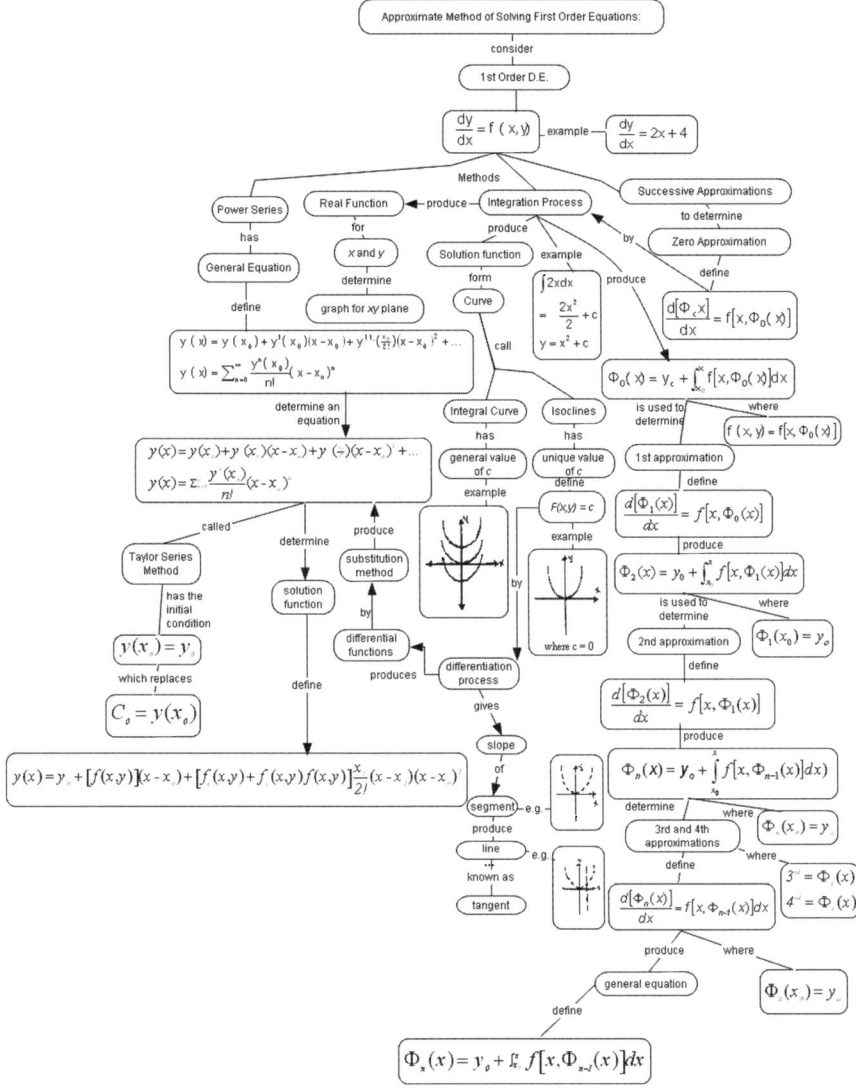

Fig. 12.4 Student 5's final concept map

sub-branch. Through critiques and subsequent revisions, his final map evolved into one that was more differentiated and enriching with substantial increases in sub-branches (from 4 to 19), average hierarchical levels per sub-branch (from 4 to 9), main branches (from 3 to 5), integrative cross-links (from 0 to 4), and multiple branching nodes (from 3 to 15). Overall, valid propositions increased from 14 to 47. Figure 12.4 shows Afa's final concept map. It was also the map with the highest proportion of valid propositions (96%).

Student 6's Topic – Partial Differential Equations (pdes)

Les' first concept map had 38 valid nodes with only 2 invalid ones due to redundant entries. The map differentiated between first and second order pdes with further differentiation at lower levels into homogenous and non-homogeneous types, and had 40 valid propositions with only 5 invalid ones due to incorrect/vague linking words and inappropriate nodes.

With further critiques and subsequent revisions, Les' final map eventually evolved into one that was substantially more complex with increases in sub-branches (from 14 to 32), average hierarchical levels per sub-branch (from 10 to 15), multiple branching nodes (from 10 to 16) and integrative cross-links (from 7 to 18). Valid propositions had also increased from 40 to 87. However, the higher number of invalid propositions (from 5 to 28) is due to an increased number of inappropriate nodes resulting from inappropriate (procedural, redundant and linking-word-type) entries and missing linking words.

Les created additional sub-branches in the final concept map to provide conceptual bases for his Vee diagram problems. He wrote in his final report: "I myself understand fully the path from one concept to another and how a conclusion can be obtained because I created the concept maps." He continued on to state that "From my experience in laying out my concept map I have learnt that differentiating first order and its special cases and second order and its special cases avoids confusion. It helps me to classify each pde I come across so that I could see the big picture." Figure 12.5 shows part of Les's final map. It was also the map with the most cross-links and multiple branching nodes.

Vee Map Data

Vee diagrams are assessed qualitatively in terms of one overall criteria and a more specific one. Specifically, the *overall criteria* assesses the appropriateness of entries in each section according to the guiding questions in Fig. 12.1 and the given problem whilst the *specific criteria* refers to the extent of integration and correspondence between listed principles and displayed main steps of the solutions. The emphasis is on the relevance, appropriateness and completeness of listed *Principles* in relation to methods and procedures displayed under *Transformations*. Rather than present each student's case, general themes emanating from their work are.

Overall Criteria

The overall criteria assess whether students had satisfactory entries for the sections *Theories*, *Concept*, and *Records* as these are basically extracted and inferred from the problem statements in accordance with the guiding questions (Fig. 12.1). Also because they were free to select their problems, obtaining the correct answers (*Knowledge Claims*) was not problematic. However, what caused a lot of critical comments and numerous revisions were the inappropriate entries for the sections,

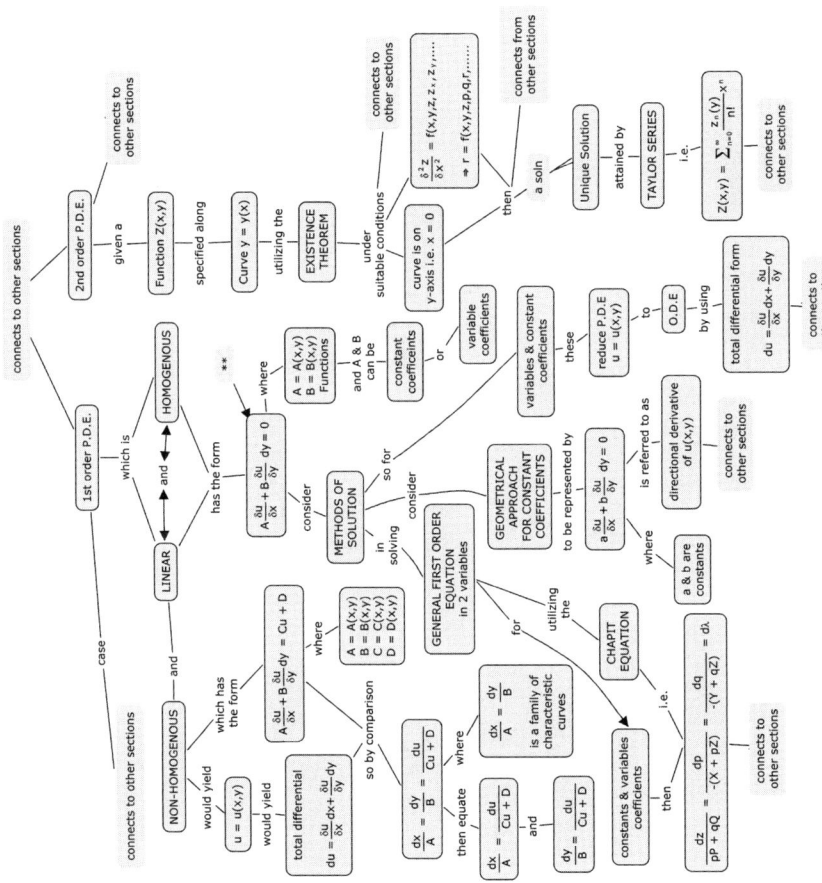

Fig. 12.5 Student 6's partial final concept map

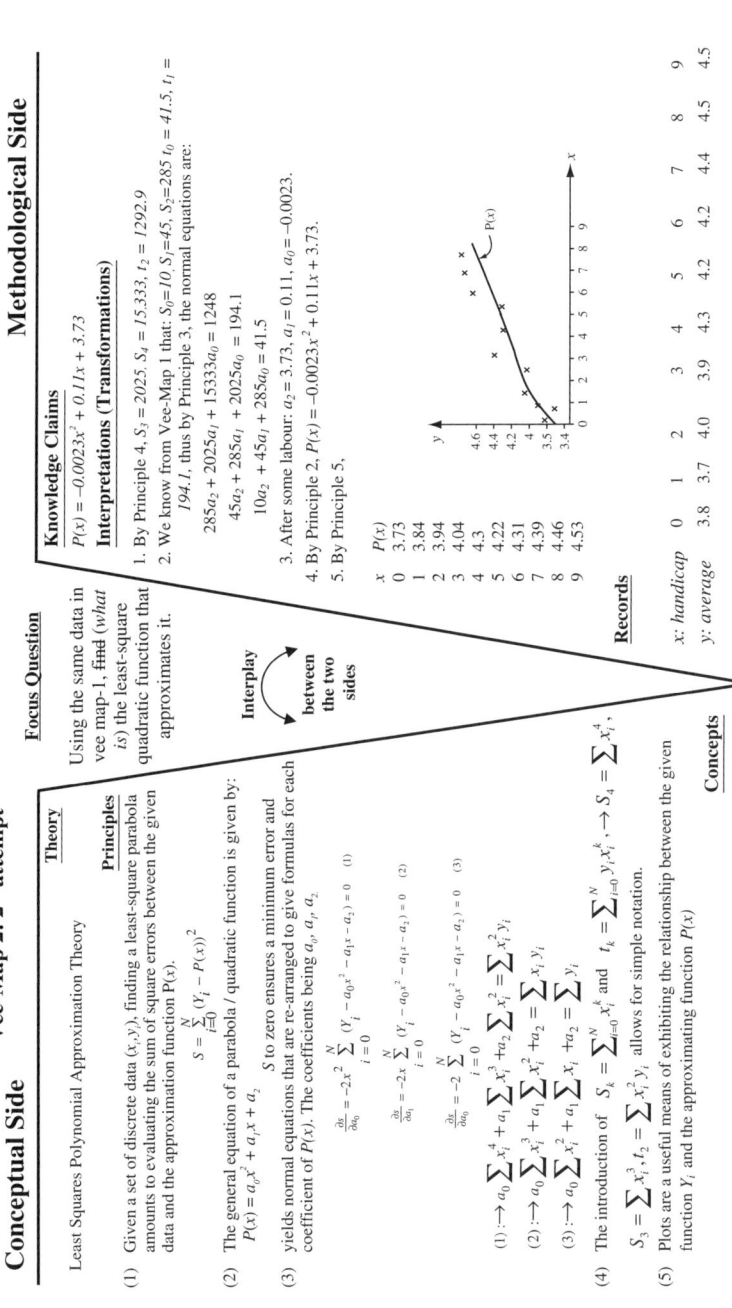

Conceptual Side   Vee-Map 2: 2nd attempt   **Methodological Side**

Theory   **Focus Question**

Principles

Least Squares Polynomial Approximation Theory

(1) Given a set of discrete data (x_i, y_i), finding a least-square parabola amounts to evaluating the sum of square errors between the given data and the approximation function $P(x)$.

$$S = \sum_{i=0}^{N} (Y_i - P(x))^2$$

(2) The general equation of a parabola / quadratic function is given by: $P(x) = a_0 x^2 + a_1 x + a_2$

(3) S to zero ensures a minimum error and yields normal equations that are re-arranged to give formulas for each coefficient of $P(x)$. The coefficients being a_0, a_1, a_2.

$$\frac{\partial s}{\partial a_0} = -2x^2 \sum_{i=0}^{N} (Y_i - a_0 x^2 - a_1 x - a_2) = 0 \quad (1)$$

$$\frac{\partial s}{\partial a_1} = -2x \sum_{i=0}^{N} (Y_i - a_0 x^2 - a_1 x - a_2) = 0 \quad (2)$$

$$\frac{\partial s}{\partial a_0} = -2 \sum_{i=0}^{N} (Y_i - a_0 x^2 - a_1 x - a_2) = 0 \quad (3)$$

$(1) :\rightarrow a_0 \sum x_i^4 + a_1 \sum x_i^3 + a_2 \sum x_i^2 = \sum x_i^2 y_i$

$(2) :\rightarrow a_0 \sum x_i^3 + a_1 \sum x_i^2 + a_2 = \sum x_i y_i$

$(3) :\rightarrow a_0 \sum x_i^2 + a_1 \sum x_i + a_2 = \sum y_i$

(4) The introduction of $S_k = \sum_{i=0}^{N} x_i^k$ and $t_k = \sum_{i=0}^{N} y_i x_i^k$, $\rightarrow S_4 = \sum x_i^4$, $S_3 = \sum x_i^3$, $t_2 = \sum x_i^2 y_i$, allows for simple notation.

(5) Plots are a useful means of exhibiting the relationship between the given function Y_i and the approximating function $P(x)$

Concepts

Discrete data, sum of square errors, equation of a parabola, coefficients, normal equations, plots, graphs

---

Using the same data in vee map-1, ~~find~~ (what is) the least-square quadratic function that approximates it.

Interplay
between the two sides

Knowledge Claims

$P(x) = -0.0023x^2 + 0.11x + 3.73$

Interpretations (Transformations)

1. By Principle 4, $S_3 = 2025$, $S_4 = 15,333$, $t_2 = 1292.9$

2. We know from Vee-Map 1 that: $S_0 = 10$, $S_1 = 45$, $S_2 = 285$, $t_0 = 41.5$, $t_1 = 194.1$, thus by Principle 3, the normal equations are:

   $285a_2 + 2025a_1 + 15333a_0 = 1248$

   $45a_2 + 285a_1 + 2025a_0 = 194.1$

   $10a_2 + 45a_1 + 285a_0 = 41.5$

3. After some labour: $a_2 = 3.73$, $a_1 = 0.11$, $a_0 = -0.0023$.

4. By Principle 2, $P(x) = -0.0023x^2 + 0.11x + 3.73$.

5. By Principle 5,

x	$P(x)$
0	3.73
1	3.84
2	3.94
3	4.04
4	4.3
5	4.22
6	4.31
7	4.39
8	4.46
9	4.53

Records

x: handicap	0	1	2	3	4	5	6	7	8	9
y: average	3.8	3.7	4.0	3.9	4.3	4.0	3.9	4.2	4.4	4.5	4.5

Object

To find the least-square parabola for the given discrete data (x_i, y_i).

Fig. 12.6 One of Lou's Vee diagram of a problem

Principles, and *Transformations*. The general weakness with the former is the language used to describe principles. For example, the intention is clear but wording were initially too procedural in contrast to theoretical statements of principles, general rules and formal definitions. The tendency was to provide only formulas without clarifications subsequently leading to ambiguities. With transformations, listed main steps did not always have supporting principles on the Vee.

Specific Criteria

The specific criteria assess the degree of correspondence between the listed principles and displayed methods. Most of the students scored low in their initial maps. However, with critiques, evolving comprehensive concept maps, and subsequent revisions over the semester, students' listed principles improved to become more conceptual statements, and the principle lists expanded to include key principles that support listed main steps in the transformation section. As one of the students wrote in the final report: "the principles section required much thought and reorganising ... my struggle was to ensure the principles were general statements and formula that became tools for solving the given problem."

Figure 12.6 shows an example of one of Lou's revised Vee diagram. Feedback from the critique concerned the need to "discuss the significance of the quadratic function approximation in relation to the linear function in Vee-map 1." Subsequently, the next revision (Vee-map 2–3rd attempt) showed (i) the addition of "Principle 6. Comparing the least-square functions for the same data values enables one to see which one approximates the given data better." and (ii) an elaborated knowledge claim "The lease-squares quadratic function or parabola for the given data is $P(x) = -0.0023x^2 + 0.11x + 3.73$ since the plot for this $P(x)$ is almost the same as the least square line, we can say that the least-square parabola for this data is unnecessary" to replace the previous one-line equation whilst the rest of the Vee remained as shown in Fig. 12.6.

Discussion

Learning a new topic and learning to use concept maps and Vee diagrams were big demands of students as Lou puts it: "I began my semester of reading a page over and over again, looking at examples and reading the same page one more time, only to realise that I had to reorganize my concepts again. This became my routine for the study of concept mapping: reading, checking, writing, organizing, ... , reading, checking, and onwards I went. " However, by the end of the semester, Lou wrote: "it would have been impossible to reach a more comprehensive map without the input from the class and lecturer." All six students found that to construct a map that made sense to the critics, they had to research more, continually revise and re-organize the concept hierarchies.

Furthermore, the construction of the Vee diagrams was greatly facilitated when based on a comprehensive integrated and differentiated concept map as evidenced by the creation of additional branches on concept maps to illustrate guiding principles for a method on a Vee diagram. In doing these activities, students learnt more about the conceptual structure of their topics in more meaningful ways and at a deeper level as well becoming proficient with the relevant methods and procedures. As Lou sums up her experiences: "When I presented my last concept map to class, it dawned on me that I had finally understood what I was struggling to know since the beginning of semester. The words, 'Least square polynomial approximations' no longer threatened me. I could close my eyes and summarize this topic to someone else without a doubt in my head that what I would be saying made sense."

The qualitative analyses of the maps/diagrams demonstrated that the concurrent use of the two tools in learning about a new topic contributed significantly in highlighting the close correspondence between the conceptual structure of a mathematics topic and its methods. For example, a student wrote: "With the help of constructive comments from critiques, I was able to work on appropriate Vee maps that elaborated on the concept map. This was the fundamental role of the Vee diagrams – to elaborate on the concepts shown by the concept maps. With this elaboration, I was able to understand the topic even better." That is, possessing only a procedural and algorithmic view of mathematics is limiting. Instead, an enriched knowledge of the conceptual bases of methods, and in-depth knowledge of the conceptual structure can motivate students to learn more about their topic. For example, Lou wrote: "Making sense out of a difficult topic through concept mapping was the miracle that I was enlightened with. In addition to this awesome discovery, I realised that the miracle was endless. That is, I could go on learning more about least square polynomial approximations because there is always more concepts waiting to be discovered, analysed and revised. So concept mapping is also a tool for extending one's knowledge."

Conclusions and Implications

Students' progressive concept maps and Vee diagrams showed improvement over time as a consequence of group presentations, individual work, peer critique and one-on-one consultations. That is, students' concept maps evolved into maps that were more meaningfully integrated and differentiated and more enriching in its conceptual structure. Their Vee diagrams showed growth in their correspondence between methods of solutions and listed principles and enhancement of the conceptual integrity of listed principles from predominantly procedural statements and bold formulas to more conceptual statements, general mathematical principles, and definitions. The increased structural and conceptual complexity reflected the growth in the extent and depth of students' understanding of the links between theoretical principles and methods of solutions of the selected topic. The established socio-mathematical norms of critiques and presentations contributed significantly to the developing quality and refinement of students' evolving understanding of their topics. The act of talking aloud (presenting and justifying to peers) required a level of reflection that aided in the problem solving process. In the study, talking aloud had the power to change students' performance (Richards, 1991, p. 37) as evident by the dynamic social interactions during critiques and demonstrated by the progressively complex concept maps and integrated conceptual and methodological Vee diagrams of their selected topic.

One of the value claims from students' perspectives is the self-realization that the construction of maps/diagrams requires and demands a much deeper understanding of interconnections than simply knowing what the main concepts and formulas are. Although time consuming, the construction of maps/diagrams facilitates learning the structure of a topic in more meaningful ways. Furthermore, students realized that the communication of their understanding is more effective if concepts are arranged in a hierarchical order complete with appropriate labels, meaningful links with concise and suitable linking words. Another value claim of the study is the potential of applying the meta-cognitive tools to other subject areas by the same students. This is succinctly captured by Lou's comments in her final report: "I was able to apply the theory of concept mapping to my other subjects and found that I became relaxed when confronted with a difficult topic. Then I was rewarded with good marks. Before I learnt of concept maps, my initial response to a difficult subject would be to panic. Then I would try to break the problem down, read, research, memorize, and do all the things an average student does before understanding some of the topic being studied. Now I wish that our high school teachers had taught us about concept mapping. It would have done wonders for me."

There are still problematic areas that need attention mainly due to the newness of the tool which students need to overcome with more practice and more time. As one of the students noted, collecting a list of relevant concepts and formulas is one thing but actually figuring out how they should all be interconnected is another. That is, the task of determining the most appropriate linking words to concisely describe the nature of the interconnection still requires further improvement. From this study, the 6 students appreciated the utility of the maps/diagrams as means of illustrating conceptual interconnections within a topic and highlighting connection between principles and procedural steps. Students also appreciated the value of the tools in mapping their growing understanding and as means of communicating that understanding to others in a social setting. Findings from this cohort suggest that concept maps and Vee diagrams are potentially viable tools for developing a deeper understanding of the structure of mathematics and facilitative tools to guide and regulate mathematical discussions and dialogues in mathematics classrooms. Using these tools as part of normal classroom routine and practices is an area worthy of further research.

References

Afamasaga-Fuata'i, K. (1998). *Learning to solve mathematics problems through concept mapping and Vee mapping.* Samoa: National University of Samoa.

Afamasaga-Fuata'i, K. (2002). A Samoan perspective on Pacific mathematics education. Keynote Address. In B. Barton, K. C. Irwin, M. Pfannkuch, & M. O. J. Thomas (Eds.), *Mathematics education in the South Pacific*. Proceedings of the 25th annual conference of the Mathematics Education Research Group of Australasia (MERGA-25), July 7–10, 2002 (pp. 1–13). University of Auckland: New Zealand.

Afamasaga-Fuata'i, K. (2005a). Mathematics education in Samoa: From past and current situations to future directions. *Journal of Samoan Studies*, 1, 125–140. Institute of Samoan Studies, National University of Samoa.

Afamasaga-Fuata'i, K. (2005b). Students' conceptual understanding and critical thinking? A case for concept maps and Vee diagrams in mathematics problem solving. In M. Coupland, J. Anderson, & T. Spencer (Eds.), *Making mathematics vital*. Proceedings of the Twentieth Biennial Conference of the Australian Association of Mathematics Teachers (AAMT), January 17–21, 2005 (pp. 43–52). University of Technology, Sydney, Australia: AAMT.

Afamasaga-Fuata'i, K. (2006). Developing a more conceptual understanding of matrices and systems of linear equations through concept mapping. *Focus on Learning Problems in Mathematics, 28*(3 & 4), 58–89.

Afamasaga-Fuata'i, K. (2007). Using concept maps and Vee diagrams to interpret "area" syllabus outcomes and problems. In K. Milton, H. Reeves, & T. Spencer (Eds.), *Mathematics essential for learning, essential for life*. Proceedings of the 21st biennial conference of the Australian Association of Mathematics Teachers, Inc. (pp. 102–111). University of Tasmania, Australia: AAMT.

Afamasaga-Fuata'i, K., Meyer, P., & Falo, N. (2007). Primary students' diagnosed mathematical competence in semester one of their studies. In J. Watson & K. Beswick (Eds.), *Mathematics: Essential research, essential practice*. Proceedings of the 30th Annual Conference of the Mathematics Education Research Group of Australasia (Vol. 1, pp. 83–92). University of Tasmania, Australia: MERGA.

Afamasaga-Fuata'i, K., Meyer, P., Falo, N., & Sufia, P. (2007). Future teachers' developing numeracy and mathematical competence as assessed by two diagnostic tests. Published on AARE's website: http://www.aare.edu.au/06pap/afa06011.pdf

Ausubel, D. P. (2000). *The acquisition and retention of knowledge: A cognitive view*. Dordrecht: Kluwer Academic.

Ball, D. (1993). With an eye on the mathematical horizon: Dilemmas of teaching elementary school mathematics. *The Elementary School Journal, 93*(4), 373–397.

Balacheff, N. (November, 1990). Beyond a psychological approach: The psychology of mathematics education. *For the Learning of Mathematics, 10*(3), 2–8.

Barton, B. (2001). How healthy is mathematics? *Mathematics Education Research Journal, 13*(3), 163–164.

Ernest, P. (1999). Forms of knowledge in mathematics and mathematics education: Philosophical and rhetorical perspectives. *Educational Studies in Mathematics, 38*, 67–83.

Gowin, D. B. (1981). *Educating*. Ithaca, NY: Cornell University Press.

Knuth, E., & Peressini, D. (2001). A theoretical framework for examining discourse in mathematics classrooms. *Focus on Learning Problems in Mathematics, 23*(2 & 3), 5–22.

Lesh, R. (2000). Beyond constructivism: identifying mathematical abilities that are most needed for success beyond school in an age of information. *Mathematics Education Research Journal, 12*(3), 177–195.

National Council of Teachers of Mathematics (NCTM) (2007). *2000 principles and standards*. Retrieved on July 28, 2007 from http://my.nctm.org/standards/document/index.htm

New South Wales Board of Studies (NSW BOS) (2002). *K-10 Mathematics Syllabus*. Sydney, Australia: NSW BOS

Novak, J. D. (1985). Metalearning and metaknowledge strategies to help students learn how to learn. In L. H. West & A. L. Pines (Eds.), *Cognitive structure and conceptual change* (pp. 189–209). Orlando, FL: Academic Press.

Novak, J. D. (1998). *Learning, creating, and using knowledge: Concept maps as facilitative tools in schools and corporations*. Mahwah, NJ: Academic Press.

Novak, J. (2002). Meaningful learning: the essential factor for conceptual change in limited or appropriate propositional hierarchies (LIPHs) leading to empowerment of learners. *Science Education, 86*(4), 548–571.

Novak, J. D. (2004). A science education research program that led to the development of the concept mapping tool and new model for education. In A. J. Canãs, J. D. Novak, & F. M. Gonázales (Eds.), *Concept maps: Theory, methodology, technology*. Proceedings of the First International Conference on Concept Mapping September 14–17, 2004 (pp. 457–467). Dirección de Publicaciones de la Universidad Pública de Navarra: Spain.

Novak, J. D., & Cañas, A. J. (2006). The theory underlying concept maps and how to construct them. Technical Report IHMC Cmap Tools 2006-01, Florida Institute for Human and Machine Cognition, 2006, available at: http://cmap.ihmc.us/publications/ResearchPapers/TheoryUnderlyingConceptMaps.pdf

Novak, J. D., & Gowin, D. B. (1984). *Learning how to learn*. Cambridge: Cambridge University Press.

Pfannkuch, M. (2001). Assessment of school mathematics: Teachers' perceptions and practices. *Mathematics Education Research Journal, 13*(3), 185–203.

Richards, J. (1991). Mathematical discussions. In E. von Glaserfeld (Ed.), *Radical constructivism in mathematics education* (pp. 13–51). London: Kluwer Academic Publishers.

Schell, V. (2001). Introduction: Language issues in the learning of mathematics. *Focus on Learning Problems in Mathematics, 23*(2 & 3), 1–4.

Thompson, A. G. (1984). The relationship of teachers' conceptions of mathematics and mathematics teaching to instructional practice. *Educational Studies in Mathematics, 15*:105–127.

Schoenfeld, A. H. (1996). In fostering communities of inquiry, must it matter that the teacher knows "the answer". *For the Learning of Mathematics, 16*(3), 569–600.

Steffe, L. P., & D'Ambrosio, B. S. (1996). Using teaching experiments to enhance understanding of students' mathematics. In D. F. Treagust, R. Duit, & B. F. Fraser (Eds.), *Improving teaching and learning in science and mathematics* (pp. 65–76). New York: Teachers College Press, Columbia University.

The Australian Association of Mathematics Teachers (AAMT) (2007). The AAMT standards for excellence in teaching mathematics in Australian schools. Retrieved from http://www.aamt.edu.au/standards, on July 18, 2007.

Schoenfeld, A. H. (1991). What's all the fuss about problem solving? *Zentrallblatt Für Didaktik der Mathematik, 91*(1), 4–8.

Wood, T. (1999). Creating a context for argument in mathematics class. *Journal for Research in Mathematics Education, 30*(2), 171–191.

Chapter 13
Concept Mapping: An Important Guide for the Mathematics Teaching Process

Rafael Pérez Flores

The chapter deals with a particular way of using concept maps to contribute to the learning of mathematics by students who begin their formation in engineering at university level. The focus is on a particular process of teaching. When the expert in mathematics analyses in detail the material of his assignment, he is able to realise that within the structure of knowledge itself, a group of concept maps are found in an implicit way. The maps made by the professor are of great importance when he uses them as a guide for his performance in the classroom. When the teacher utilises in the classroom a guided way of performing with the maps, he implements a didactic strategy and he helps students to develop skills of thought to achieve meaningful learning. This way of performing allows the development of the processes of basic thinking in mathematics; it facilitates the understanding of concepts and the application of them within the process of solving problems.

Introduction and Antecedents

When one looks at the different paradigms about education frequently encountered today, such as the cognitive, humanistic and socio-cultural paradigms, in general terms it can be observed that learning is understood as a development of both cognitive as well as affective elements.

Resulting from an analysis of the principle aspects of each of the educational paradigms, Román and Díez (1999) put forward the idea of a paradigm or integrating model, a socio-cognitive paradigm, consisting of the fusion of two: the cognitive and the socio-cultural, an integration resulting from a search for complementary models. The authors point out that together the cognitive and the socio-cultural paradigms can complement each other to give meaning to what is learnt. This complementary model finds support in the affirmation by Vygotsky (1979), namely, the potential of learning (i.e., the cognitive dimension) is developed by means of contextualized socialization (i.e., the cultural dimension).

R.P. Flores (✉)
Universidad Autónoma Metropolitana, Mexico City, México
e-mail: pfr@correo.azc.uam.mx

K. Afamasaga-Fuata'i (ed.), *Concept Mapping in Mathematics*,
DOI 10.1007/978-0-387-89194-1_13, © Springer Science+Business Media, LLC 2009

Román and Díez (1999) point out that certain important aspects must be taken into consideration, such as: how does the student learn, how does the teacher teach and what is the atmosphere and life in the classroom. The authors, from this perspective, emphasize that one can and should complement both paradigms. While the cognitive paradigm is more individualistic, the socio-cultural paradigm is more socializing. The first centers around the individual, the thought processes both of the teacher and of the student, and the second centers around the context-group-individual interaction and vice versa. Following these ideas, learning mathematics may be understood as a development of capacities (i.e., cognitive elements the student develops while relating with the mathematical content during the learning process) as well as attitudes (i.e., affective elements developed in the context of the classroom as a product of the interactions with the teacher and companions).

As is well known, concept maps are an important teaching tool to assist in the learning of different contents (Novak and Gowin, 1988). But one must not forget that this in its turn implies starting up thought processes that favor the development of a cognitive structure and in addition offer a great opportunity, depending on the way the teaching activities are organized, for the development of affective aspects in the context of collaborative learning.

In a great number of curricula models rooted in different educational paradigms, reasons can be found for the use of concept maps in the process of teaching and learning. But also by analyzing educational panoramas in different regions of the world, we find other reasons of great importance that justify the study and introduction of novel teaching methodologies that include concept maps. Considering what today is known as, namely, "the society of knowledge" or "the society of information", the development of cognitive and affective elements, also known as the development of competence, has become the leader in educational systems, from the early levels up to university, in many nations.

It is important to mention the process of agreement in the European Area of Higher Education and the development of competence, (Hernández, Martínez, Da Fonseca, and Rubio, 2005). The Universities in the Europe of Knowledge play a very important role in social and human growth/development. The universities have the task of affording the citizens the necessary competence to face the challenges of the new millennium, as well as contributing towards the consolidation and enrichment of the citizens of Europe. The declaration made in 1988 at the Sorbonne emphasizes the central role of the Universities in the development of European cultural dimensions and the creation of the European Area of Higher Education in order to favor the mobility of the citizens and the ability to obtain employment for the general development of the continent.

The systems of higher education and research of all the nations, not only in Europe, must be continually adapted to the ever changing needs and demands of society. The transformations that constantly appear in societies evoke reflections in the organizations about its role, commitment and especially the way to confront the new needs and demands of society. In the case of educational institutions reflection in the pedagogical world is important in order to be able to attend to and offer a better formation of the students in accord with the characteristics of the context.

The commonly evident characteristics in present day societies of almost all the countries force reflection and discussion about the role of the universities and their commitment to society. In the same way today, to speak of changes to aid the formation of the student demands special attention to the process of learning and teaching in order to create ways of acting in the classroom different to the usual or classical ones. In other words, it is paramount not to forget the importance of the role of didactics when carrying out actions in the classrooms to expedite the cognitive activity of the student allowing him to develop his thought potential which in its turn spurs him on to learn. As Román (2005) has described, it is a question of achieving in the student a meta-cognitive autonomy to facilitate his integration in a present day social context characterized by information, knowledge and a great diversity of activities.

When one looks at the educational institutions, both at the basic level and at the university level, one can appreciate and distinguish difficult situations. Nowadays there are throughout the world problems related to the teaching and learning of mathematics. Certainly, this reality in education has increased an interest to learn about, from different standpoints, studies that offer an alternative or guide for teaching and learning of this content. Mathematics is of great importance, among other aspects, for its application in science, technology and even in the context of daily life, but, also, it has great educational value at all levels since it permits the development of thought De Guzmán (1999).

Something even more important, pointed out by different thinkers like Casassus (2003), is that the changes that imply new methods (methods of teaching and learning depending on the development of competence) that permits the citizen to integrate in the activities of society, are changes or transformations that aim to a great extent at social equity, as well as improving higher education.

With regard to the social commitment of the universities there are weighty aspects in the proposals (for the development of competence) that are primordial for the student's learning by taking into consideration the cognitive and affective dimension of the learner. As explained by Casassus (2003), the variables within the educational institutions (internal variables) have greater weight than the variables outside the institutions (external variables) such as the families, the social milieu and what this implies, for instance: the previous academic formation of our students. The variable within the institution: "teaching without regard for learning", in other words without attention to the development of competences, allows certain external variables to enter into action leading to, as a result, drop outs or failure at school.

Whatever strategy eases the learning process of mathematics is important for its potential contribution to the development of competence and all that it implies. All the work towards achieving proper formation in the students begins in the classroom with different ways of acting by the teachers, reflecting upon the attitudes and current didactic tools. This chapter proposes that concept maps represent a didactic tool that contributes to learning of quality; the latter is defined as learning that is constructive, meaningful and by experimentation. Within the framework of the socio-cognitive paradigm quality learning is understood to be the development of cognition and affectivity.

Concept Maps and The Learning Process

One of the fundamental ideas about concept maps and mathematics is that the concept maps are implicit in the content of mathematics itself, (see Fig. 13.1). For example, a group of particular elements conform generalities that can be called "low level generalities". At the same time, a group of "low level generalities" form "mid level generalities". Finally a group of "mid level generalities" form "high level generalities". Depending upon the characteristics of the contents themselves there can be few or many generalities. The conformation or grouping of information represents a structure that is implicit in the knowledge of mathematics. This structure or network can be noted when the mathematical content is analyzed or weighed up. But also, these structures represent concept maps. Novak (1998) points out that concept maps represent knowledge. That knowledge can be concepts and can be propositions as well, depending on the context. Concept maps, as a representation of knowledge, serve as support for teaching and learning.

As can be seen in Fig. 13.2, a concept map combines particularities in a generality and also combines low level generalities in a generality of a higher level. Concept maps are structures characterized by very particular elements in the lower part and general elements or very general elements in the upper half.

The teacher interested in finding different teaching methods can construct his own concept maps once he has carried out a profound analysis of the structure of knowledge he wishes to teach. These maps represent for the teacher a guide, letting him know how to act in the classroom to achieve adequate learning. The maps show him what information he can present in order to respect the stages present in

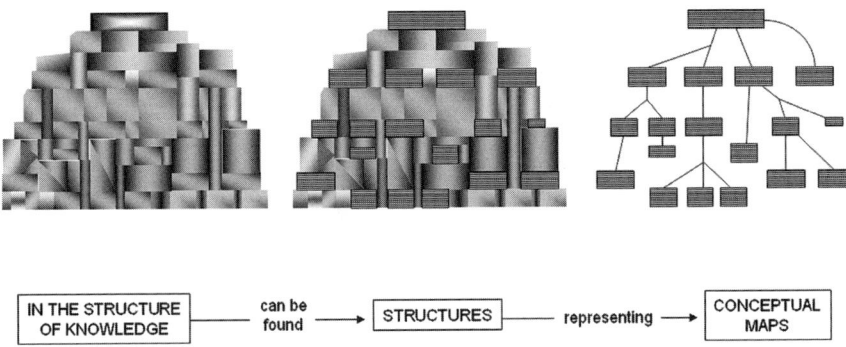

Fig. 13.1 Structure of knowledge and concept maps

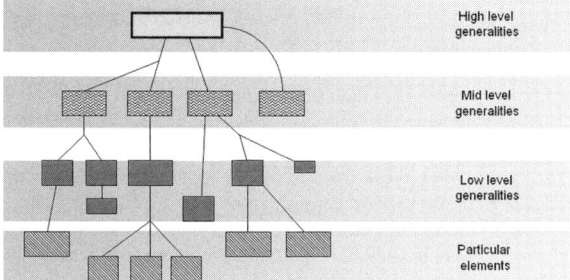

Fig. 13.2 Levels of generality

the learning process (e.g., perception, representation and conceptualization) by contemplating the different levels of generalities and promoting the starting up of basic operations of thought (such as induction and deduction) favoring the development of competences in the student (i.e., cognitive development).

The stages present in the process of conformation or learning of knowledge: perception, representation and conceptualization are, according to Román and Díez (1998), stages or basic levels of learning. The authors consider that learning therefore implies articulating in the classroom adequately these three basic levels of learning. For them learning supposes the initiation of a cyclical process which first must be inductive (from perception to conceptualization) and later deductive (from concepts to facts). For the authors this learning is considered a Cyclical Process of Scientific Learning, (see Fig. 13.3). They recommend that teaching professionals must be respectful of this cyclical learning process, since in this way the students will make sense of what they are learning. Added to this is the idea that the student constructs knowledge by induction (from the particular to the general) and deduction (from the general to the particular).

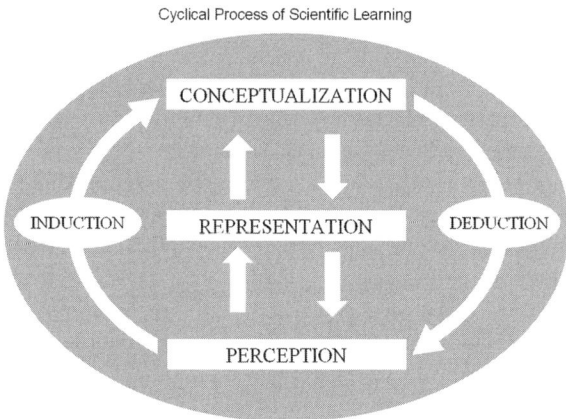

Fig. 13.3 Cyclical process of scientific learning

The following are a series of ideas through which Román and Díez (1998) described the three basic levels of learning.

Perception The basis of learning is in the sensorial perception of facts, examples and experiences captures preferably from the context or milieu in which the learner lives. The basis of perception is above all auditive and visual since today we move in an audiovisual culture. The latest theories of short and long term memory try to revalue perception as a basis of the active and constructive memory. Short term memory acts by codifying and storing information of a perceptive origin and constructing a data base (facts, examples, experiences, perceptions and concrete sensations) that later will have to be re-elaborated in order to constitute a knowledge base through representations and conceptualization. To construct a base of knowledge without empirical data and without contrasting with reality will be problematic for the learner.

Representation Facts, examples and experiences capture by sensations and perceptions are converted into representations and mental images, more less organized, usually associated with other representations. In this way what has been perceived is converted into images. According to Aristotle, images presuppose perception and sensation by which the human being conserves the content. The image becomes the data or what has been given and as such is not a creative act. It is a generic representation of what has been given (facts, examples, experiences, . . .) possessing a high intelligible potential for the creation and development of concepts. The concept is supported by images of mental representations – the image in this way becomes the raw material for the concepts and acts as a potentially intelligible object.

Conceptualization To learn is to conceptualize or symbolize concepts, theories, principles, conceptual systems, hypotheses and laws. It implies managing concepts and symbols adequately and interrelating them. From this point of view, the data base, captured by perception and semi-organized by representation, becomes a base of knowledge structured in the form of concepts and symbols interrelated adequately. The authors mention that Aristotle denominates concepts as universal and considers they are products of intelligence and thought (abstraction) (pp. 178–183).

Basic Processes and The Learning of Mathematics

There are some authors who have pointed out important aspects about the basic thought processes in the context of the teaching and learning of mathematics. For example, Usabiaga, Fernandez, and Cerezuela (1984) mentions that the interchange of inductive and deductive proceedings can help the student to ways of acting that complement each other and definitely this interchange allows the student to have available mental elements and creative resources to resolve difficulties. As to the relation induction-deduction, the author says it is not right to drop the inductive elements too soon in the cognitive activity of the student in the classroom. On the one hand, it is not correct to assimilate in a reduced way the scientific method, in other words, only employ the deductive part, or only use the hypothetical-deductive method. The lack of inductive processes (the abuse of deductive processes) limits the options of personal research important for the development of thought. That does not mean that as the student reaches formal levels he leaves aside the deductive processes. On the way to high levels of abstraction the student must count, in his personal experience, on a certain prior training in inductive processes. This implicitly suggests a balance between inductive and deductive processes during the learning of mathematics. When designing the didactic strategies this harmony must be sought without falling into an inductive extreme or a deductive extreme.

While insisting on the ideas, one must remember as a fundamental element in the didactic strategies to be designed, from the position of Scientific Learning, for the teaching of mathematics, the teaching activity, implies propitiating an inductive process by analyzing the contents and demonstrating the facts, examples and experiences, accessible to the intellect of the students, to be perceived by them. From the perception of this information (i.e., starting from the particular), the teacher in charge will guide the students, without forgetting the processes of representation (i.e., imagination) and with visual support (De Guzmán, 1996), towards conceptualization (i.e., construction of concepts; understanding of ideas; arriving at the general). The other way round, to promote the deductive process, the teacher will guide the students so that, starting from concepts (i.e., generalities) and through images the student arrives at facts and examples, at the understandable (i.e., the particular). The concept maps created by the teacher are an important guide for this kind of teaching task.

This method, as a didactic or learning strategy is not only relevant to mathematics or physics but it was also the way the thinkers and scientists such as Aristotle, Galileo and Newton among others, reached knowledge. It is a way that represents the form in which such knowledge was elaborated. Teaching methods are not something foreign to the knowledge that is mathematics. As González and Díez (2004) points out, the didactic of mathematics is something implicit in its very construction. The route followed by the scientist which led him to knowledge was also the way he learnt.

It is very common to find in the everyday teaching world methods that mainly follow deductive processes. An inductive process is not employed first so that later the student can be taken through a deductive process. Generally, perhaps more often at the university, a formal conceptual system is introduced by means of symbols that include principles, definitions, theorems, and laws; that is to say, they introduce theory. This is more notable in what is known as master classes, commonly found in university lecture halls (De Miguel Díaz et al., 2006). Since the student is not in a condition to assimilate such a system, audiovisual methods are used so they understand. From a theoretical framework, they wish to pass to representation. Afterwards, when they realize the student is not able to apply the concepts, even with the help of audiovisual methods, they teach him the applications in reality. So finally they reach the reality (the particular) where they should have started. Thus in classical and present day teaching, the order of learning activities is not respected. Learning of quality implies teaching that procures a balance between the thought processes that conform abstraction.

To sum up, (see Fig. 13.4), it is suggested that concept maps are to be found in the structure of knowledge of mathematics. The concept maps are a guide to teaching that respects the basic levels of learning and that develop competence. In this sense, as the title of this chapter states, concept maps are an important guide for the teacher, for the process of teaching mathematics.

Fig. 13.4 Concept map: "A guide for the process of teaching"

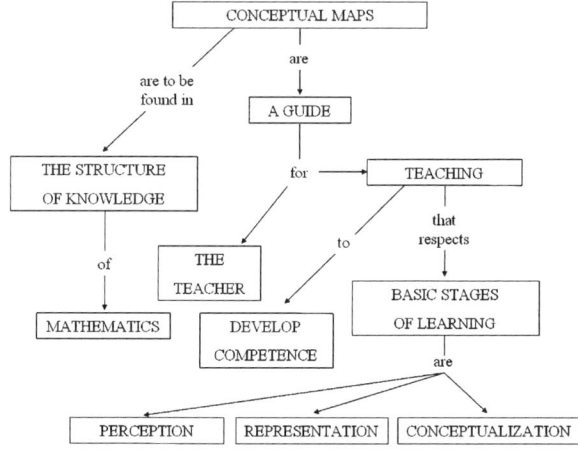

Concept Mapping for Mathematics Teaching Process: Some Examples

The following are a collection of concept maps. They have been selected with the aim of elaborating the subject further as a guide for the teacher during the teaching process. Their characteristics will be explained and the possible ways of acting in the lecture room are described to illustrate their potential to support the development of thought processes. Teachers who wish to carry out their teaching with these ideas, can make their own adaptations, based on their experiences, for the important and detailed task of teaching.

"Critical Point of a Function" Concept Map

If one reads down from top to bottom the concept map entitled *"Critical point of a function"* presented in Fig. 13.5, the following appears: *a critical point in a function* can be *an extreme point* (extreme value) or *a non extreme point* (non extreme value). *An extreme point* can be *a maximum point* or *a minimum point*. Also *a maximum point* can be a point with a *horizontal tangent line* (a tangent line with an inclination equal to zero) or a point without a *tangent line*, that is, a point where *the derivative* is not defined (does not exist). *A minimum point* can be a point with a *horizontal tangent line* or a point where *the derivative* is not defined. In addition, a *non-extreme point* can be the one that has *a horizontal tangent line* or a point that has a *vertical tangent line*. In the concept map *"Critical point of a function"* one can note in the lower part there is particular information specified in images (i.e., graphs).

These images that can be shown to the student represent a support leading to comprehension and the learning of concepts at a low level of generality. In this concept map the concepts *extreme point* or *non-extreme point* can be considered as

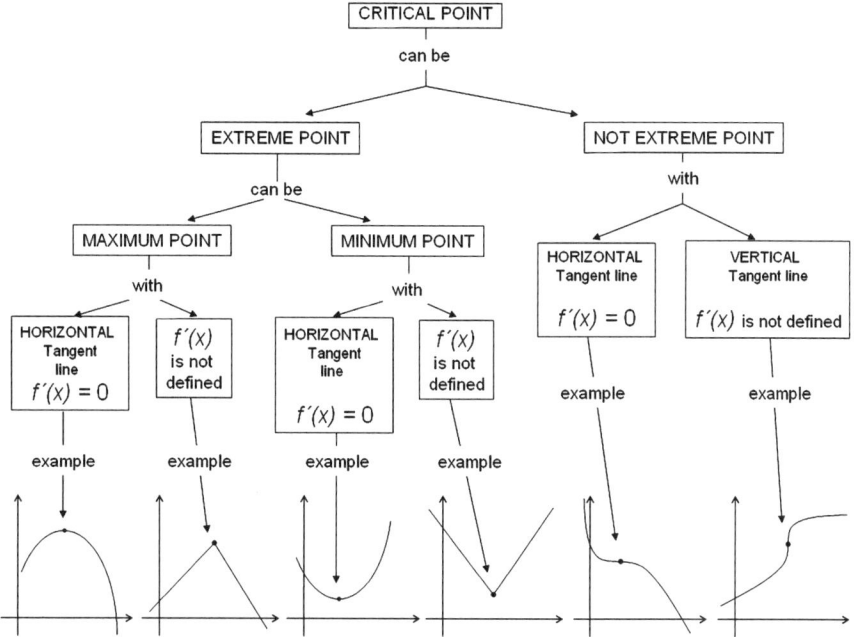

Fig. 13.5 Concept map: "Critical point of a function"

concepts at a medium level of generality. The concept *critical point of a function* is at the highest level of generality in the map. As Ausubel explains, the inductive and deductive thought processes allow the disposition of information respecting conceptual hierarchies, achieving subordinate and supra-ordered learning, moving from the particular to the general and vice versa (Ausubel, 1976).

The concept map *Critical point of a function* represents a guide for the teacher to offer information going from the particular to the general promoting an inductive process. The information in this map is related to the applications of the concept of the *derivative* and is a very important topic to help resolve problems in real life engineering. For each one of these examples, the teacher can demonstrate by applying the mathematical tools relating to the *derivative* to obtain the value of the critical points and then classify them. A wide battery of functions can be shown that have *critical points* to demonstrate the different types, to move toward the more general concept. Various functions can show *maximum and minimum points* as *extreme values* and other functions can show just *non-extreme values*. This can all come together as *critical points of a function*.

It is useful for the student to bear in mind the characteristics of each of the *critical points*. This will help him realize a part of the analysis of a function and will be very useful information to obtain a graph. The deductive process appears when concepts that were understood previously are compared with facts (the information obtained from the analysis of a function) in order to conclude with the *critical points of a function* and its classification. In other words, after the student has determined the *critical points* he can classify them into *extreme or non-extreme values*. If it is a question of *extreme values* he can classify them into *maximum or minimum points*. If it is a question of *non-extreme values* he can classify them in points with a *vertical tangent line* or a *horizontal tangent line*. As a particular interpretation of Piaget's ideas, contrasting facts with concepts and concepts with facts is considered as leading to inductive and deductive processes thus contributing to constructive learning (Piaget, 1979). It is worth indicating that the study of *critical points of a function* should have previously assimilated the definition of *the derivative*.

It is very important to point out that the teaching of mathematics from this perspective is not merely presenting the students with a concept map elaborated by the teacher. They are only personal materials used by the teacher. Perhaps if the teacher wishes, he can show these maps at the end of the classes. It is also useful for learning and to reinforce the development of cognitive capacities that the students try to construct their own maps guided by the teacher. This activity, which can be carried out in groups, requires special sessions during a course where collaborative learning is undertaken, creating in the classroom, among other aspects, important affective elements for the student in the learning process. Some of the previous ideas can be summed up in the following manner: in the dynamics of collaboration, learning happens through the interactions in the social context of the classroom.

"Extreme Point" Concept Map

To further examine the topic of critical points, another map can be made showing other concepts. In the concept map *"Extreme point"* of Fig. 13.6 the following can be read: an *extreme point* can be a *maximum point* or a *minimum point*. In turn, a *maximum point*, with a *horizontal tangent line*, can be an *absolute or relative maximum point*. On the other hand, a *minimum point* with a *horizontal tangent line* can be an *absolute or relative minimum point*. It can be seen that this map concentrates on the extreme points with a horizontal tangent line. This kind of critical points are those that appear frequently in the functions that model a great number of situations in real life proper to the context of engineering. In the lower part of the map a collection of graphs can be seen that represent visual support for the comprehension of concepts. It is information that is particular and accessible at the level of students in the course of calculus.

Through the visual supports common aspects can be determined in order to arrive at the concept of the more generalized map: *an extreme point* (Fig. 13.6).The mathematics teacher must realize he can construct many concept maps according to the different contents of his course. Some will be more specific than others but they will always represent an important tool for his work and for stimulating the thought processes.

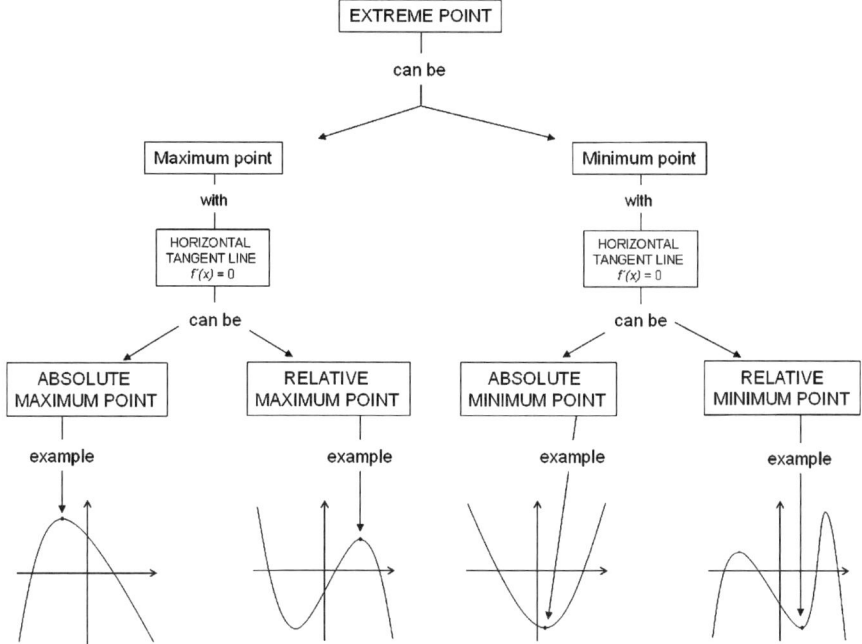

Fig. 13.6 Concept map: "Extreme point"

"Real Numbers" Concept Map

Figure 13.7 gives a useful map entitled "*Real numbers*". In the lower part there is the more particular, different classes of numbers that can be placed in a horizontal line as visual support in order to understand the great concept of the map: *real numbers*. The teacher can play with information to show that a *natural number* is a*positive integer number* and a *integer number* either positive or negative is a *rational number*. He can also play with the information to show how certain *irrational numbers* can be placed in the horizontal line, for example $\sqrt{2}$. What is important is for the student to initiate learning by perceiving particular information accessible to his intellect. During the teacher's classes or in the sessions to foster collaborative learning (at the moment of elaborating maps in a groups with the help of the teacher) different systems implicit in learning are set in motion.

From the point of view of Bruner and the theory of learning by discovery, starting from an enactive system (facts, examples and experiences) up to a symbolic system (concepts) passing through an iconic system (images) learning by discovery happens permitting the development of inductive processes (Bruner, 1988). The question of *real numbers* is important for the study of calculus. The concept map the student can create on this topic, after hearing the presentation and explanations of the teacher, in addition to setting in motion thought processes, represents an image that can be stored easily by the student to which he can return when necessary during his studies for the solution of problems.

Fig. 13.7 Concept map:
"Real numbers" – Taken
from
http://canek.azc.uam.mx

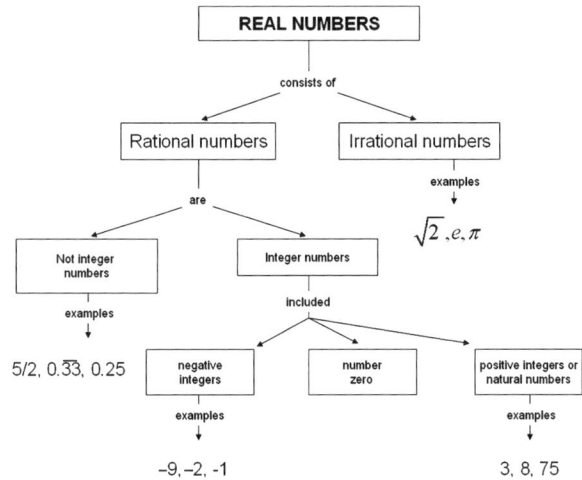

"Displacement of Functions" Concept Map

On occasions, presentations or university courses begin from generalities. An example is when initiating the subject of transformations of function by presenting the expression $g(x) = k f(a x - b) + c$ and explaining the effects of the constants k, a, b y c on the function $f(x)$. This procedure in the lecture room forces a deductive process without any previous inductive process and motivates memorization rather than the comprehension of the ideas. The concept map in Fig. 13.8 presents two concrete cases of displacement, a horizontal displacement and a vertical one of a particular function. These two cases, can be presented first by the teacher, contain appropriate images that allow the comprehension of the effect of the constants b and c on a function. Various concept maps can be constructed to foster thought processes during the explanation of the displacements such as the lengthening, the compressions and reflections of functions, leading finally to link all this information with the expression $g(x) = k f(a x - b) + c$. As has been said before, at later stages of the explanations, the students, with the advice of the teacher, can elaborate their own maps. The proper construction of these maps represents satisfactory learning in the subjects of calculus.

Mathematics Problem Concept Maps

But in mathematics there are not only concepts. The exercises and problems that naturally imply the handling of concepts are a part of the mathematical content too. Following the ideas and characteristics of concept maps, structures and networks can be constructed that contain information on the procedures to resolve exercises or part of them, where generalities or concepts appear as previously explained and learnt (see Figs. 13.9 and 13.10). It is common that at times one should wish to

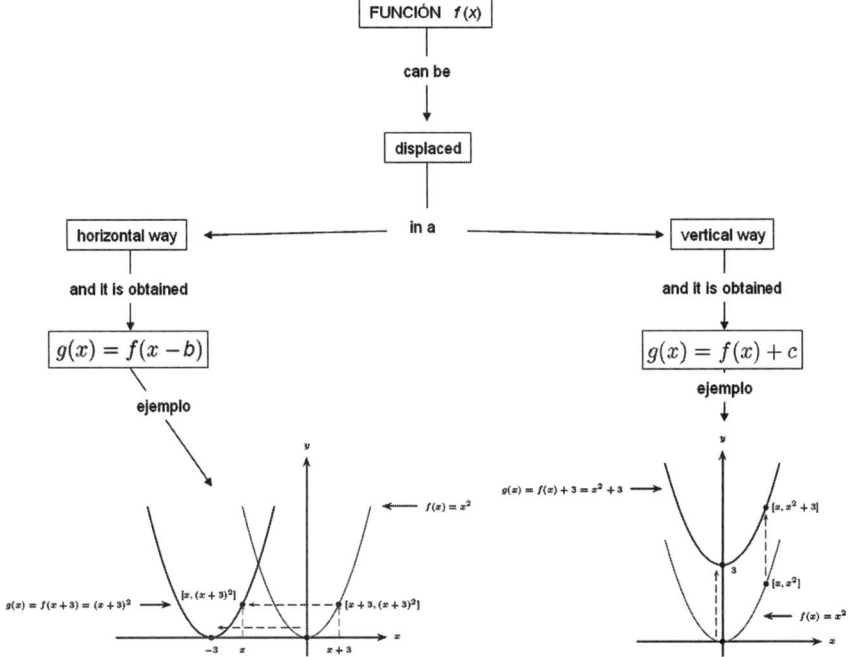

Fig. 13.8 Concept map: "Displacements of functions" – graphs taken from http://canek.azc.uam.mx

proceed in the lecture hall in one way or another, emphasizing certain aspects, in regard to the experiences and particular difficulties appreciated during the resolution of problems and exercises.

"Solution of Certain Inequalities" Concept Map

The concept map of Fig. 13.9 shows two structures that link a generality ($|x–b|< M$ & $|x–b| > M$) with a collection of exercises as examples (or particular exercises). It is a way of showing that generality can be applied to resolve inequalities of a particular kind.

The teacher can show these or other exercises as examples and explain the application of the general expression to reach the particular solution. Of course, the idea of generality is as explained earlier. Naturally these exercises are simple for the teachers but not for the students.

Simple concept maps like these, when prepared by the students, represent images easy to recall and useful to resolve exercises appropriately.

"Elements of a Function" Concept Map

The concept map presented in Fig. 13.10 gives elements of a function and how to obtain them. This map shows applications of the *concept of the derivate*. The

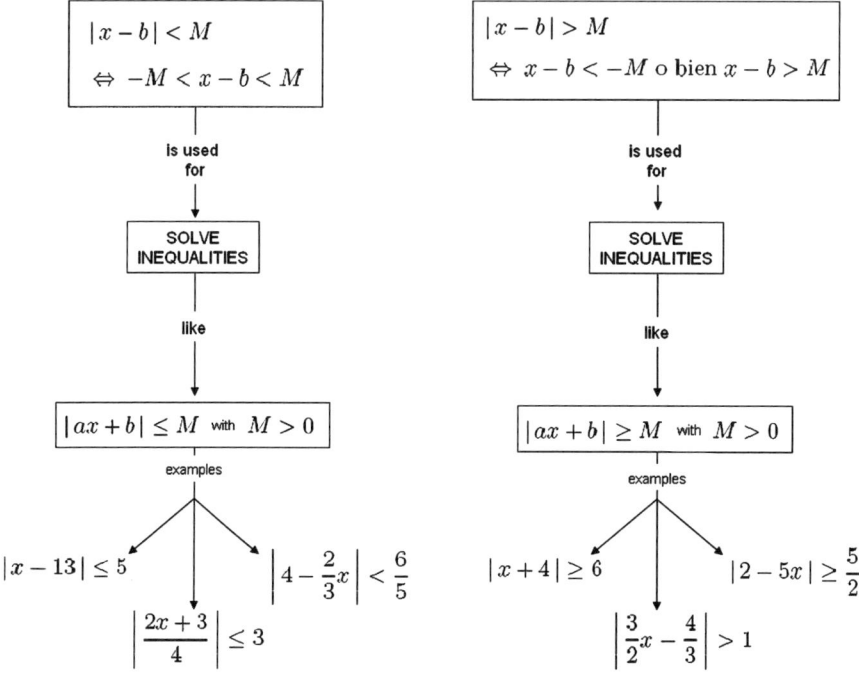

Fig. 13.9 Concept map: "Solution of certain inequalities"

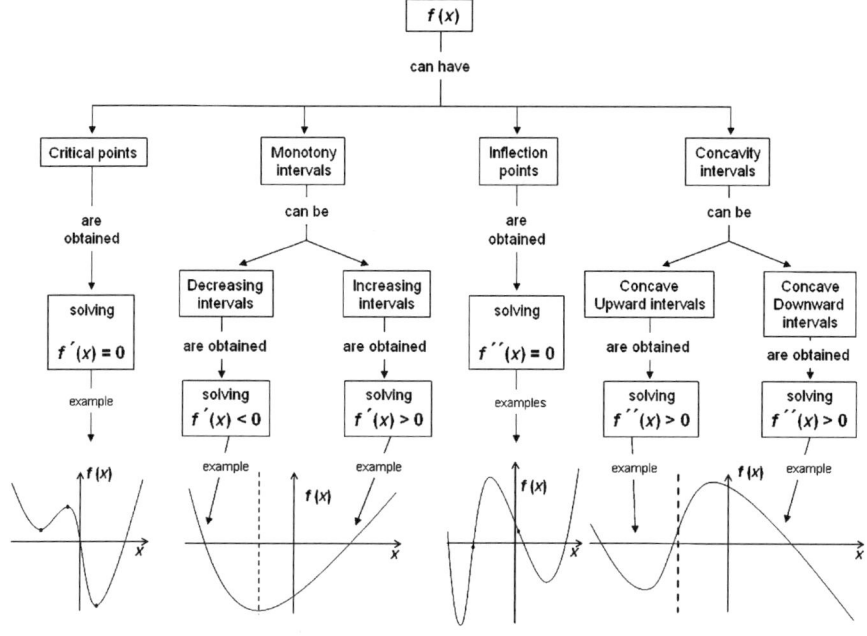

Fig. 13.10 Concept map: "Elements of a function"

structure presents a series of graphs showing *critical points, intervals of growth and decrease, points of inflection and intervals of concavity and convexity.* Obtaining all this information about a function in particular is necessary to be able to obtain a graph with a certain precision for the said function. A map such as this suggests the information to be presented for the student to reach conclusions and to carry out processes of abstraction. A series of functions can be given to show that by resolving $f'(x) = 0$, $f'(x) > 0$, $f'(x) < 0$, $f''(x) = 0$, $f''(x) > 0$ and/or $f''(x) < 0$ a series of characteristics are obtained about $f(x)$. After this handling of particular information a formal mathematical explanation must be given about the application of the derivate. One must always remember that previous work with particular information permits a satisfactory assimilation of the later explanation about mathematical generalities.

Cognitive Development of the Learner

By using concept maps in the classroom or lecture room with a series of strategies (as previously described) favors, in general terms, the cognitive development of the learner. In particular, with this method, the aim is to seek the development of: (1) the skills of induction and deduction considered as part of the capacity for logical reasoning and (2) the skills of situating, localizing and expressing graphically, considered as part of the capacity for spatial orientation. Both capacities form, among others, the operations of thinking about thought (i.e., cognition). When developing the cognitive processes implicit in mathematics, thought is developed (De Guzmán, 1999). Developing the skills and capacities during the teaching process also implies the development of student competence.

There are a great number of ways of acting in the classroom for the teaching of mathematics and perchance many of them have good results for learning. The use of concept maps as a guide for the teacher is yet another way of acting and it is certain that other ways of acting have in them the aspects presented above. A synthesis of these ideas for the teacher might be: show the information feasible for the intellect of the student, let him play with this information, offer images that complement and nourish their mental representations, so that the way toward generalities may be an educational truth followed with pleasure. Miguel De Guzmán, a great mathematician and educator insisted:

> Playing and beauty are the origin of a great part of mathematics. If the mathematicians of all times have had such fun playing and contemplating their game and their science, why not try to learn it and communicate it through games and beauty.
>
> (cited in Sanz, n.d.)

Concept Map and Students: Important Considerations

To speak of concept maps as a guide for the teacher is to speak of maps as important elements for intervention. As has been mentioned before, concept maps contribute to learning, being understood as the development of elements in the cognition of the student. But this development of the cognitive sphere is not isolated from the development of the affective sphere. The development of thought processes affects the affectivity understood as a grouping of attitudes. At the same time one can speak of the opposite, that is to say, the development of attitudes has influence on the development of thought processes. In this last section various important effects are presented as the effects of the concept maps on the students such as effects on the development of cognitive capacities and the development of affective elements.

The Intervention and Its Methodology

As part of the educational experiences in the Universidad Autónoma Metropolitana of Mexico City concept maps have been used. These have been a substantial part of a Program of Intervention for a Calculus course to help in the learning of mathematics. Two groups were formed: the control group and the experimental group, each one with 20 students. The control group began an ordinary Calculus course and the experimental group began the same course but with the Program of Intervention. During this investigation both quantitative and qualitative instruments were used in order to carry out an evaluation of the teaching process with teaching guided by concept maps. The instruments employed to carry out the qualitative register were; a diary of each of the students, a diary by the teacher and the partial evaluations of the content carried out during the course. The information offered by the students in these evaluations reflected qualitative aspects of academic benefits understood as the level of learning obtained by the students from the contents of the subject and about the development of capacities and skills. To carry out the quantitative register the following instruments were used: Test of differential aptitudes DAT-5 (Differential Aptitude Test, Fifth Edition. Authors: George K. Bennett, Harold G. Seashore and Alexander G. Wesman) measuring Numerical Reasoning, Abstract Reasoning and Spatial Relationships.

Results

The information gathered from the students revealed qualitative aspects of the academic advances. The students frequently expressed the fact that they understood the initial information very well, which in its turn helped them understand the concepts. It is a question of the emphasis made in the presentation of particular information (examples) accessible to their intellect that supports the introduction of the concepts. Also the students commented that the images and graphic representations for particular information is an effective aid to the comprehension of more complex concepts. They evaluated this methodology as appropriate since it contributes to the learning of the theory of calculus in a way that does not imply memorizing. They referred to the fact that they comprehended the theory and had not memorized it.

On the other hand the students considered it important to know the concept maps prepared by the teacher once the practical and theoretical aspects had been presented. They said this activity motivated them to construct their own maps on the content of the course, thus developing their thought processes.

The results obtained from the DAT-5 tests represent evidence of the favorable effects of the didactic strategies guided by concept maps. Having processed the information from Dat-5 measuring Numerical Reasoning of a control group and an experimental group, statistically significant differences were observed only in the experimental group with a confidence level of 99% between before and after. The processing of information from DAT-5 measuring Abstract Reasoning and Spatial Relationships revealed, in addition, statistically significant differences in the experimental group with a confidence level of 99% between before and after. Consequently, it is obvious there is a significant evolution of Numerical Reasoning, Abstract Reasoning and Spatial Relationships in the experimental group. This increase can be explained principally from the effect of integrating into the teaching-learning process for a Calculus course, of concept maps as tools for the development of cognitive skills and capacities in the students. It is important to emphasize the theoretical ideas supporting the concept maps, mainly those implicit in the way of employing them, which can plausibly explain the changes experienced by the students.

Discussion and Implications

Hernández et al. (2005) underlines two points of view about learning. He points out that Marton and Säljo (1976a, 1976b) were the first to coin the terms *deep focus* and *superficial focus* of learning to refer to the two differentiated ways of processing information. In general terms, *deep focus* depends upon the student centering his attention on the content of the learning material and is orientated toward the message this contains, toward the meaning. *Superficial focus* is characterized by the fact that the student directs his attention to the text itself, toward the sign, which indicates a reproductive conception of learning. What is important in *deep focus* is the comprehension and in *superficial*, memorization.

Hernández et al. (2005) place the emphasis on the fact that the primordial ideal in regard to the methodology in the classroom is not to try to change the subject (i.e., student) in order to improve learning. Rather it is to change his learning experience or his concept of learning. The inclusion of concept maps as a teaching guide directs the new experiences toward a *deep focus* for learning.

Non intellectual elements suffer modifications during the learning experience. The development of these elements implies the development of motivational factors and the development of affective factors. Motivational factors are the expectations, the aims, the interests and the attitudes. The development of affective factors refers to the development of a feeling of security, a feeling of independence and development of the concept of oneself (*auto-concept*).

To speak of the *auto-concept* is to speak of the image the subject has of himself. The auto-concept is a product of many factors; among them are the learning experiences. One must not forget that in the classroom or the lecture hall, a consideration of emotional and cognitive situations, favor a learning atmosphere and collaboration in the resolution of problems, which reflects in the cognitive and non cognitive competence and in the *auto-concept* of the students. Polya (1981) asserts it would be a mistake to consider the solution to a problem is a purely intellectual matter; determination and emotions play an important part. Auto-concept is a basic factor in learning, which can be modified or developed in the classroom with repercussion in better learning. Undertaking collaborative activities in the classroom for the creation of concept maps has a strong effect upon motivational and affective factors in the students.

Teachers through strategies in the classroom have a great task so that the students learn to control feelings such as anxiety, fear, despair, and perplexity that limit the cognitive processes. Developing *auto-concept* can be understood as developing the capacity of a student for knowledge, control and gestation of feelings. Anxiety and despair can appear at a moment of cognitive conflict, which presents the student with two options: control of his feelings or abandoning the task or activity. Every effort related to improving the learning processes represents a way of achieving educational equity.

The university domain needs new research that contemplates the learning and teaching processes. Research for the enhancement of pedagogical knowledge includes: students, teachers contents and technology. Currently in the Universidad Autónoma Metropolitana of Mexico City, Mathematics teachers are constructing, backed by the courses, a mathematics portal in Internet called "CANEK" to be found at http://canek.uam.mx. There is material online for which concept maps are being created and gradually introduced. At present theory and problems can be found in the page offering learning support; in the future it is hoped that all this online material with its concept maps will represent an element for further educational research to contribute to pedagogical knowledge in the field of the technology of information.

References

Ausubel, D. (1976). *Psicologia educativa. Un punto de vista cognoscitivo*. México: Trillas.

Bruner, J. (1988). *Desarrollo cognitivo y educación*. Madrid: Morata.

Casassus, J. (2003). *La escuela y la (des)igualdad*. Santiago de Chile: LOM Ediciones.

De Guzmán, M. (1996). *El rincón de la pizarra*. Madrid: Pirámide.

De Guzmán, M. (1999). *Para pensar mejor*. Madrid: Pirámide.

De Miguel Díaz, M. (coord), Alfaro Rocher, I. J., Apodaca Urquijo, P. M., Arias Blanco, J. M., García Jiménez, E., Lobato Fraile, C., et al. (2006). *Metodología de Enseñanza y Aprendizaje para el Desarrollo de Competencias*. Madrid: Alianza.

Hernández, P., Martínez, C., Da Fonseca, R., & Rubio, E. (2005). *Aprendizaje, Competencias y Rendimiento en Educación Superior*. Madrid: La Muralla.

González, J., & Díez, B. (2004). Las didácticas específicas: Consideraciones sobre principios y actividades. *Revista Complutense de Educación, 15*(1), 253–286.

Marton, & Säljo. (1976a). On qualitative differences in learning I – outcomes and processes. *British Journal of Educational Psychology, 46,* 4–11.

Marton, & Säljo. (1976b). On qualitative differences in learning II – Outcome as a function of the learner's conception of the task. *British Journal of Educational Psychology, 46,* 115–127.

Novak, J. D. (1998). *Conocimiento y aprendizaje.* Madrid: Alianza.

Novak, J. D., & Gowin, D. B. (1988). *Aprender a aprender.* Barcelona: Martínez Roca.

Piaget, J. (1979). *Tratado de Lógica y conocimiento científico. Epistemologia de la matemática.* Buenos Aires: Paidos.

Polya, G. (1981). *Cómo plantear y resolver problemas.* México: Ed. Trillas.

Román, M. (2005). *Sociedad del conocimiento y refundación de la escuela desde el aula.* Madrid: EOS.

Román, M., & Díez, E. (1998). *Aprendizaje y currículum. Diseños curriculares aplicados.* Santiago de Chile: FIDE.

Román, M., & Díez, E. (1999). *Currículum y evaluación: Diseños curriculares aplicados.* Madrid: Complutense.

Sanz, A. P. (n.d.) *In memoriam Miguel de Guzmán. El último pitagórico.* Retrieved on November 29, 2007 from http://platea.pntic.mec.es/aperez4/miguel/Miguel%20de%20Guzm%E1n.htm

Usabiaga, C., Fernandez, J., & Cerezuela, M. (1984). *Aproximación didáctica al método científico.* Madrid: Narcea

Vygotsky, L. S. (1979). *El desarrollo de los procesos psicológicos superiores.* Barcelona: Crítica.

Chapter 14
Concept Mapping and Vee Diagramming "Differential Equations"

Karoline Afamasaga-Fuata'i

The chapter presents a case study of a student's (Nat's) developing understanding of differential equations as reflected through his progressive concept maps and Vee diagrams (maps/diagrams). Concept mapping and Vee diagramming made Nat realized that there was a need to deeply reflect on *what* he really knows, determine *how* to use what he knows, identify *when* to use *which* knowledge, and be able to justify *why* using valid mathematical arguments. Simply knowing formal definitions and mathematical principles verbatim did not necessarily guarantee an in-depth understanding of the complexity of inter-connections between mathematical concepts and procedures. The presentations of his ideas and understanding of differential equations, Nat found, was greatly facilitated by using his individually constructed concept maps and Vee diagrams. The external projection of his ideas visually on maps/diagrams also facilitated social critiques and mathematical communication during seminar presentations and one-on-one consultations with the researcher. The chapter discusses some implications for teaching mathematics.

Introduction

Distinctions between *doing* mathematics and *meaningfully learning* mathematics are often drawn to highlight the difference between the algorithmic proficiency of simply substituting values into formulas to get an answer and that which values, in addition to this procedural proficiency, a deeper conceptual understanding that seeks to make sense of connections between principles, concepts and methods where concepts exist in and are characterized by systems of relations to other concepts, each concept serving as an axis around which other forms of knowledge can be organized (Ernest, 1999). When mathematical understanding is conceived as a personally-constructed system of relations among concepts, symbols, formulas, methods, and objects or situations, then valid assessment of this understanding

K. Afamasaga-Fuata'i (✉)
University of New England, Armidale, Australia
e-mail: kafamasa@une.edu.au

K. Afamasaga-Fuata'i (ed.), *Concept Mapping in Mathematics*,
DOI 10.1007/978-0-387-89194-1_14, © Springer Science+Business Media, LLC 2009

requires that we obtain information about how students view these as systems of interconnecting concepts and formulas and multiplicity of methods as a demonstration of students' connected understanding and ability to apply this understanding to different situations and contexts. This chapter proposes that the construction of hierarchical concept maps and Vee diagrams enables an individual to demonstrate visually and publicly his/her knowledge and understanding of a mathematical domain's conceptual structure and different methods.

Hierarchical concept maps are graphs of interconnecting concepts (i.e. nodes) representing the key concepts of a domain, arranged hierarchically with linking words on interconnecting links to describe the nature of the relationships between connected nodes. The basic semantic unit is a proposition made up of a triad or strings of triads comprising linked nodes with linking words. For example, two nodes: "*Graphs*" and "*Parabolas*" may be interconnected and described with linking words: "may be" on the link, to form the proposition "*Graphs*" may be "*Parabolas*." A Vee diagram on the other hand is a Vee with its tip situated in the problem to be solved with its two sides representing the conceptual information (i.e. theory, principles and concepts) on the left and the methodological information (given information, transformations and knowledge claims) on the right. An active interplay between the two sides ensures cohesiveness of the displayed information to generate a summative overview of the theoretical principles and concepts informing and guiding the methods of transforming the given information to answer some focus questions (Novak & Gowin, 1984). Examples of concept maps and Vee diagrams are provided later.

The use of concept maps and/or Vee diagrams in learning subject matter more meaningfully and more effectively has been the focus of numerous research in the sciences (Mintzes, Wandersee, & Novak, 1998; Novak, 2002, 1998). Others have researched the use of concept maps in mathematics, as a tool to illustrate conceptual understanding of a topic (Williams, 1998), and pedagogical content knowledge of teachers (Liyanage & Thomas, 2002; Hansson, 2005; Brahier, 2005). The author also investigated the use of both concept maps and Vee diagrams as tools to facilitate the meaningful learning of, and problem solving in, mathematics by secondary students (Afamasaga-Fuata'i, 1998, 2002), undergraduate mathematics students (Afamasaga-Fuata'i, 2002, 2001, 2004, 2005, 2006a, 2007a), practicing teachers (Afamasaga-Fuata'i, 1999) and preservice teachers (Afamasaga-Fuata'i, 2006b, 2007b; Afamasaga-Fuata'i & Cambridge, 2007). Relevant to this chapter are the studies involving undergraduate mathematics students and university mathematics, briefly described next.

Concept Mapping and Vee Diagram Studies

The undergraduate concept map and Vee diagram studies (mapping studies) evolved out of a need to seek innovative ways in which Samoan students' ways of learning mathematics and solving problem could be improved beyond their technical proficiency in applying known procedures and algorithms to solve familiar problems.

The predominantly rote memorization style of learning in schools often means students "learn" mathematics by simply applying formulas and memorising facts to be applied to given problems whether or not they are appropriate (Afamasaga-Fuata'i, 2002).

The mapping studies, involving different cohorts of undergraduate mathematics students, explored and investigated the impact of using concept maps and Vee diagrams on students' understanding of the structure of mathematics and the consequential effect of this understanding on their efficiency and effectiveness in solving mathematics problems (Afamasaga-Fuata'i, 2002). Different cohorts, over a decade, chose mathematics topics that varied in the recency of their formal study of it. "Formal study" is defined as the study of the topic as part of a traditional semester-long, undergraduate mathematics course while "recency" is characterised three ways, namely, *recent past*, *concurrent*, or *new*. "Recent past" refers to formal study in the previous one or two semesters before the research while "concurrent" means the student is formally studying the topic in a separate course whilst also participating in the research. "New" topic means students have not yet formally studied it in a semester-long mathematics course.

The data that is presented in this chapter is that of a student (Nat) who chose a topic he had studied in the recent past. The case study documents Nat's journey as he learnt to use the two meta-cognitive tools of concept maps and Vee diagrams (maps/diagrams) to make explicit his understanding of the topic of "*differential equations.*" He was part of a cohort of students who learnt to use maps/diagrams to communicate their knowledge and understanding of selected mathematics topics over a semester. Unlike Nat, the rest of the cohort chose topics that were new (see Chapter 12 for the cohort results). Before presenting Nat's case study, the theoretical principles driving the mapping studies are presented next.

Theoretical Perspectives

To explain the processes of knowledge construction and development of conceptual understanding, Ausubel–Novak's theory of meaningful learning was used as a guiding theoretical framework. In particular, its principles of assimilation and integration of new and old knowledge into existing knowledge structures assisted in making sense of the processes of meaningful learning. It is argued that when learning new knowledge, students may begin by trying to decide which established ideas within their cognitive structures (or patterns of meanings) are most relevant to it. The deliberate linking of concepts to relevant existing concepts may take place by progressive differentiation and/or integrative reconciliation (Ausubel, Hanesian, & Novak, 1998; Ausubel, 2000). To illustrate these processes, one of Nat's concept maps is used. Figure 14.1 shows a partial view (left vertical segment) of Nat's first concept map in which the concept "$a_0 \, dy/dx + a_1 \, y = Q(x)$" is progressively differentiated into two links to connect to concepts "$Q(x) = 0$ and $Q(x) \neq 0$." In

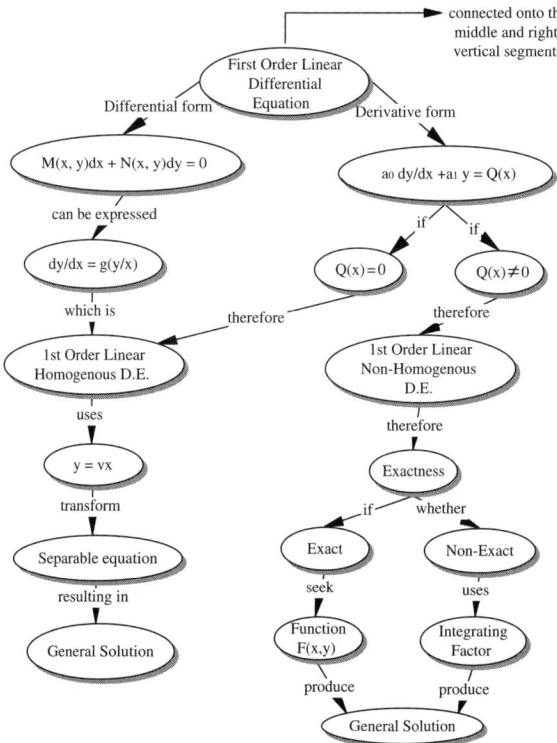

Fig. 14.1 Left vertical segment – Nat's first concept map

contrast, the two concepts *"Function (x, y)"* and *"Integrating Factor"* are integratively reconciled with the less general concept *"General Solution."*

According to Ausubel–Novak's theory, students' cognitive structure should be hierarchically organized with more inclusive, more general concepts and proposition superordinate to less inclusive, more specific concepts and propositions to facilitate assimilation and retention of new knowledge (Ausubel et al., 1978; Ausubel, 2000; Novak, 1998). Knowledge about knowledge or meta-knowledge learning is viewed "as a form of meta-cognition, where learners acquire an understanding of the nature of concepts and concept formation and the processes of knowledge creation" (Novak, 1985). That people remember better, longer, and in more detail if they understand, actively organize what they are learning, connect new knowledge to prior knowledge, and elaborate is also supported by Bransford, Brown, and Cocking (2000). Students will remember procedures better, longer, and in more detail if they actively make sense of procedures, connect procedures to other procedures, and connect procedures to concepts and representations. These perspectives collectively recommend that the best way to remember is to understand, elaborate, and organize what you know thus providing support for the principles of meaningful learning and the use of maps/diagrams as one way of visually explicating the interconnectedness of the main concepts of a domain and its methods.

The social constructivist approach to learning provides the opportunity for the development of students' meta-cognitive skills within a classroom setting particularly as through interactions with others, students actively construct mathematical knowledge as their ideas are challenged causing them to reflect more on their thinking and reasoning. Hence, through public presentations, students necessarily seek to negotiate meaning, justify and argue the validity of their work. For the social interactions and negotiations to be mathematically relevant, the substance of the communication needs to be focused on important mathematical ideas. Engaging students in evidence-based arguments as part of classroom communication, by focusing on explanations, arguments, and justifications, builds conceptual understanding. Communication should include multiple modes (talking, listening, writing, drawing, etc.) because making connections among multiple ways of representing mathematical concepts is central to developing conceptual understanding (Moschkovitch, 2004). This chapter suggests an innovative, alternative mode of communication, namely, concept mapping and Vee diagramming in a mathematics classroom.

The socio-linguistic view promoted by Richards (1991), that mathematics learning is "a task practiced by communities in terms of linguistic discourse enacted by its members," facilitates distinction between prevailing mathematics learning in schools (i.e. learning mathematics as a collection of facts and procedures) and the ideal, which is inquiry mathematics (i.e. students think, discuss and act mathematically).

Collectively these (cognitive, social constructivist and socio-linguistic) perspectives highlight the fundamental principle that "learning mathematics involves being initiated into mathematical ways of knowing, ideas and practices of the mathematical community and making these ideas and practices meaningful at an individual level" (Ernest, 1999) as well as having the ability to maintain and conduct a mathematical discussion (Richards, 1991) and communicating more effectively through multiple modes within a classroom setting (Moschkovitch, 2004).

Nat's Case Study

As previously mentioned (in Chapter 12), the mapping studies were qualitative, exploratory teaching experiments conducted over a semester of 14 weeks with different cohorts of mathematics students enrolled in a research course. A teaching experiment is defined as "an exploratory tool ... aimed at providing understanding of what might go on in (students') heads as they engage in mathematical activity." (Steffe & D'Ambrosio, 1996). The course introduced students to the meta-cognitive tools as means of learning mathematics more meaningfully and solving problems more effectively. The selection of procedures and activities were guided by the principles of social constructivist and socio-linguistic epistemologies namely that of building upon students' prior knowledge, group work, negotiation of meanings, consensus and provision of time-in-class to allow students to reflect on their own understanding. For example, the mapping studies included a familiarization phase, which introduced students to maps/diagrams and the new socio-mathematical norms of group and one-on-one (1:1) critiques including the expectations that each student should be prepared to justify and address critical comments from peers and researcher, and then later on critique peers' presented work. Time was set aside in between critiques for students to revise and modify their maps/diagrams where appropriate in readiness for the next critique. Students underwent 3 cycles of group and 1:1 critiques before completing a final report.

Topics selected by Nat's cohort of 7 students included *partial differential equations, approximation methods for first order differential equations, multiple variable functions, Laplace's transform, least square polynomial approximations*, and *trigonometric approximations*. This chapter reports data from Nat's *differential equation* concept maps and Vee diagrams.

Nat's Data and Analysis

Data collected included Nat's progressive concept maps (4 versions) and progressive Vee diagrams of 4 problems (2 versions each), and final reports. Analysis of the maps/diagrams, are in accordance with the Ausubel–Novak theory of meaningful learning particularly in terms of how Nat organized his knowledge. Nat's perceptions of the value of maps/diagrams as meta-cognitive tools were obtained through written responses to two questions on the advantages and disadvantages of using the tools in learning mathematics. The written responses and all versions of maps/diagrams formed part of the final reports. In the following sections, the concept map data is presented first followed by those for the Vee diagrams. Excerpts from Nat's final report are used, where appropriate, to support the discussion of the results.

Concept Map Data Analysis

Concept Map Criteria

A review of the literature shows different ways of assessing and scoring concept maps (Novak & Gowin, 1984; Ruiz-Primo & Shavelson, 1996; Liyanage & Thomas, 2002) however for this case study, a qualitative approach is adopted using only counts of occurrences of each criterion. Nat's concept maps are assessed mainly in terms of the *complexity of the network structure* (*structure*) of concepts, *nature of the contents* (*contents*) within concept boxes and *valid propositions* (*propositions*). The structure criteria indicate the depth of differentiation and extent of integration between concepts, whilst the contents criteria reflect the nature of Nat's perceptions of what constitutes mathematical knowledge. Tables 14.1 and 14.2 list the relevant breakdown within each category (*conceptual elements, inappropriate entries,* and

Table 14.1 Contents and propositions criteria of Nat's concept maps

	First map Count	Final map Count	% Increase
Criteria: Contents			
A: Conceptual contents			
Concept names/labels	31	27	−13
Concept symbols/expressions	4	13	225
Symbols	0	4	
Mathematical statements/expressions	8	13	63
Names of methods	2	6	200
General formulas/expressions	0	8	800
Formula concepts	0	1	100
Total A – Conceptual contents	45 (86.5%)	72 (88.9%)	2.4
B: Inappropriate entries			
Procedures	0	1	
Linking words used in concept boxes	0	2	
Redundant entries	7	2	
Total B – Inappropriate entries	7 (13.5%)	5 (6.2%)	−7.3
C: Definitional entries			
Parts of definitions	0	3	
TOTAL C – definitional entries	0	3	
D: Examples			
Examples	0	1	
Total D – Examples	0	1	
Totals A+B+C+D	52	81	55.8
Criteria: Propositions			
Valid propositions	33	54	63.6
Invalid propositions	16	19	18.8

definitional entries for contents; *valid and invalid propositions* for propositions; and *main branches, sub-branches, average hierarchical levels per sub-branch, integrative cross-links* and *multiple branching nodes* for structure).

Valid propositions are mathematically meaningful statements formed by connecting valid concept contents with suitable linking words correctly describing the nature of the interrelationship between the connected-concepts.

The occurrences of lengthy definitional phrases (*definitional entries*) although conceptual indicate students have not completely analyzed them to identify major concepts. Other types of entries are generally categorized as *inappropriate entries* since they require further modifications and revisions from their existing forms to make them more suitable. For example, *procedural* entries include those that specifically describe a procedural step, which are more appropriate on Vee diagrams. *Redundant* entries indicate students' tendency to learn information verbatim, and/or in isolation instead of seeking out meaningful integrative cross-links with the first occurrence of the concept elsewhere in the map. *Inappropriate linking word entries* are those that contain words more suitable as linking words; these indicate students' difficulties to distinguish between mathematical concepts and descriptive phrases.

Table 14.2 Structural criteria of Nat's concept maps

	First map Count	Final map Count	% Increase
Criteria: Structure			
Main branches	8	8	0
Sub-branches	13	17	31
Average hierarchical levels per sub-branch	7	11	57
Integrative cross-links			
Between (sub-)branches at same level	2	2	
Between (sub-)branches at different levels	1	12	
Total – *Cross-links*	3	14	366
Multiple branching nodes			
Progressive differentiation links from nodes at:			
Level 2	5	2	
Level 3	2	3	
Level 4	2,2,2	3	
Level 5		5,2,2	
Level 6	2,2	2,2	
Level 7	2,2	7	
Level 8		2	
Level 9		2	
Level 10		2,3	
14		2	
TOTAL *Multiple branching nodes*	9	14	56

Concept Map Data

The data summarized in Tables 14.1 and 14.2 and partial views of Nat's concept maps in Figs. 14.1–14.4 show that there was an overall 2.4% increase (86.5–88.9%) in the number of nodes with valid conceptual entries compared to an associated 7.3% (13.5–6.2%) decrease in inappropriate entries due to inclusion of procedural steps, linking words or redundant entries.

Specific significant increases were noted with the sub-criteria: *concept symbols/expressions* (225%), *mathematical statements/expressions* (63%), *general formulas/expressions* (800%) and *names of relevant methods* (200%) (see Table 14.1) as Nat tried to enhance the meaningfulness of various concept hierarchies (branches/sub-branches) by progressively differentiating concepts (56% increase in nodes with multiple branching (i.e. more than 1-outgoing link), Table 14.2) resulting in increased hierarchical levels (by 200% from Level 7 to Level 14) and increased integration across (sub-)branches (366% increase in valid cross-links, Table 14.2).

Figure 14.1 shows the left vertical segments of Nat's first map that attracted various critical comments during the first group presentation. His peers felt it was not illustrating sufficient information to guide the solution of a differential equation problem. One of the comments referred to the inappropriate use of important concepts such as "differential form" and "derivative form" as linking words

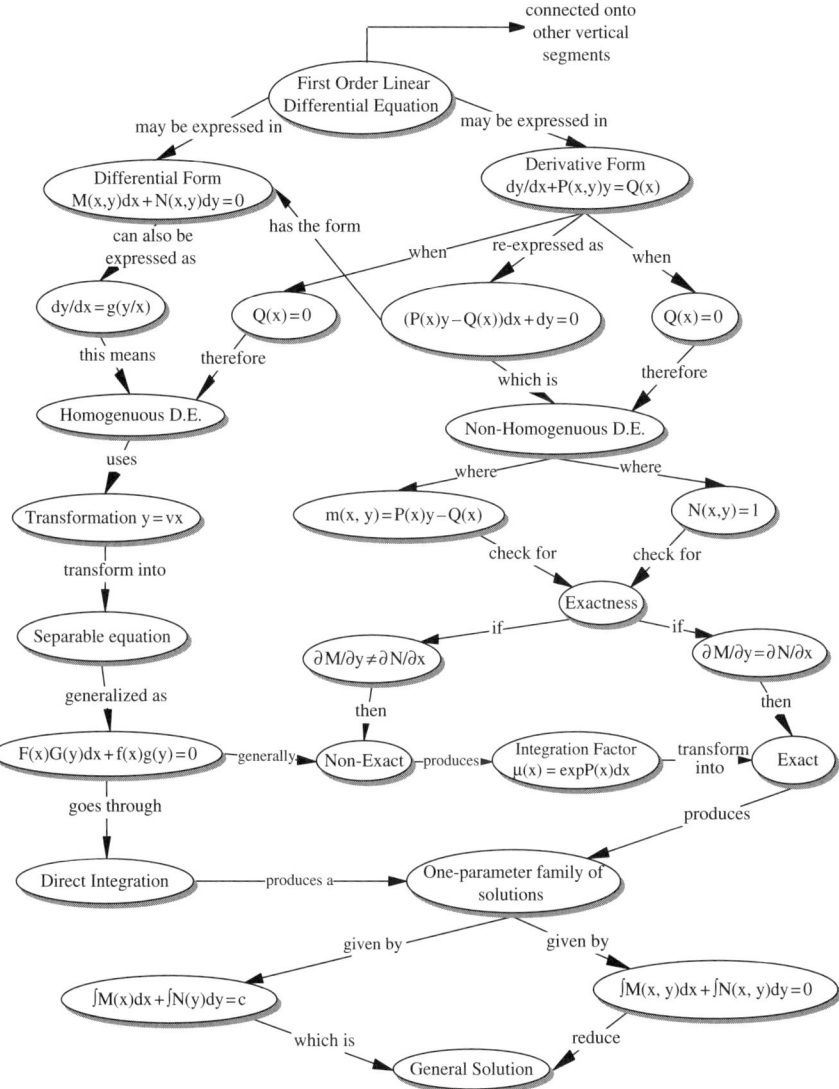

Fig. 14.2 Left vertical segments – Nat's final concept map

in the propositions: "*First Order Linear Differential Equation*" differential form "*M(x,y)dx + N(x,y)dy = 0*" and "*First Order Linear Differential Equation*" derivative form "*a_0 dy/dx + a_1 y = Q(x).*"

Over the semester, as a consequence of critiques and feedback from his peers and researcher, these segments (Fig. 14.1) evolved to its final form shown in Fig. 14.2 in

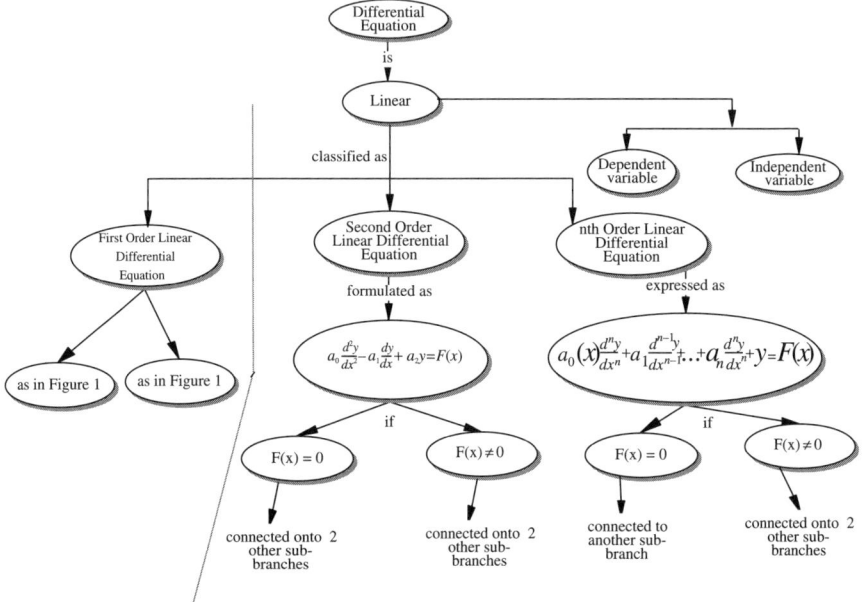

Fig. 14.3 Middle & right vertical segments – Nat's first concept map

which the above propositions were subsequently revised to read: "*First Order Linear Differential Equations*" may be expressed in "*Derivative Form* $\frac{dy}{dx} + P(x, y)y = Q(x)$" and "*First Order Linear Differential Equations*" may be expressed in "*Differential Form: M(x, y)dx + N(x, y)dy = 0*" Fig. 14.2 also shows a more integrated and differentiated network structure that included 9 more new nodes, much higher average number of hierarchical levels and twice as many multiple branching nodes. In terms of the structure criteria, data and partial views in Figs. 14.2 and 14.4 illustrate that the structural complexity of Nat's final map had changed significantly compared to the first one.

In spite of the unchanged number of main branches, the final map was significantly more complex. For example, there was a substantial increase (366%) in the number of integrative cross-links, predominantly more progressive differentiation of nodes at various levels of the hierarchy (Levels 2–10 and Level 14, Table 14.2), a 56% increase in multiple branching nodes, a 31% increase in the number of sub-branches and a 57% increase in the average number of hierarchical levels. Table 14.1 shows that the number of valid propositions also increased significantly (63.6%) by the final map. Some of the changes (compare Figs. 14.3 and 14.4) were deletions of nodes (e.g. independent variable, dependent variable), and merging of two branches (second order linear differential equation, nth order linear differential equation) to give a more enriched, differentiated and integrated concept hierarchy as partially shown in Fig. 14.4.

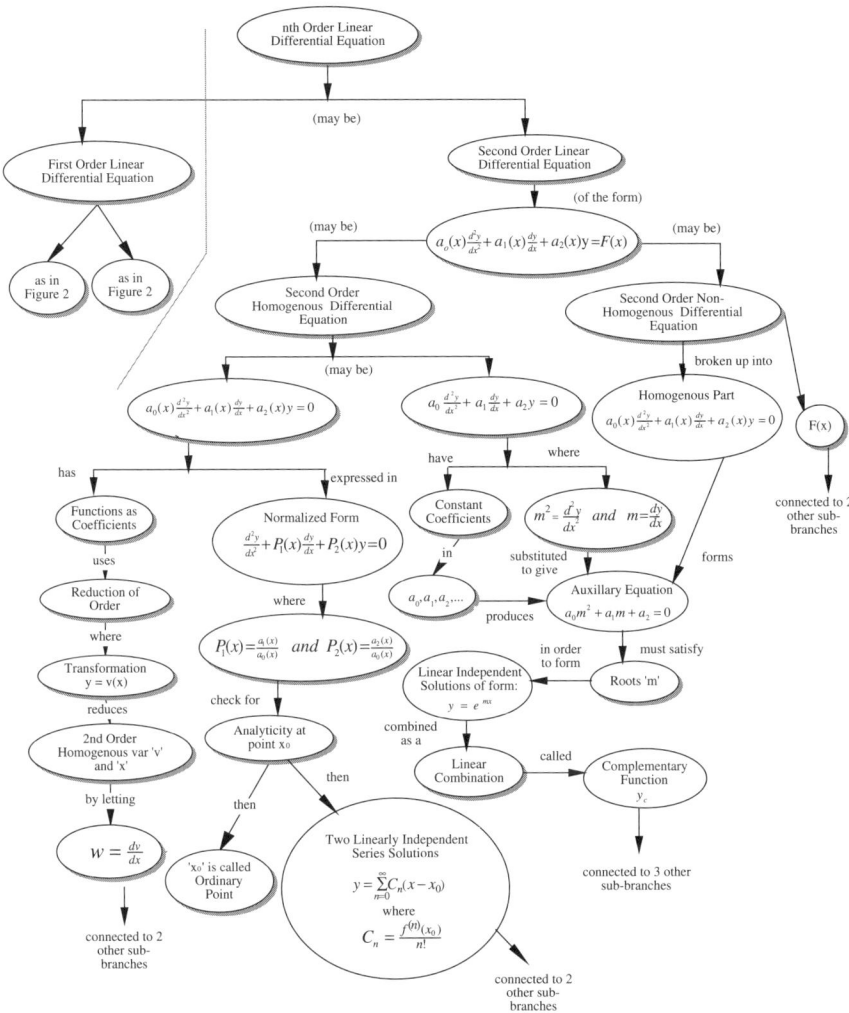

Fig. 14.4 Middle & right vertical segments – Nat's final concept map

Vee Diagram Data Analysis

The structure of the Vee (see Fig. 14.5 for one of Nat's Vee diagram) and guiding questions (see Table 14.3 below) provide a systematic guide to students as they reason from the problem context (*Event/Object*) to extract the given information (*Records*) and identify relevant principles, theorems, formal definitions and major rules (*Principles*) and *Concepts* which could guide the development of appropriate methods and procedures (*Transformations*) to find an answer (*Knowledge Claim*) to the *Focus Question*. Engaging in this reasoning process enables the students to

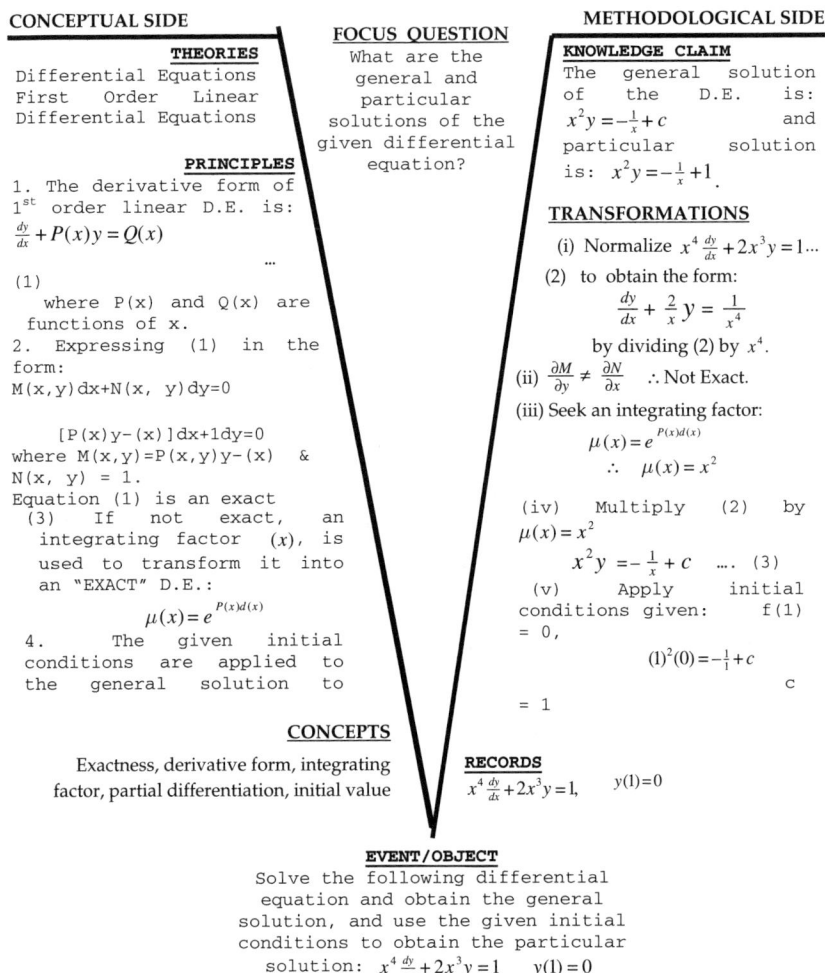

CONCEPTUAL SIDE

FOCUS QUESTION
What are the general and particular solutions of the given differential equation?

METHODOLOGICAL SIDE

THEORIES
Differential Equations
First Order Linear Differential Equations

KNOWLEDGE CLAIM
The general solution of the D.E. is:
$x^2y = -\frac{1}{x} + c$ and particular solution is: $x^2y = -\frac{1}{x} + 1$.

PRINCIPLES
1. The derivative form of 1st order linear D.E. is:
$\frac{dy}{dx} + P(x)y = Q(x)$
...
(1)
    where P(x) and Q(x) are functions of x.
2. Expressing (1) in the form:
M(x,y)dx+N(x, y)dy=0

    [P(x)y-(x)]dx+1dy=0
where M(x,y)=P(x,y)y-(x) &
N(x, y) = 1.
Equation (1) is an exact
    (3)   If   not   exact,   an
    integrating factor (x), is
    used to transform it into
    an "EXACT" D.E.:
        $\mu(x)=e^{P(x)d(x)}$
4.   The   given   initial
conditions   are   applied   to
the   general   solution   to

TRANSFORMATIONS
(i) Normalize $x^4\frac{dy}{dx} + 2x^3y = 1$...
(2) to obtain the form:
$\frac{dy}{dx} + \frac{2}{x}y = \frac{1}{x^4}$
    by dividing (2) by x^4.
(ii) $\frac{\partial M}{\partial y} \neq \frac{\partial N}{\partial x}$ ∴ Not Exact.
(iii) Seek an integrating factor:
    $\mu(x)=e^{P(x)d(x)}$
    ∴ $\mu(x) = x^2$
(iv)   Multiply   (2)   by
$\mu(x) = x^2$
    $x^2y = -\frac{1}{x} + c$ .... (3)
(v)   Apply   initial
conditions given: f(1)
= 0,
        $(1)^2(0) = -\frac{1}{1} + c$
                                    c
= 1

CONCEPTS
Exactness, derivative form, integrating factor, partial differentiation, initial value

RECORDS
$x^4\frac{dy}{dx} + 2x^3y = 1,$ $y(1)=0$

EVENT/OBJECT
Solve the following differential equation and obtain the general solution, and use the given initial conditions to obtain the particular solution: $x^4\frac{dy}{dx} + 2x^3y = 1$ $y(1) = 0$

Fig. 14.5 Nat's Vee diagram – Problem 1 (P1)

consider not only the methods of solving the problem but including as well the articulation of the conceptual bases of these methods.

As Gowin (1981) explains: "The structure of knowledge may be characterized (in any field ...) by its telling questions, key concepts and conceptual systems; by its reliable methods and techniques of work ..." (pp. 87–88). Therefore, a Vee diagram is a potentially useful tool for not only assessing students' proficiency in solving a problem but it can assess the depth and extent of students' conceptual understanding of mathematical principles and concepts underpinning the methods. A completed Vee diagram, consequently, provides a record of both the conceptual and methodological information involved in solving a problem.

Table 14.3 Guiding questions for Vee diagrams of mathematics problems

Sections	Guiding Questions
Theory	What theory(ies), major principles govern the methods?
Principles	How are the concepts related? What general rule, principle, formula do we need to use?
Concepts	What are the concepts used in the problem statement? Relevant concepts required to solve problem?
Event/Object	What is the problem statement?
Records	What are the "givens" (information) in the problem?
Transformations	How can we make use of the theories/principles/concepts/records to determine a method?
Knowledge claims	What is the answer to the focus question given the event?
Focus question	What is the problem asking about?

For assessment, a completed Vee diagram may be evaluated qualitatively in terms of (i) *Overall Criteria,* which consider the appropriateness of section entries against the guiding questions (Table 14.3) in the context of the given problem, and (ii) more *Specific Criteria*, which assess the extent of integration and correspondence between listed principles and listed main steps. Nat's Vee diagrams were drawn after one cycle of group and 1:1 critiques, which means that he was able to use his first revised concept map to guide the completion of his Vee diagrams. The first problem (P1) (Fig. 14.5) was on first order whilst the other 3 were all on second order differential equations (D.E.) of type non-homogenous with constant coefficients (second problem P2), homogenous with constant coefficients (third problem P3, Fig. 14.6) and homogenous with variable coefficients (fourth problem P4).

Overall Criteria

In all four Vee diagrams, the common entry under *Theory* was "differential equations" with a second one reflecting the general order ($n = 1$ or 2) of the D.E. A third entry of "homogenous with constant coefficients" was included for P3, and for P4, additional entries were "homogenous" and "power series."

For the sections, *Event/Object*, *Focus Questions*, and *Records* all entries were appropriate and extracted directly from the problem statements and in accordance with the guiding questions of Table 14.3. In contrast, Nat's selections of entries under *Concepts* included others not explicitly in the problem statement but were considered relevant. This means that, Nat had already recognized potential underlying principles, and guided by his revised concept map, he selected the most "suitable" principles, and subsequently relevant concepts, for each Vee diagram. For example, in Fig. 14.5, his concept list was: {exactness, derivative, integrating factor, partial differentiation, *initial value* problem, *general solution*, *particular solution*} where 3 of the 7 concepts listed were explicitly stated in the problem statement whilst the rest were inferred as being relevant; similarly, for the other 3 Vee diagrams.

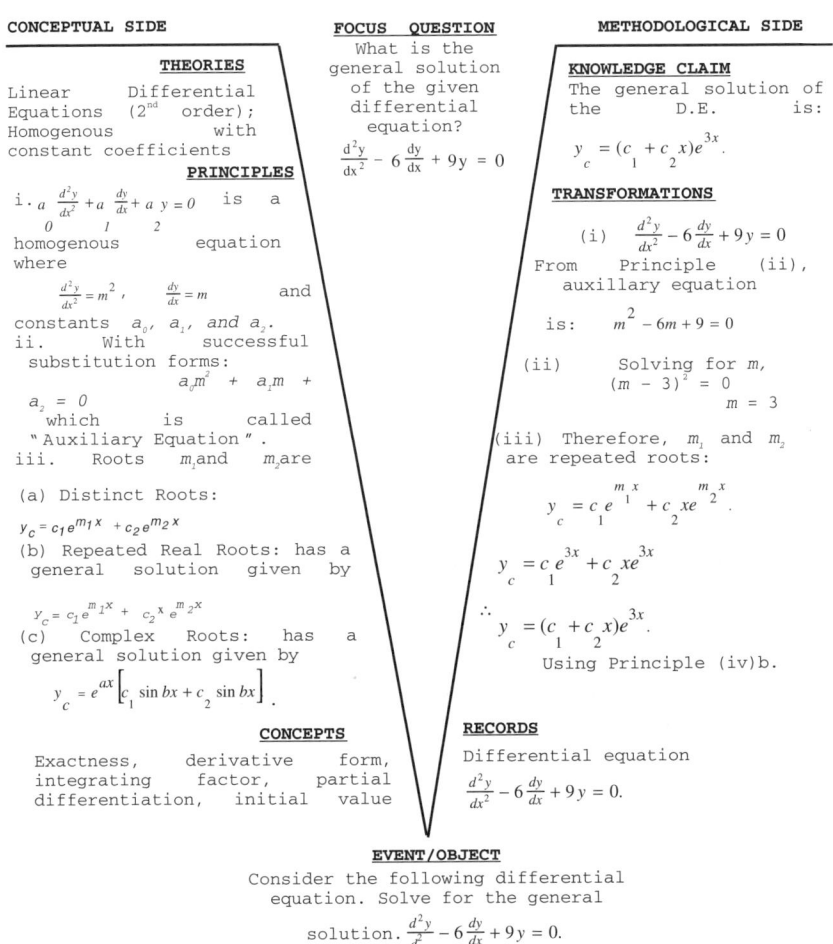

Fig. 14.6 Nat's Vee diagram – Problem 3 (P3)

However, entries for *Principles* required some reflection and consideration. For the first problem, his initial diagram only included 3 of the listed principles. A fourth one was added in the final version when he realized that none of the 3 listed principles justified part (v) under Transformations. This was a positive self-realization indicative of a growing confidence in his skills to complete the Vee diagram. However, he did not pick up the missing principle underlying the normalization step in part (i) of the transformations.

The rest of the main steps had corresponding principles supporting their transformations. As a presenter of his own work and critic of peers' maps, he was aware that maintaining a close correspondence between principles and main steps of the solution was a critical aspect of, and a common problematic area when constructing a Vee diagram.

Specific Criteria

The specific criteria of a tighter integration and correspondence between main steps and principles calls into consideration the inclusion of relevant principles and the "statement," or "wording" of identified principles. That is, are the listed principles stated in theoretical terms (i.e. formal definitions or general rules) and not as procedural instructions?

An inspection of Nat's Vee diagrams shows a number of listed principles have the ingredient concepts; however the statements are phrased in procedural or algorithmic terms. For example, in the Vee diagram for P3 (Fig. 14.6), Nat listed principles (iii) and (iv) as: "*With successful substitution forms: $a_0m^2 + a_1m + a_2 = 0$ which is called 'Auxiliary Equation'*" and "*Roots m_1 and m_2 are obtained after solving the auxiliary equation for m*" respectively. These statements are more procedurally worded (*substitution* and *obtained after solving*) than conceptual. This algorithmic nature is also evident in the choice of linking words (*substituted to give, must satisfy,* and *check for*) used in the partial segment shown in Fig. 14.4.

It appears that his background in computer programming and experience with flow charts are influencing his perceptions of what constitutes appropriate "linking words" and "principles." Alternatively, this procedural view of mathematics suggests that his perceptions of "mathematical knowledge" up to this point may have been predominantly as "a collection of methods and procedures." Thus, in spite of the improvement in aligning the listed principles with main steps of solution in Vee diagrams, Nat still needs to rephrase his "procedural" principles to be more theoretical.

Discussion

Both Nat's final concept maps and final Vee diagrams showed significant changes by the end of the study compared to his initial attempts. For example, the final map was structurally more complex with increased integrative cross-links, more progressive differentiation links, additional concepts, and increased number of valid propositions. These increased criteria provided empirical evidence that Nat's understanding of differential equations had become more structured, better organized and more enriched to the point that it greatly facilitated the construction of his Vee diagrams.

For example, he wrote: "the Vee diagram consists of the theoretical and practical senses of the event/object ... directed by the concept map flow (right segment, Fig. 14.4) ... these senses help guide the transformations." That is, he was beginning to realize for himself the value of a clearly organized concept map as a guide to make decisions about effectively solving a problem. Since his Vee diagrams were constructed after the first cycle of critiques, they attracted relatively less criticism during critiques unlike his concept maps. From the socio-cultural perspective, Nat's conceptual, more structured and better organized understanding of D.E. was partially a result of sustained social interactions, and critiques from his peers and researcher throughout the semester. For example, Nat wrote: "Due to questions raised in class, on what 'auxiliary equations' represented and how it's formed. Therefore, more details were presented …" and "the whole concept of reduction of order needed clarification in all senses due to its shortened presentation before." When there was no criticism, he tended to leave the maps/diagrams unchanged. For example, he said: "I encountered very little or no critical comment during class presentations … (so). … I'm left with no urge to make any further modifications … not criticized at all, therefore no changes." That is, his determination to minimize critical comments from others motivated Nat to continually develop a more complete and comprehensive map, which accommodated critical comments of the previous critique to ensure minimal criticism in the next critique. For example, he added branches to clarify the concepts: *auxiliary equation, reduction of order* in his second map, and added a new branch in the final map to accommodate *power series solutions* (see Fig. 14.4).

From the perspectives of Richards (1991) and Knuth and Peressini (2001) Nat and his peers engaged in inquiry mathematics where classroom discourse was both "to convey meaning" and "to generate meaning." Nat, through the use of maps/diagrams engaged in mathematical discussions, dialogues and critiques with his peers and researcher. His fluency with the language of D.E. and the communication of this mathematical understanding publicly was made more effective and efficient with maps/diagrams involving important mathematical ideas. Substantive mathematical communication and multiple communication modes to support the development of conceptual understanding in mathematics, as emphasized by Moschkovitch (2004) for bilingual learners, were evident in Nat's case study. Further, noticeable growth in Nat's in-depth understanding of D.E. was indicated by the increased number of valid propositions and structural complexity of conceptual networks in his final map, and improved correspondence between listed principles and solutions on Vee diagrams. However, because of his tendency to view interconnections and principles with a procedural bias, Nat needs to continuously revise his maps/diagrams with appropriate critiques and feedback from other mathematics people to ensure that the conceptual interconnections are made more explicit on links on concept maps and the principles of Vee diagrams are more theoretically stated and less procedural.

Finally, the established socio-mathematical norms, and use of maps/diagrams appeared to promote a classroom environment that was alive with meaningful discussions as students engage in the critiquing and justification processes. Whilst it could be argued that this would happen irrespective of the type of meta-cognitive tools used, the author proposes that the unique visual structures of the maps/diagrams were pivotal in facilitating and promoting dialogues and critiques. It appeared then from the empirical evidence that the established socio-mathematical norms and use of the meta-cognitive tools had substantially influenced Nat's perceptions of mathematics as reflected by the progressive improvement in his maps/diagrams and his responses in the final report. Clearly, with the right supportive classroom environment, students can be encouraged to dialogue, discuss and communicate mathematically. In so doing, students begin to realize that learning mathematics meaningfully involves much more than implementing a sequence of steps. Findings from this case study suggest that students are amenable to changes in classroom practices, and with their cooperation, and using appropriate meta-cognitive tools, mathematics learning can be rendered more enriching and meaningful at an individual level.

Implications

The results of the case study suggest that active organization of one's knowledge and understanding and visually presenting and communicating this understanding publicly through individually-constructed concept maps and Vee diagrams has the potential to make mathematics learning and problem solving more than simply "doing" mathematics (i.e. applying, and substituting values in, formulas). By constructing maps/diagrams, students can be challenged to represent and organize their mathematics knowledge and to demonstrate their understanding of the structure of the mathematics embedded in a problem or relevant to a topic. Socially interacting and negotiating meanings with others has the potential to influence one's thinking and reasoning such that it may result in consolidating one's understanding or signaling the need for further reflection and revisions. Such a process of reorganization, revision and reflection is necessarily cyclic and interactive, which can ultimately result in the development of one's conceptual understanding and meaningful learning of mathematics. Further research is recommended to investigate these ideas further in mathematical classrooms at all levels.

References

Afamasaga-Fuata'i, K. (1998). *Learning to solve mathematics problems through concept mapping and Vee mapping.* Samoa: National University of Samoa.

Afamasaga-Fuata'i, K. (1999). Teaching mathematics and science using the strategies of concept mapping and Vee mapping. *Problems, Research, and Issues in Science, Mathematics, Computing and Statistics, 2*(1), 1–53. Journal of the Science Faculty at the National University of Samoa.

Afamasaga-Fuata'i, K. (2001). *Enhancing students' understanding of mathematics using concept maps & Vee diagrams.* Paper presented at the International Conference on Mathematics Education (ICME), Northeast Normal University of China, Changchun, China, August 16–22, 2001.

Afamasaga-Fuata'i, K. (2002). *A Samoan perspective on Pacific mathematics education.* Keynote Address. Proceedings of the 25th annual conference of the Mathematics Education Research Group of Australasia (MERGA-25), July 7–10, 2002 (pp. 1–13). New Zealand: University of Auckland.

Afamasaga-Fuata'i, K. (2004). Concept maps and Vee diagrams as tools for learning new mathematics topics. In A. J. Canãs, J. D. Novak, & F. M. Gonãzales (Eds.), *Concept maps: Theory, methodology, technology.* Proceedings of the First International Conference on Concept Mapping (Vol. 1, pp. 13–20). Navarra, Spain: Dirección de Publicaciones de la Universidad Pública de Navarra.

Afamasaga-Fuata'i, K. (2005). Students' conceptual understanding and critical thinking? A case for concept maps and Vee diagrams in mathematics problem solving. In M. Coupland, J. Anderson, & T. Spencer (Eds.), *Making mathematics vital*. Proceedings of the Twentieth Biennial Conference of the Australian Association of Mathematics Teachers (AAMT) (Vol. 1, pp. 43–52). Sydney, Australia: University of Technology.

Afamasaga-Fuata'i, K. (2006a). Developing a more conceptual understanding of matrices and systems of linear equations through concept mapping. *Focus on Learning Problems in Mathematics*, *28*(3 & 4), 58–89.

Afamasaga-Fuata'i, K. (2006b). Innovatively developing a teaching sequence using concept maps. In A. Canas & J. Novak (Eds.), *Concept maps: Theory, methodology, technology*. Proceedings of the Second International Conference on Concept Mapping (Vol. 1, pp. 272–279). San Jose, Costa Rica: Universidad de Costa Rica.

Afamasaga-Fuata'i, K. (2007a). Communicating students' understanding of undergraduate mathematics using concept maps. In J. Watson, & K. Beswick (Eds.), *Mathematics: Essential research, essential practice*. Proceedings of the 30th Annual Conference of the Mathematics Education Research Group of Australasia (Vol. 1, pp. 73–82). University of Tasmania, Australia: MERGA.

Afamasaga-Fuata'i, K. (2007b). Using concept maps and Vee diagrams to interpret "area" syllabus outcomes and problems. In K. Milton, H. Reeves, & T. Spencer (Eds.), *Mathematics essential for learning, essential for life*. Proceedings of the 21st biennial conference of the Australian Association of Mathematics Teachers, Inc. (pp. 102–111). University of Tasmania, Australia: AAMT.

Afamasaga-Fuata'i, K., & Cambridge, L. (2007). Concept maps and Vee diagrams as tools to understand better the "division" concept in primary mathematics. In K. Milton, H. Reeves, & T. Spencer (Eds.), *Mathematics essential for learning, essential for life*. Proceedings of the 21st biennial conference of the Australian Association of Mathematics Teachers, Inc. (pp. 112–120). University of Tasmania, Australia: AAMT.

Ausubel, D. P. (2000). *The acquisition and retention of knowledge: A cognitive view*. Dordrecht: Kluwer Academic.

Ausubel, D. P., Novak, J. D., & Hanesian, H. (1978). *Educational psychology: A cognitive view*. New York: Holt, Rhinehart and Winston. Reprinted 1986, New York: Werbel and Peck.

Brahier, D. J. (2005). *Teaching secondary and middle school mathematics* (2nd ed.). New York: Pearson Education, Inc.

Bransford, J., Brown, A., & Cocking, R. (2000). *How people learn: Brain, mind, experience, and school*. Washington, DC: National Academy Press.

Ernest, P. (1999). Forms of knowledge in mathematics and mathematics education: Philosophical and rhetorical perspectives. *Educational Studies in Mathematics*, *38*, 67–83.

Gowin, D. B. (1981). *Educating*. Ithaca, NY: Cornell University Press.

Hansson, O. (2005). Preservice teachers' view on $y=x+5$ and $y=\pi r^2$ expressed through the utilization of concept maps: A study of the concept of function. In H. Chick & J. L. Vincent (Eds.), *Proceedings of the 29th conference of the international group for the psychology of mathematics education* (Vol. 3, pp. 97–104). Melbourne: PME.

Knuth, E., & Peressini, D. (2001). A theoretical framework for examining discourse in mathematics classrooms. *Focus on Learning Problems in Mathematics*, 23(2 & 3), 5–22.

Liyanage, S., & Thomas, M. (2002). *Characterising secondary school mathematics lessons using teachers' pedagogical concept maps*. Proceedings of the 25th annual conference of the Mathematics Education Research Group of Australasia (MERGA-25), July 7–10, 2002 (pp. 425–432). New Zealand: University of Auckland.

Mintzes, J. J., Wandersee, J. H., & Novak, J. D. (Eds.). (1998). *Teaching science for understanding. A human constructivistic view*. San Diego, CA, London: Academic Press.

Moschkovitch, J. (2004). Using two languages when learning mathematics. Retrieved on February 20, 2008 from http://math.arizona.edu/~cemela/spanish/content/workingpapers/UsingTwoLanguages.pdf

Novak, J. D. (1985). Metalearning and metaknowledge strategies to help students learn how to learn. In L. H. West & A. L. Pines (Eds.), *Cognitive structure and conceptual change* (pp. 189–209). Orlando, FL: Academic Press.

Novak, J. D. (1998). *Learning, creating, and using knowledge: Concept maps as facilitative tools in schools and corporations.* Mahwah, NJ: Academic Press.

Novak, J. (2002). Meaningful learning: the essential factor for conceptual change in limited or appropriate propositional hierarchies (LIPHs) leading to empowerment of learners. *Science Education, 86*(4), 548–571.

Novak, J. D., & Gowin, D. B. (1984). *Learning how to learn.* Cambridge: Cambridge University Press.

Richards, J. (1991). Mathematical discussions. In E. von Glaserfeld (Ed.), *Radical constructivism in mathematics education* (pp. 13–51). London: Kluwer Academic Publishers.

Ruiz-Primo, M. A., & Shavelson, R. J. (1996). Problems and issues in concept maps in science assessment. *Journal of Research in Science Teaching, 33*(6), 569–600.

Steffe, L. P., & D'Ambrosio, B. S. (1996). Using teaching experiments to enhance understanding of students' mathematics. In D. F. Treagust, R. Duit, & B. F. Fraser (Eds.), *Improving teaching and learning in science and mathematics* (pp. 65–76). New York: Teachers College Press, Columbia University.

Williams, C. G. (1998). Using concept maps to access conceptual knowledge of function. *Journal for Research in Mathematics Education, 29*(4), 414–421.

Chapter 15
Using Concept Maps to Mediate Meaning in Undergraduate Mathematics

Karoline Afamasaga-Fuata'i

The chapter presents the concept map data from a study, which investigated the use of concept maps and Vee diagrams (maps/diagrams) to illustrate the conceptual structure of a topic, its relevant problems and common procedures. Students were required to construct comprehensive topic maps/diagrams as ongoing exercises throughout the semester and to present these for critique before individuals finalized them. With improved mapping proficiency and on-going social critiques, students' mathematical understanding deepened, becoming more conceptual as a result of continually revising their work as the validity of each map is dependent on how effective it illustrated the intended meanings and correct mathematics structure. Students also developed an appreciation of the crucial inter-linkages between mathematical principles, common procedures and formulas, and how all of these mutually reinforce each other conceptually and methodologically. Incorporating concept mapping as a normal mathematical practice in classrooms can potentially alter the learning of mathematics, making it more meaningful and conceptual to supplement the predominantly procedural proficiency practised in many mathematics classrooms.

Introduction

It is not enough just to be able to solve problems. Students must be able to reason out connections between the underlying mathematics in the context of a problem or topic and they must be able to make conjectures, justify methods, and communicate solutions and strategies. These are the core ideas embedded in various curriculum standards (e.g., NCTM, 2007; AAMT, 2007). For many undergraduate students,

K. Afamasaga-Fuata'i (✉)
University of New England, Armidale, Australia
e-mail: kafamasa@une.edu.au

A shorter version of this paper was presented at the 30th Annual Meeting of the Mathematics Education Research Group of Australasia (MERGA), July 4–6, 2007, University of Hobart, Tasmania and published as Afamasaga-Fuata'i (2007b) in the proceedings.

explaining why their methods work is difficult, mainly because they do not possess the language with which to explain their solutions. Alternatively, explaining and justifying their methods have not been normal practices in their classrooms (Pratt & Kelly, 2005). Richards (1991) describes this situation as a communication problem resulting from students' inability to "understand the appropriate language ... of a mathematical discussion, and no sense of making conjectures or evaluating mathematical assumptions". Instead, answers are often in terms of methodological issues or sequential relations rather than a conceptual framework, which indicate that a student's proficiency and adeptness in executing a procedure may not necessarily reflect deep, conceptual understanding of the mathematics involved. To challenge the depth of student understanding in any domain, they may be asked to effectively represent and communicate that understanding for public scrutiny.

Mathematics Education in Samoa

From a socio-cultural perspective, students' answers typically reflect the norms, discourse and practices of their cumulative mathematics experiences (Pratt & Kelly, 2005). In Samoa, the fragmented, and narrow view of mathematics most undergraduate students have, reflect their secondary mathematics experiences. While most students are proficient in applying formulas to familiar problems (i.e., routine expertise (Hatano, 2003)), there is a remarkable lack of critical thinking and analysis especially evident when they are given novel problems to solve. More particularly, students may have the knowledge of relevant content areas and procedures but are often unable to, independently and flexibly, apply what they know to problems unless substantial guidance is provided (Afamasaga-Fuata'i, 2005a, 2002; Afamasaga-Fuata'i, Meyer, Falo, & Sufia, 2007). Such manifestations (a) characterize students who learn mathematics by memorising collections of facts and procedures with little effort in making connections between topics, and (b) typically reflect prevailing socio-cultural practices of their school mathematics classrooms. According to Hiebert and Carpenter (1992), understanding takes place when students develop relationships and connections in their mathematical knowledge whereas adaptive expertise (Hatano, 2003), in contrast to routine expertise, is meaningful (deep conceptual and procedural) knowledge that can be applied creatively, flexibly, and appropriately to new, as well as familiar tasks.

At the national level, failure rates in secondary mathematics examinations are fairly high up to 70% (Afamasaga-Fuata'i, 2002, 2001). Despite the restricted entry into the National University of Samoa (NUS) University Preparatory Year (UPY)[1] program, mathematics failure rates are still high at up to 65% a semester. Although UPY students represent the top 10% of those that passed the Pacific Senior Secondary Certificate (PSSC) examinations in the previous year at the end of secondary level, they nonetheless consistently struggle with applications

---

[1] The UPY program subsequently changed its name to Foundation program in 2004.

of basic mathematics principles to solve problems that require investigation and interpretation (Afamasaga-Fuata'i, Meyer, & Falo, 2007), demonstrating that UPY students and those that continue onto undergraduate studies consistently perceive mathematics learning narrowly as basically manipulating symbols and substituting values into formulas with a tendency to use any procedure to get an answer to a word problem without really checking whether the algorithm is suitable to the problem (Schoenfeld, 1996). If and when explanations are given, they are "reflective of precedent-thriven comprehension with no logical basis for a priori reasoning" (Baroody, Feil, & Johnson, 2007), indicative of previous socio-mathematical practices in classrooms where mathematical dialogues and discussion of mathematical ideas are neither supported nor encouraged (Richards, 1991). Further exacerbating this problem is the examination-driven teaching of secondary school mathematics, which naturally inculcates a narrow view of mathematics. Against this general background, a series of concept mapping and Vee diagrams studies (mapping studies) with different cohorts of undergraduate students was initiated (Afamasaga-Fuata'i, 2005b, 2002, 2001) primarily to explore different ways in which mathematics learning could be made more meaningful and more conceptual. Therefore, the research questions for this chapter are: (1) *How can hierarchical concept maps illustrate improvements in students' understanding of mathematics topics?* (2) *In what ways do social interactions influence students' developing understanding?*

The chapter presents data from one of the cohorts that selected topics they recently or concurrently studied at the time, as content for the application of two meta-cognitive tools (concept maps and Vee diagrams). An overview of the relevant theoretical framework guiding the studies is examined next followed by a review of relevant studies.

Literature Review

Theoretical Framework

The two main theories guiding the studies are Ausubel's theory of meaningful learning and Vygotsky's theory of development. The fundamental idea in Ausubel's cognitive psychology is that learning takes place by the assimilation of new concepts and propositions into existing concept and propositional frameworks (i.e., cognitive structures) held by the learner (Novak & Cañas, 2006). The assimilation of new knowledge into students' existing patterns of meanings takes place through the process of progressive differentiation and/or integrative reconciliation where new meanings are acquired by the interaction of new, potentially meaningful ideas with what is previously learnt. This interactional process results in a modification of both the potential meaning of new information and the meaning of the knowledge structure to which it becomes connected to (Ausubel, 2000).

The process of progressive differentiation involves the consequent refinement of meanings of a more general more inclusive concept in terms of less general and more specific concepts. In contrast, integrative reconciliation of ideas involves the synthesis of a group of coherent concepts with another concept or concepts not initially connected to them. According to Ausubel's theory of meaningful learning, meaning is created through the deliberate connection of concepts and in how this interrelationship is interpreted and internalised by the learner. This meaningful learning is "a constructive process involving both our knowledge and our emotions or the drive to create new meanings and new ways to represent these meanings" (Novak & Cañas, 2006).

To provide evidence of students' developing understanding of a domain, concept maps and Vee diagrams may be constructed (Novak & Gowin, 1984). Concept maps are hierarchical networks of interconnecting concepts (nodes) with linking words describing the meanings of the interrelationships. Strings of triads of valid concept nodes, linking words and concept nodes form meaningful statements called propositions (examples of concept maps are presented later). Vee diagrams on the other hand is a Vee situated in the problem or activity to be examined with its left hand side displaying the conceptual information required to interpret the problem/activity with the methodological information on the right hand side to illustrate how the conceptual information is applied to the given information to generate answers to the focus questions (see Chapter 12 for an example).

Whilst Ausubel's cognitive theory of meaningful learning defines the construction of knowledge and meaning by a learner, Vygotsky's zone of proximal development describes how a learner, through interactions, and discussion with help from adults or more capable peers, master concepts and ideas that they cannot understand on their own, in order to, through the process of mediation, become more knowledgeable and expert in a particular domain or more proficient with a particular skill. The emphasis here is that the learner reaches "cognitive maturation" (Vygotsky, 1978) in his/her cognitive development. Social interactions or external stimuli are consequential to a learner only to the degree that he/she can assimilate them by means of his/her cognitive structures (Piaget, 1969). Hence, while social interactions and negotiations are influential to a student's thinking and reasoning that he/she has reached cognitive maturation (i.e., at the peak of his/her zone of proximal development) would be manifested by what he/she does with directed thought. The latter is internal and occurs without communication with another as social thought that is increasingly influenced by experience and logic and no longer tied to an immediate social context. Consequently, higher mental processes engage the "social" interactions of inner speech to create new knowledge (Marsh & Ketterer, 2005). Furthermore, Vygotsky proposed that human beings achieve control over natural mental functions by bringing socio-culturally formed mediating artifacts into thinking activity while cultural artifacts, situated within a human activity system, mediate human activity (Marsh & Ketterer, 2005). Again the emphasis falls on learners actively constructing knowledge and meaning through participating in activities and challenges, with the added emphasis on the interaction between learners and facilitators in order to arrive at a higher level of truth (Sternberg & Williams, 1998). In problem solving, having information about which to think, analytically, creatively, or practically is as important as the thinking process itself. In other words, thinking requires information to analyze, creative thinking, to go beyond the given, requires knowledge of the given, and practical thinking must make use of knowledge of the situation (Sternberg & Williams, 1998).

Relevant Studies

Numerous studies investigated the use of concept maps and Vee diagrams (maps/diagrams) as assessment tools of students' conceptual understanding over time in the sciences (Novak & Cañas, 2006; Novak, 2002; Mintzes, Wandersee, & Novak, 2000), and mathematics (Afamasaga-Fuata'i, 2005b, 2007). Research in secondary (Afamasaga-Fuata'i, 1998, 2002) and university mathematics (Afamasaga-Fuata'i, 2007b, 2004, 2006b) found students' conceptual understanding of mapped topics was further enhanced after a semester of concept mapping. Research with preservice teachers showed maps were useful pedagogical planning tools (Afamasaga-Fuata'i, 2006a, 2007a; Brahier, 2005) and means of examining students' conceptual understanding of mathematical concepts (Baroody & Bartel, 2000; Williams, 1998; Hansson, 2005; Swarthout, 2001). Workshops with science and mathematics specialists and teachers found maps/diagrams have potential as teaching, learning and assessment tools (Afamasaga-Fuata'i, 2005b, 2002, 1999).

Collectively, these studies provide empirical evidence of the value of concept maps as a way of illustrating, representing, and organizing one's understanding of the interconnections of a group of coherent concepts, and as methodological tools to organise the analysis of interview protocols into schematic categories that are systematically interrelated. Both these uses emphasize the importance of valid interconnections and meaningful hierarchical organization of concepts. Vee diagrams on the other hand, have been used to unpack participants' epistemological beliefs and perceptions of how knowledge is produced and structured in a discipline as contextualized in a problem/activity and as a pedagogical planning tool for the design of learning activities.

Concept Mapping and Vee Diagram Studies

The mapping studies conducted at NUS in Samoa, investigated the dual impact of the *deliberate* construction of maps/diagrams and *altered* socio-mathematical classroom norms have on students' understanding of mathematics particularly in terms of (i) their fluency with the language of mathematics, (ii) their effectiveness in articulating and communicating their understanding to others, and (iii) their perceptions of mathematics. The newly established socio-mathematical norms required students to conduct conceptual analyses of a topic and its problems and to display the results using maps/diagrams in order to publicly communicate their understanding for critique in a social setting, comprising their peers and the researcher, during individual presentations to the group and one-on-one consultations with the researcher.

Preparatory work for maps/diagrams involved the analyses of a topic to identify its relevant key concepts, principles, formal definitions, theorems, and formulas in addition to being proficient in executing its methods and procedures.

Methodology of the Study

The study reported here was an exploratory teaching experiment to investigate students' developing understanding of particular topics (Steffe & D'Ambrosio, 1996). Students met twice a week for 50 minutes each time over 14 weeks. Maps/diagrams were introduced as means of learning mathematics more meaningfully and solving problems more effectively. The content included the topics: *limits and continuity, indeterminate forms, numerical methods, differentiation, integration, motion, multiple integrals, infinite series, normal distributions* and *complex analysis*. The epistemological principles, namely, building upon students' prior knowledge, negotiation of meanings, consensus and provision of time-in-class for student reflections, guided classroom practices. Hence, the study included a familiarization phase, which introduced the new socio-mathematical norms of students presenting and justifying their work publicly, addressing critical comments, and then later on critiquing peers' presented work. Time was set aside between critiques to revise maps/diagrams. The cyclic process of: presenting (to peers or researcher) → critiquing → revising → presenting underpinned the study. Of the 13 students, 3 chose topics outside of mathematics (computer programming, cell biology, and organic chemistry). This paper reports the data from the mathematics maps only.

Analysis of Concept Maps

Whilst the literature documents a variety of assessment/scoring techniques (Novak & Gowin, 1984; Ruiz-Primo, 2004; Liyanage & Thomas, 2002), a modified version of the Novak scheme was adopted using counts of a criterion particularly as the depth of a student's understanding is determined by the number, accuracy, and quality of connections (Hiebert & Carpenter, 1992; Baroody et al., 2007). The three criteria are the *structural* (complexity of the hierarchical structure of concepts), *contents* (nature of the contents or entries in the concept nodes) and *propositions criteria* (valid propositions).

The structural criteria are in terms of integrative *cross-links* between concept hierarchies (integrative reconciliation) signaled by nodes with multiple incoming links, progressive differentiation evidenced by nodes with *multiple branching* (more than one outgoing link) (which create *main branches* and *sub-branches*), and *average number of hierarchical levels per sub-branch*. The contents criteria indicate students' perceptions of mathematical *concepts* in terms of suitable labels and illustrative *examples*. *Inappropriate* entries included those describing procedural steps (more appropriate on diagrams), redundant entries (indicating the need for a reorganization of concepts), and linking words as concept labels (linking-word-type).

The definitional-phrase invalid node, although conceptual is too lengthy, its presence signals the need for further analysis to identify "*concepts*" as distinct from "linking words". The propositions criteria define *valid propositions* as those formed by valid triads (i.e., "valid node- valid-linking words → valid node"). Propositions are meaningful statements made up of triads of valid nodes and valid linking words.

Concept Maps Collected

The data collected consisted of students' progressive maps (4 versions) and progressive Vee diagrams of 3 problems (at least 2 versions per problem), and final reports. Only the data from maps are presented here. The three criteria were used to assess students' first and final maps, to identify any changes. The map data for Students 1–10 are in Tables 15.1 and 15.2. The results show that 7 out of 10 students displayed increases in valid propositions by the final map with five students showing increases in at least 4 of the 5 structural sub-criteria.

In this chapter, a second level analysis was conducted to extend beyond that presented in Afamasaga-Fuata'i (2007b), to determine an *overall rating* for each final cmap, by averaging the ratings based on the two criteria (1) *valid propositions* and (2) *structural complexity*. This extended analysis is to provide a more in-depth investigation of features that distinguish between different types of maps. Implicit in the criteria: *valid propositions* is the count of valid nodes and valid linking words to form a valid proposition while the *structural complexity* criteria is in terms of whether or not there were increases (↑) with its 5 sub-criteria.

For *valid propositions* (VP), a rating scale of 1–5 is defined based on the percentage of valid propositions in the final cmap. That is, rating 1 means at least 80% VP, 2 (70% ≤ VP < 80%), 3 (60% ≤ VP < 70%), 4 (50% ≤ VP < 60%), and 5 (VP < 50%). For *structural complexity*, a rating scale of 1–5 indicates the number of increased sub-criteria, that is, a rating of 1 is for 5 increased sub-criteria (↑), 2 for 4↑, 3 (3↑), 4 (2↑), and 5 (1↑).

Within the *overall rating framework*, a rating of 1 signifies a map that is conceptually meaningful, evident by its high percentage of valid propositions, and structurally complex, reflected by increased integrative reconciliation between concept hierarchies and progressive differentiation of concepts. In contrast, an overall rating of 5 indicates a map that has a high proportion of invalid proportions and structurally less complex with predominantly linear links that are sparsely, or, not, connected to each other or fragmented hierarchies. A summary of overall ratings for the 10 final maps are shown in Table 15.3 with a further categorization to describe types, namely, *good*, *above average*, *average*, *below average*, and *poor*.

The *good* maps (top-3) indicate high proportions of valid propositions (rating 1) with 4 increased structural sub-criteria out of 5 (rating 2) while the bottom map (rating of 4) had a high proportion of invalid propositions and only 3 increased structural sub-criteria. The *below-average* maps (rating of 3.5) were mainly due to few (1 or 2) increased structural sub-criteria with less than 70% valid propositions.

Table 15.1 Concept map data for students 1–3

Student	1	1	2	2	3	3	4	4	5	5
Cmap Criteria	First	Final	First	Final	First	Final	First	Final	First	Final
1. Contents										
Valid nodes										
– *Concepts*	35 (67)	30 (65)	17 (44)	32 (56)	73 (99)	83 (83)	40 (59)	66 (99)	44 (86)	50 (74)
– *Examples*	5 (10)	1 (2)	19 (49)	16 (28)	0 (0)	6 (6)	10 (15)	0 (0)	0 (0)	0 (0)
Invalid nodes										
– *Definitional*	4 (8)	14 (30)	1 (3)	8 (14)	1 (1)	8 (8)	12 (18)	1 (1)	1 (2)	15 (22)
– *Inappropriate*	8 (15)	1 (2)	2 (5)	1 (2)	0 (0)	3 (3)	6 (9)	0 (0)	6 (12)	3 (4)
Total nodes	52	46	39	57	74	100	68	67	51	68
2. Propositions										
Valid propositions	27 (52)	26 (44↓)	25 (69)	29 (49↓)	77 (96)	106 (88↓)	32 (51)	85 (97↑)	35 (66)	54 (67↑)
Invalid propositions	25 (48)	33 (56)	11 (31)	30 (51)	3 (4)	14 (12)	31 (49)	3 (3)	18 (34)	27 (33)
Total propositions	52	59	36	59	80	120	63	88	53	81
3. Structural										
Cross-links	3	10 (↑)	0	6 (↑)	9	10 (↑)	4	17 (↑)	6	22 (↑)
Sub-branches	17	14 (↓)	9	19 (↑)	26	33 (↑)	22	19 (↓)	9	32 (↑)
Average H/levels per sub-branch	6	8 (↑)	6	8 (↑)	10	9 (↓)	7	8 (↑)	8	7 (↓)
Main branches	6	6 (↔)	5	7 (↑)	5	8 (↑)	4	5 (↑)	6	9 (↑)
M/B nodes	8	10 (↑)	5	8 (↑)	18	19 (↑)	9	18 (↑)	9	19 (↑)

Key: H/levels, Hierarchical levels; M/B, Multiple branching; Count, (% of total number); ↓, Decrease; ↑, Increase; ↔, No change.

Table 15.2 Concept map data for students 4–6

| Student | 6 | | 7 | | 8 | | 9 | | 10 | |
Map criteria	First	Final	First	Final	First	Final	First	Final	First	Final
1. Contents										
Valid nodes										
– *Concepts*	165 (87)	159 (84)	32 (67)	30 (73)	41 (36)	52 (87)	34 (76)	63 (71)	100 (57)	79 (72)
– *Examples*	11 (6)	12 (6)	0 (0)	1 (2)	20 (18)	0 (0)	0 (0)	7 (8)	34 (19)	17 (16)
Invalid nodes										
– *Definitional*	1 (1)	2 (1)	3 (6)	3 (7)	3 (3)	7 (12)	6 (13)	15 (17)	8 (5)	6 (6)
– *Inappropriate*	12 (6)	16 (8)	13 (27)	7 (17)	50 (44)	1 (2)	5 (11)	4 (4)	33 (19)	6 (7)
Total nodes	189	189	48	41	114	60	45	82	141	92
2. Propositions										
Valid propositions	148 (74)	166 (79↑)	15 (32)	26 (62↑)	48 (40)	39 (67↑)	28 (56)	88 (81↑)	110 (60)	82 (73↑)
Invalid propositions	51 (26)	45 (21)	32 (68)	16 (38)	71 (60)	19 (33)	22 (44)	20 (19)	73 (40)	30 (27)
Total propositions	183	112	47	42	119	58	50	108	183	112
3. Structural										
Cross-links	8	8 (↔)	1	4 (↑)	11	12 (↑)	13	21 (↑)	5	6 (↑)
Sub-branches	68	66 (↓)	16	13 (↓)	19	16 (↓)	10	24 (↑)	44	35 (↓)
Average H/levels per sub-branch	6	6 (↔)	6	6 (↔)	12	9 (↓)	11	9 (↓)	9	7 (↓)
Main branches	19	19 (↔)	4	6 (↑)	12	6 (↓)	3	10 (↑)	10	10 (↔)
M/B nodes	24	34 (↑)	4	9 (↑)	13	11 (↓)	9	17 (↑)	20	16 (↓)

Key: H/levels, Hierarchical levels; M/B, Multiple branching; Count, (% of total number); ↓, Decrease; ↑, Increase; ↔, No change.

Table 15.3 Summary of overall ratings for the final concept maps

Student # – Name	Criteria		Final concept map	
	Valid Propositions	Structural Complexity	Overall Rating	Type of concept map
3 – Fia	1	2	1.5	Good
4 – Vae	1	2	1.5	Good
9 – Toa	1	2	1.5	Good
5 – Heku	3	2	2.5	Above average
2 – Loke	5	1	3	Average
7 – Fili	3	3	3	Average
8 – Pasi	3	4	3.5	Below average
6 – Santo	2	5	3.5	Below average
10 – Salo	2	5	3.5	Below average
1 – Pene	5	3	4	Poor

Of the *above-average, average,* and *below-average* types (overall ratings 2.5, 3 and 3.5 respectively), Students 2's, 6's and 10's maps stand out because of the one lowest rating (i.e., 5).

The next sections present the cases of each type of map to illustrate some of the key features distinguishing between them.

Examples of "Good" Concept Maps

Student 3: Fia – Numerical Methods

Fia's first map had a high percentage of valid propositions (96%) reflecting her careful organization of propositions. As a result of critiques, revisions and further research, the final map showed increased number of valid concept nodes (from 73 to 83) and valid propositions (from 77 to 106) but proportionally reduced (valid nodes from 99 to 89% and valid propositions from 96 to 88%) due to increased definitional-phrase and inappropriate nodes (from 1 to 11%).

Structurally, the final map expanded (increased main branches from 5 to 8), becoming more integrated (increased cross-links from 9 to 10) and more differentiated (increased multiple-branching nodes from 18 to 19 and increased sub-branches from 26 to 33) with more compact sub-branches (reduced average hierarchical levels from 10 to 9).

Figures 15.1 and 15.2 show sections of the final map, which depict general and more inclusive concepts at the top with progressively less inclusive and more specific concepts towards the bottom, examples of progressive differentiation (e.g., *algorithms, collocation polynomial*) and integrative reconciliation (e.g., *errors*).

An example of a linking-word-type of inappropriate entry (Fig. 15.1) is *"used as a measure of accuracy"* while those exemplifying definitional-type are *"Rounding up of numbers due to the given decimal place"* and *"When a number is automatically*

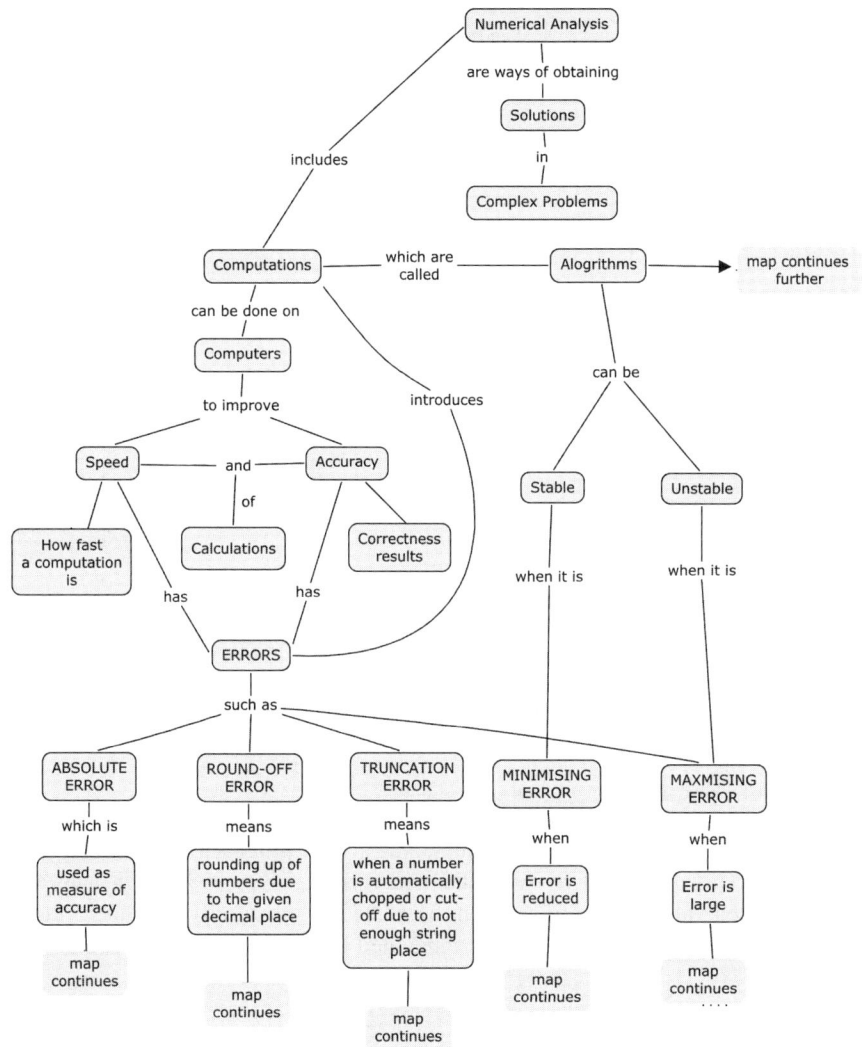

Fig. 15.1 Partial view of student 3's final concept map

chopped or cut-off due to not enough string place". The last two entries require further analysis to provide more precise concept labels. Figure 15.2 displays more examples of progressive differentiation and clear hierarchical arrangements of nodes.

Student 4: Vae – Limits and Continuity

Vae's first map showed inclusion of complete formal definitions as concept labels (see partial view, Fig. 15.3), which the first peer critique highlighted as problematic.

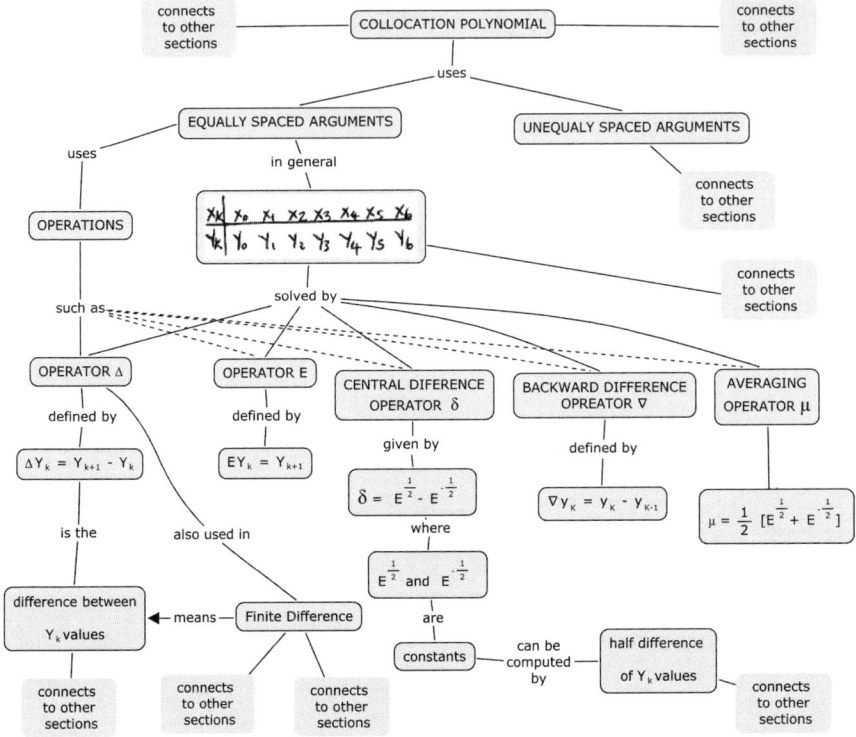

Fig. 15.2 Partial view of student 3's final concept map

As a result of revisions, and critiques, Vae's map progressively evolved into a more conceptual one (increased valid nodes from 74 to 99%) with substantially increased valid propositions (from 51 to 97%), structurally expanded (main branches increased from 4 to 5), more integrated (cross-links increased from 4 to

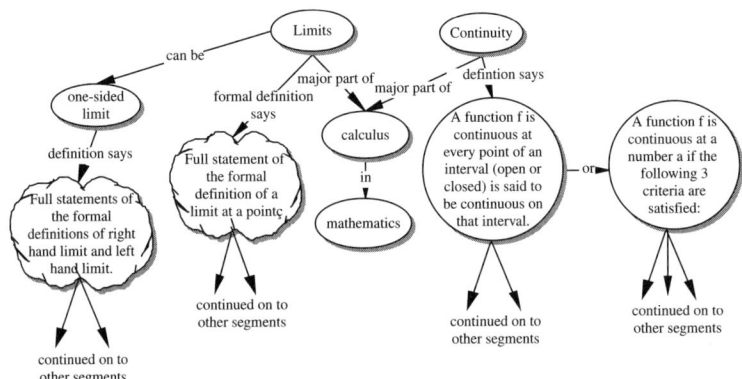

Fig. 15.3 Partial view of student 4's first concept map

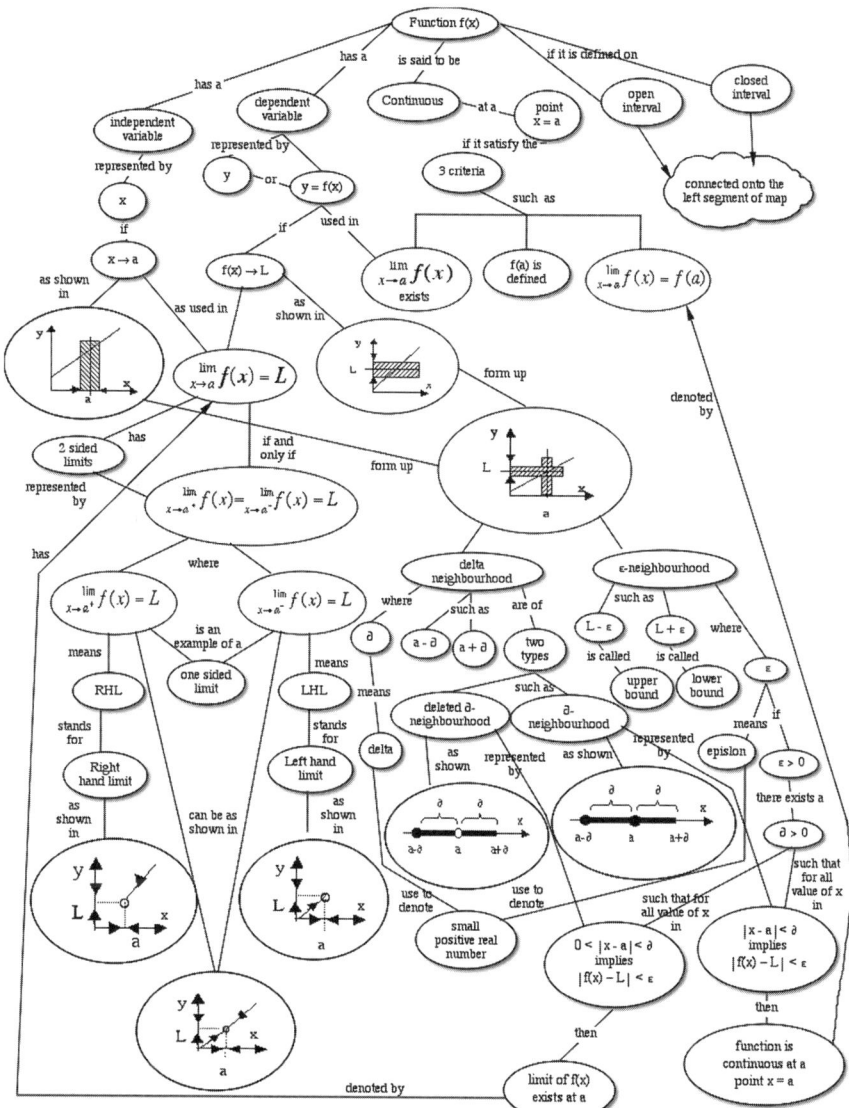

Fig. 15.4 Partial view of student 4's final concept map

17), more differentiated (increased multiple branching from 9 to 18 and increased average hierarchical levels per sub-branch from 7 to 8), and more compact (reduced sub-branches from 22 to 19) (Fig. 15.4).

Figure 15.4 displays a partial view of the final map showing a more differentiated illustration of concepts, more precise concept labels, examples of cross-links and progressive differentiation in contrast to bulk definitions of the initial cmap.

Evidently, continuous revisions enhanced the hierarchical interconnections such that formal definitions were analysed substantively for its key concepts, with concepts appropriately linked and interconnections described meaningfully.

Student 9: Toa – Normal Distributions (ND)

Toa felt challenged to construct a map that included ND, Poisson distributions (PD) and binomial distributions (BD). He wrote: *(it was) hard to think of a concept to start the map and then link the others right down to the end when it introduces (BD, PD and ND).* The first peer critique commented the map had "*too many useful concepts ... missing*", and the "*concepts used were paragraphs*", see Fig. 15.5.

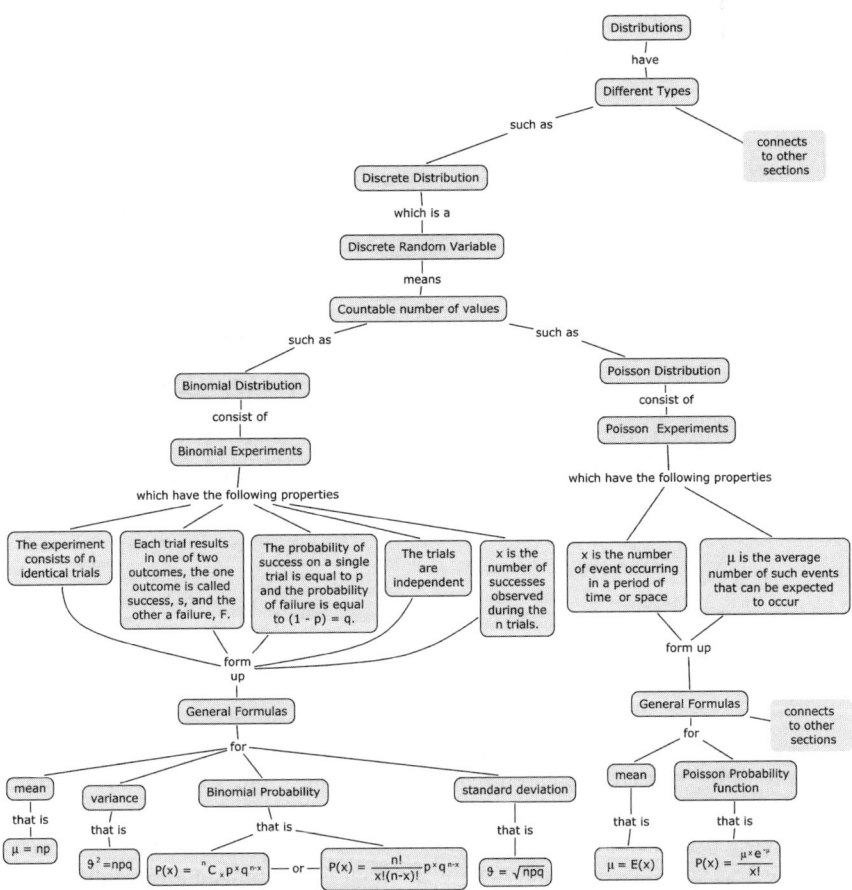

Fig. 15.5 Partial view of student 9's initial concept map

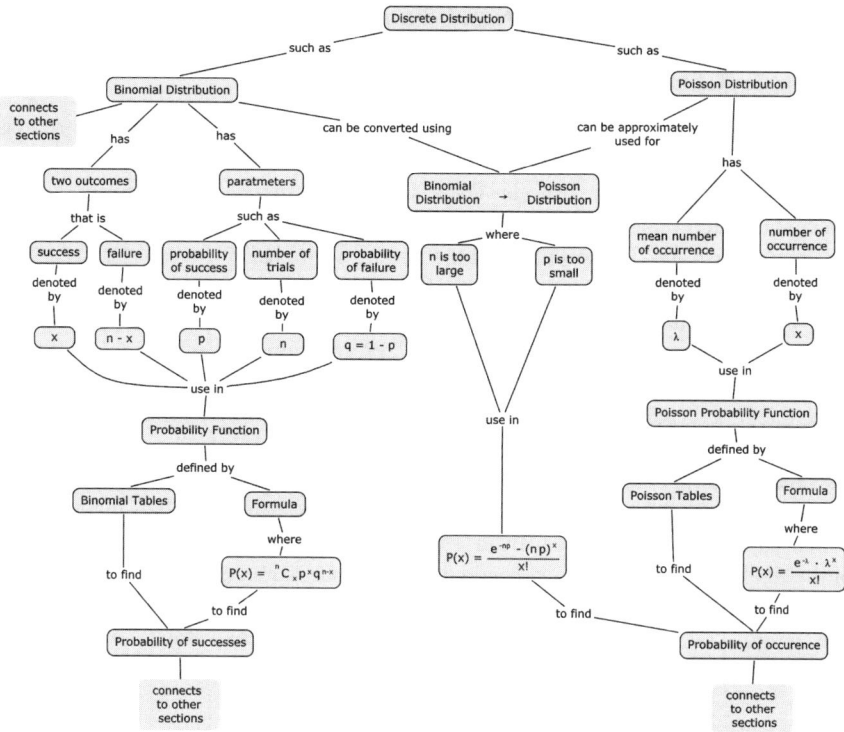

Fig. 15.6 Partial view of student 9's final concept map

In subsequent revisions, he *"tried to break down those paragraphs into one or two concept names"* and *"re-organized concept hierarchies"* eventually resulting in a final map that was more conceptual (increased valid nodes from 76 to 79%) with increased valid propositions (from 56 to 81%). Structurally, the final map (partial view in Fig. 15.6) became more expanded (increased main branches from 3 to 10), more integrated (increased cross-links from 13 to 21), more differentiated (substantial increases with multiple branching nodes from 9 to 17 and sub-branches from 10 to 24) and more compact within sub-branches (reduced average hierarchical levels from 11 to 9). Shown in Fig. 15.5 is a partial view of Toa's first map, with paragraphs as nodal entries. The corresponding revisions in the final map (Fig. 15.6) include more precise concept labels and examples of integrative cross-links between two branches (propositions *"Binomial Distribution* can be converted using *Binomial Distribution → Poisson Distribution"* and *"Poisson Distribution* can be converted using *Binomial Distribution → Poisson Distribution"*), multiple branching nodes (*Binomial Distribution, Poisson Distribution,* and *parameters*) and integrative reconciliation of a number of nodes merging into a single node (nodes x, $n - x$, p, n, $q = 1 - p$ with merging links to *Probability Function*).

Example of an "Above Average" Final Concept Map

Student 5: Heku – Motion

Half of Heku's initial map (Fig. 15.7) was mostly derivation of different formulas for speed and distance with the other half (not shown) displaying graphs of distance, speed and acceleration versus time. The first critique targeted the need to show more conceptual interconnections, less demonstration of procedural steps (more appropriate on Vee diagrams), to complement the various formulas.

As a result of critiques and revisions, Heku's final map became more conceptual with increased number of valid concept nodes (from 44 to 50 but proportionally reduced from 86 to 74%) and increased valid propositions (from 66 to 67%). Structurally, the final map was more expanded (increased main branches from 6 to 9), more integrated (increased cross-links from 6 to 22), more differentiated (increased multiple branching nodes from 9 to 19 and increased sub-branches from 9 to 32) but relatively more compact within sub-branches (reduced average hierarchical levels from 8 to 7). Increased invalid nodes (from 14 to 26%) resulted mainly from increased definitional phrases (from 2 to 22%). Provided in Fig. 15.8 is a section of the final map showing less formula derivation but more conceptual nodal entries with an improved structural hierarchical organization. Examples of definitional-type entries are: "rate in which something moves" and "rate of velocity of a moving object".

Example of an "Average" Final Concept Map

Student 2: Loke – Differentiation

Loke's first map had relatively more illustrative examples (49%) than conceptual entries (44%) (see partial view in Fig. 15.9). As a result of critiques, revisions and independent research, the final map was relatively more conceptual (increased valid concept nodes from 44 to 56% and a reduction in examples from 49 to 28%), structurally more expanded (addition of 2 more main branches), more integrated (addition of 6 new cross-links) and more differentiated (increased multiple-branching nodes from 5 to 8 and increased sub-branches from 9 to 19). However, the reduction of valid propositions (from 69 to 49%) was due mainly to increased definitional-phrase invalid nodes (from 3 to 14%).

A partial view of the final map (Fig. 15.10) shows some examples of definitional phrases are "*finding the limit of the difference quotient*", "*gradient of secant at point PQ*" and "*secant, join by point PQ*", which could be either analysed further to extract relevant concepts from linking works or revised to represent more precise concept names. Examples of cross-links and progressive differentiation are also shown. Overall, the final map was structurally more differentiated and more integrated with the quality of the revised nodes more conceptual than the first map.

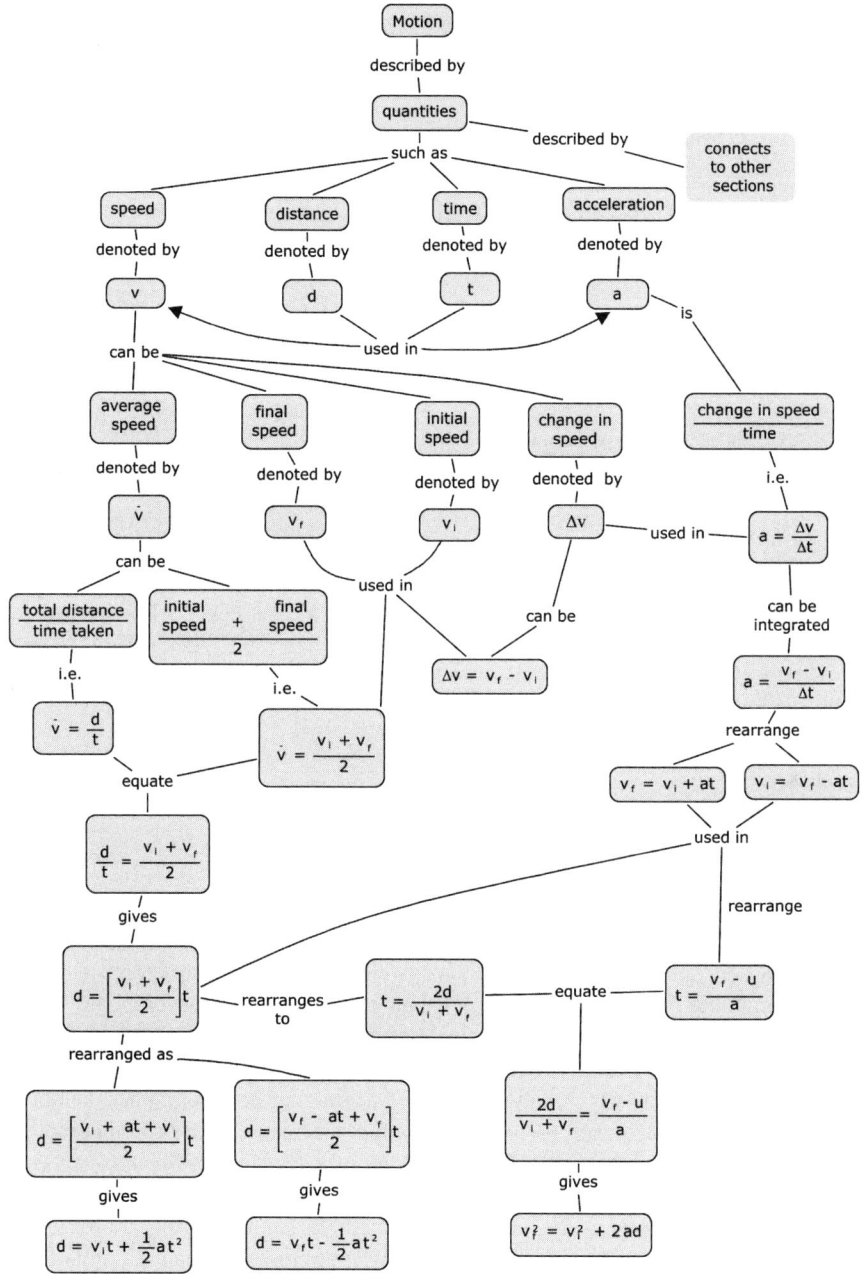

Fig. 15.7 Partial view of student 5's first concept map

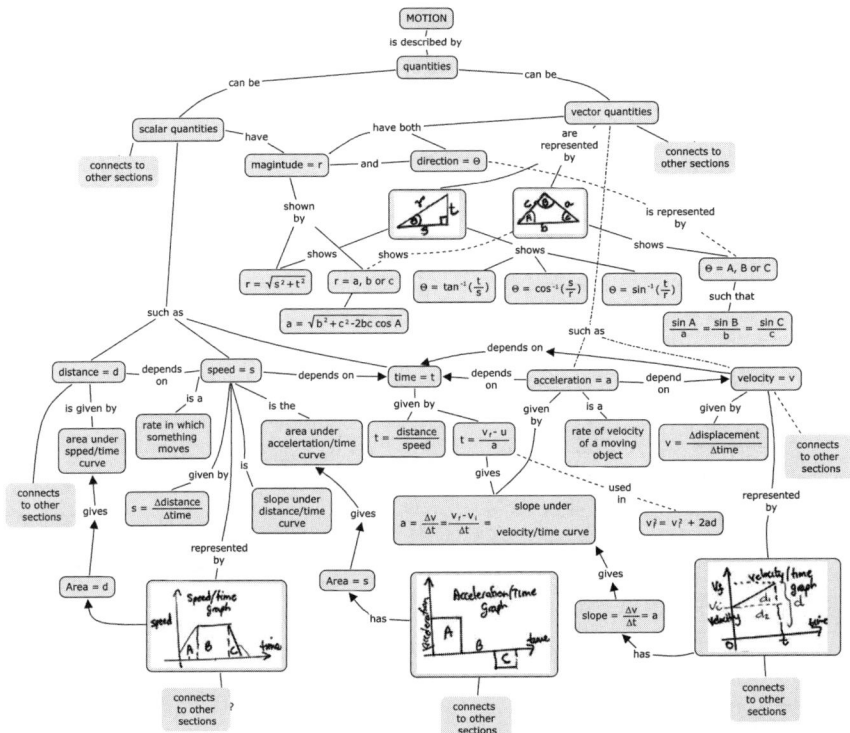

Fig. 15.8 Partial view of student 5's final concept map

Example of a "Below Average" Final Concept Map

Student 8: Pasi – Integration

Pasi's first concept map was at first glance, well organized and structured with the 12 main branches neatly arranged and consisting of a total of 19 sub-branches with a fairly high average number of hierarchical levels of 12. However, from the contents criteria, 55% (63/114) of the nodes had entries related to the illustrative examples compared to only 36% (41/114) which were classified as concept names, symbols or general formulas with 3% (3/114) with entries that were more or less complete formal definitions such as statements like *"Limit of Riemann sum (as max $\Delta x_i \rightarrow 0$ of f for interval [a,b]"* (see Fig. 15.11), which could be decomposed further into precise concepts. The inclusion of main steps of methods (more appropriate for a Vee diagram and demonstrating his procedural knowledge) resulted in a high proportion of illustrative examples.

As a consequence of the cyclic process of presenting → critiquing → revising → presenting, Pasi's final map evolved into a substantially more conceptual one (increased valid nodes from 54 to 87%) with increased valid propositions (from 40 to 67%) and a conspicuous absence of illustrative examples (of procedural steps) Fig. 15.12.

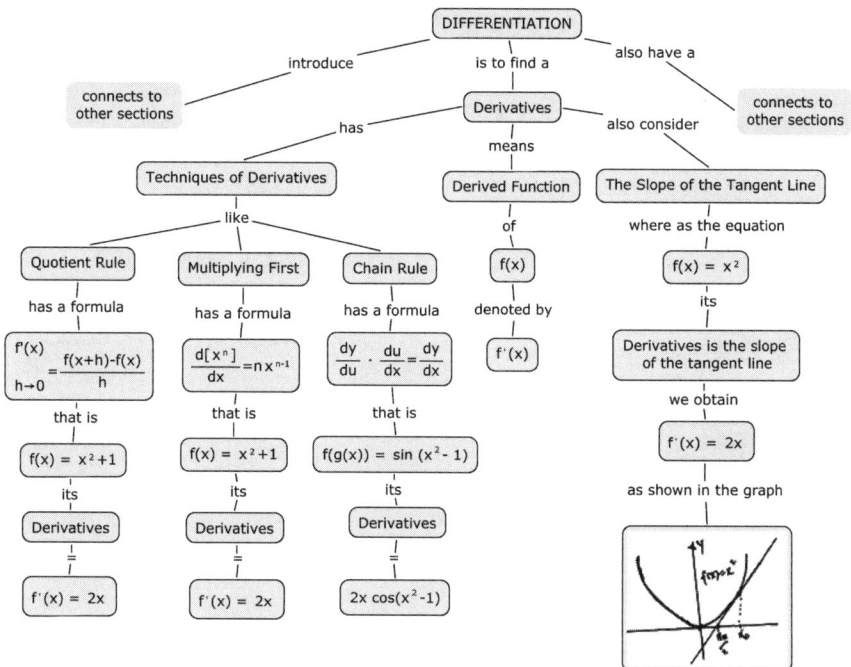

Fig. 15.9 Partial view of student 2's first concept map

Structurally, the map was more compact (reduced multiple-branching nodes (from 13 to 11), reduced sub-branches (from 19 to 16), reduced main branches (from 12 to 6), and reduced average hierarchical levels per sub-branch (from 12 to 9)). Overall, the final map was predominantly more conceptual with its concept labels and more valid propositions with a more parsimonious, compact final structure. That its structural criteria was rated 4 (resulting in an overall rating of 3.5), was caused mainly by the reduction of excessive procedural steps (and hence hierarchical levels) while the quality of propositions was relatively better than the initial map.

A partial view of Pasi's final concept map (Fig. 15.12) shows the more conceptual and expanded *numerical limit* and *area under the curve* sub-branches in contrast to the initial version in Fig. 15.11, which was predominantly an illustration of the application of the definite integral formula. Whilst the middle and right branches subsumed under the more inclusive node: *Definite Integral* are cross-linked towards the bottom (Fig. 15.12), the middle and left branches are not.

Example of a "Poor" Final Concept Map

Student 1: Pene – Indeterminate Forms

Despite encountering "indeterminate forms" in first year mathematics, Pene struggled to begin a map. The segment of his first map (see Fig. 15.13) shows, radiating

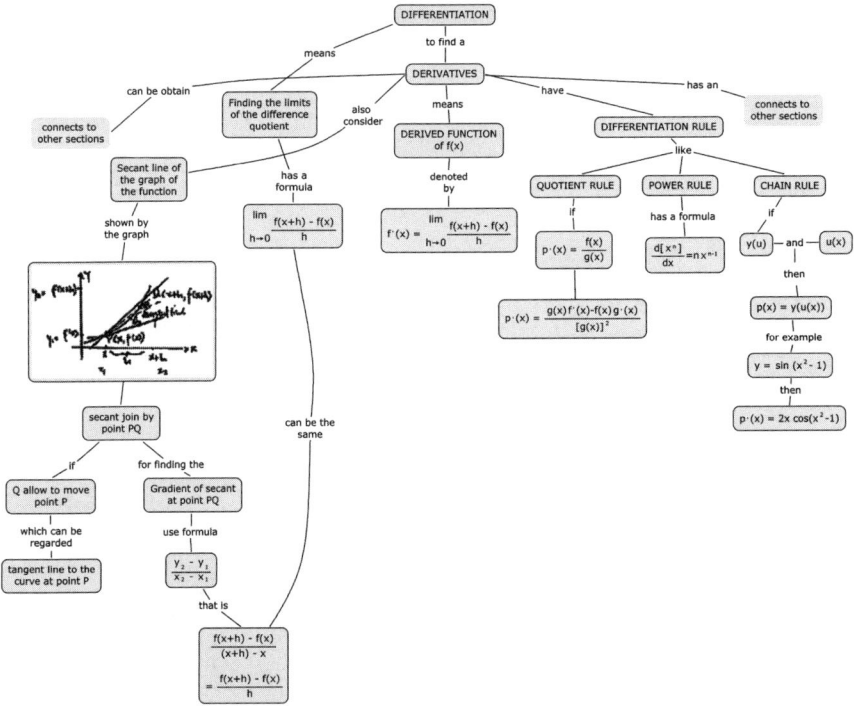

Fig. 15.10 Partial view of student 2's final concept map

from node: *Limit*, are linear, sequential relations (or strings of triads), each expressing a mathematical statement or fact with little integrative linking to adjacent concepts or other concept hierarchies. As a result of critiques, revisions and independent
research, Pene's final map became structurally more integrated (increased cross-
links from 3 to 10), more differentiated (increased multiple-branching nodes from 8
to 10 and increased average hierarchical levels per sub-branch from 6 to 8) and more
compact (decreased sub-branches from 17 to 14) with main branches remaining
unchanged (Table 15.1). However, the percentage of valid nodes (from 77 to 67%)
and valid propositions (from 52 to 44%) decreased due to increased definitional-
phrase invalid nodes (from 8 to 30%) and vague/incorrect linking words. Some
examples of definitional phrases in Fig. 15.14 are "*Ratio of two functions as* $x \rightarrow a$",
"*Quotient of the two functions*", and "*Differentiable in an open interval* $a < x <
b$ *and* $f(x) \neq 0$". Despite this, the final map was conceptually richer in its choice of
concept labels but linking words were still of poor quality and inaccurate, with a
structurally parsimonious, network of linked nodes.

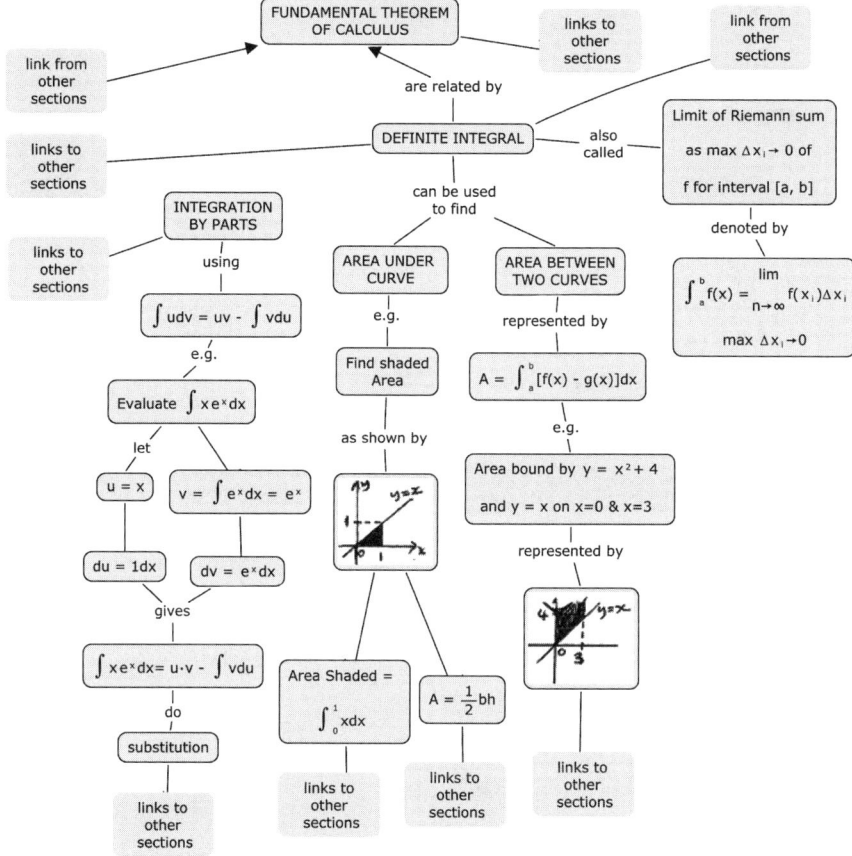

Fig. 15.11 Partial view of student 8's first concept map

Discussion

Findings suggested that students' progressive maps became integrated and differentiated as students continually strove to illustrate valid nodes and meaningful propositions, in response to specific concerns raised during social critiques and in anticipation of subsequent critiques. These progressive maps provided evidence of students' deepening conceptual understanding of the topics as indicated by the increasing complexity of the hierarchical organisation of concepts from most general to less general, progressive differentiation between more inclusive and less inclusive concepts, integrative links, and quality of propositions.

Social interactions evidently influenced students to structurally reorganize concept hierarchies and further analyse nodal entries to be more precise, less

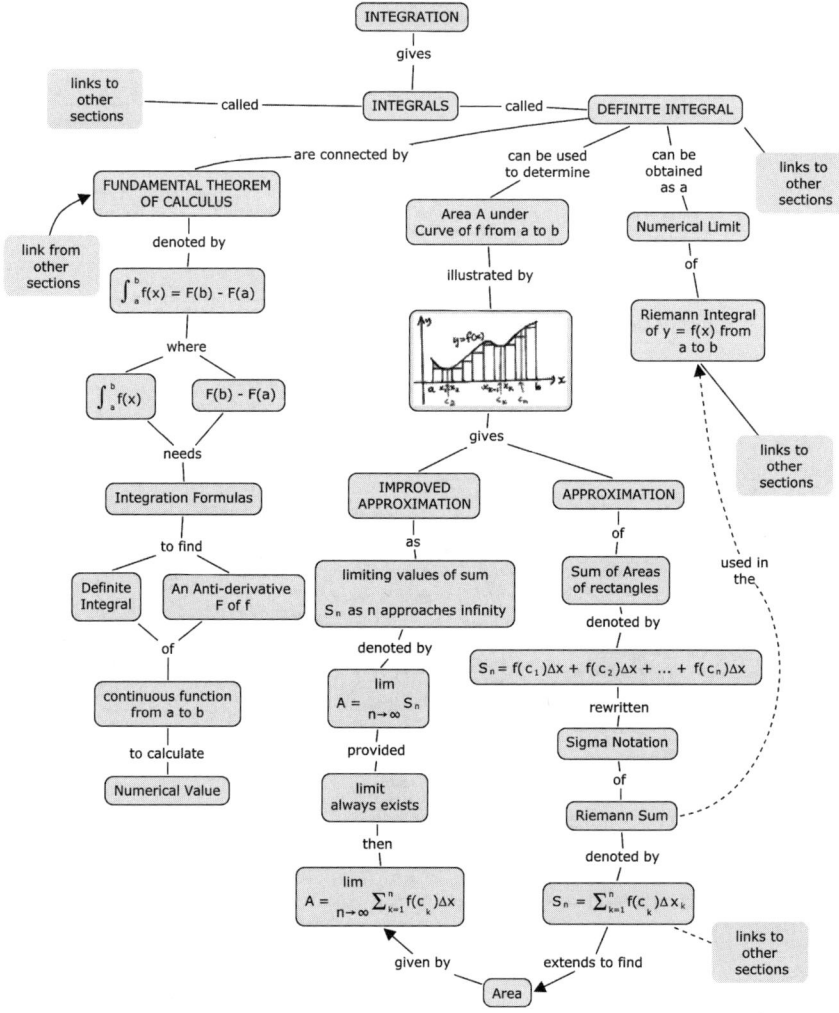

Fig. 15.12 Partial view of student 8's final concept map

definitional and non-procedural. In terms of Sternberg and William's view of problem solving, these structural and conceptual changes with the maps indicated that students were not only thinking and reasoning out the conceptual structure of their topics (analytical), but that they also attempted (a) to think creatively by going beyond the analysed information to hierarchically organize the concepts into networks of connected concepts, and (b) to think practically and flexibly in order to fit all of the above meaningfully on a map in preparation for social critiques.

Consequently, the re-definition of socio-mathematical norms appeared to substantively affect the quality of students' mathematical thinking and reasoning as

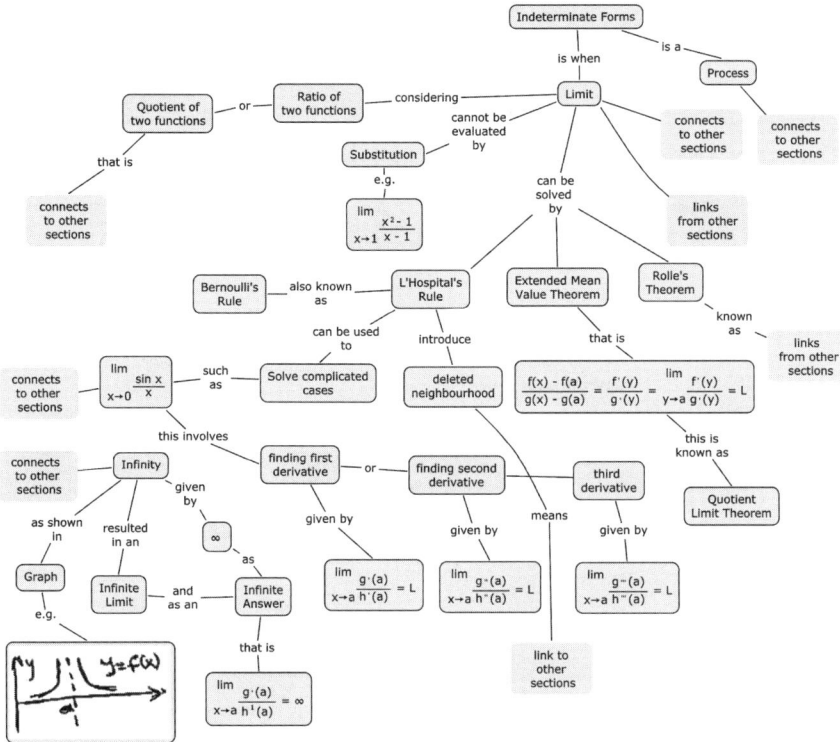

Fig. 15.13 Partial view of student 1's first concept map

reflected in the evolving quality of their progressive maps, particularly given that students were additionally required to publicly justify their displayed connections, negotiate meanings with their peers and reach a consensus to revise or not. The effects of the dialectic processes of social interactions and individual thinking and reasoning materialised as conceptual and structural changes on the maps. For example, the majority of students showed increases in valid propositions with half the students showing increased structural complexity by the final map. It appeared that the public presentation of the maps prompted the students to strive for effective communication of their ideas as manifested through improved hierarchical organization, precise nodal entries and more meaningful linking words albeit to different extents as indicated by the final maps' overall ratings.

For the students in the study, they were challenged to develop a deeper understanding of their topic, in order to address concerns raised in social critiques, through further revisions and independent research, as they internalised the meanings of concepts and being transformed by them as they communicated these internalised meanings (as their understanding) during presentations and visually through their individually constructed maps. Consequently, students constructed their own

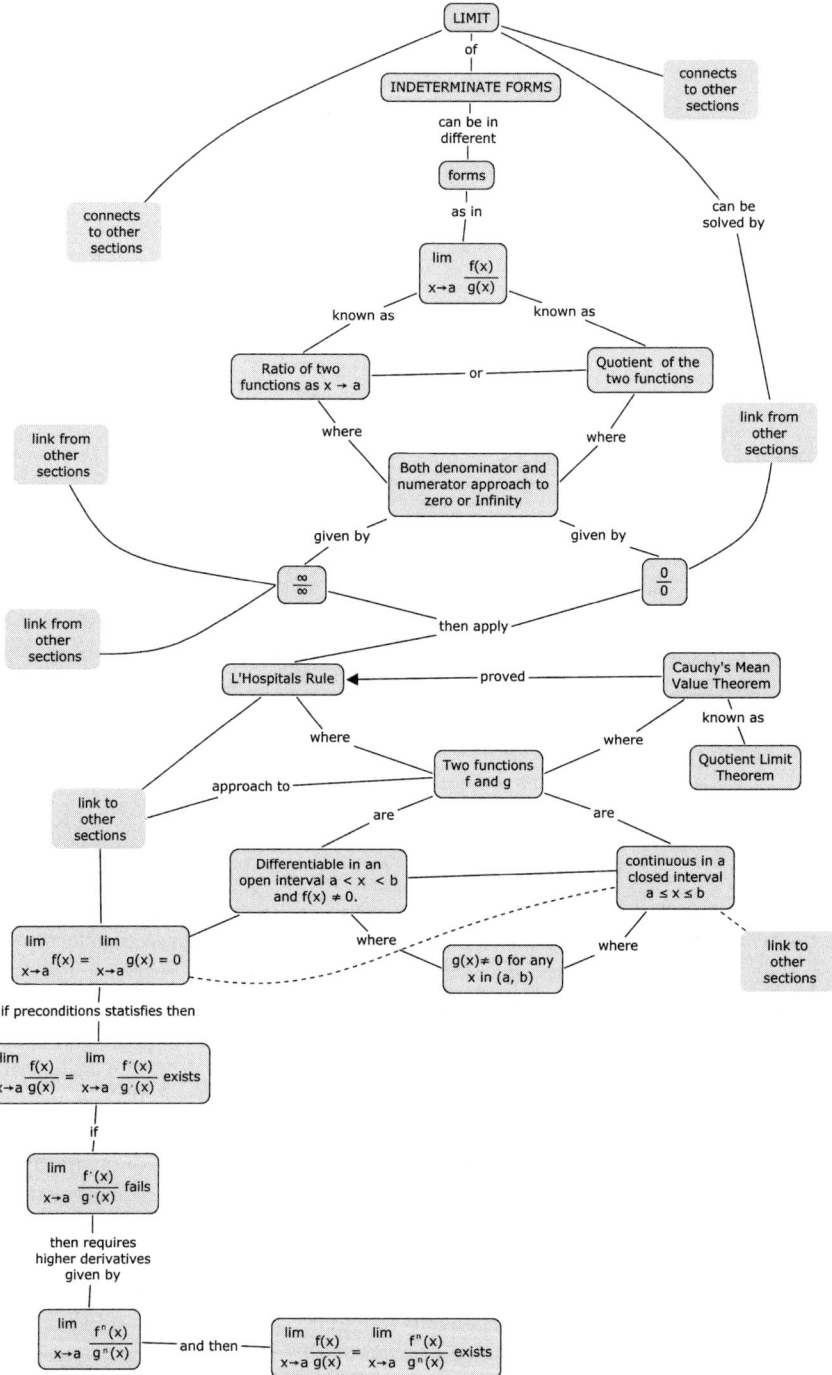

Fig. 15.14 Partial view of student 1's final concept map

knowledge and understanding (on maps) and developed mathematical meanings (as propositions) as they learnt to publicly explain and justify their thinking to others.

There was a marked shift from simply providing entire paragraphs, procedural steps, formulas, and excessive illustrative examples to seeking out more precise concept labels and valid linking words, and structurally more integrated and differentiated conceptual interconnections, which reflected the impact of the social interactions on an individual's evolving understanding, and developing directed thought and meta-cognitive skills.

Students necessarily had to reflect more deeply, as individuals, about the conceptual structure of topics than they normally did. Because of the need to communicate their understanding competently in a social setting, over time and with increased mapping proficiency, students became more analytical, flexible, creative, and precise in their selection of concept entries and more astute in describing the nature of the relationships between connecting concepts more correctly to minimize critical comments.

From students' perspectives, they realized that mathematics has a conceptual structure, the socially validated body of knowledge, which underpins its formal definitions and formulas. By searching for missing relevant concepts to make the maps more robust and comprehensive, students eventually realized that an in-depth understanding of topics required much more than re-stating a definition or formula. Instead, concept mapping invited the students to identify more inclusive and less inclusive concepts of their topics, demonstrate an integrated understanding of concepts, definitions and formulas, and visually organise this understanding as a meaningful hierarchy of interconnecting nodes with rich linking words to form valid propositions for public scrutiny and critique. In so doing, concept mapping provided one means of reinforcing a "view (of) mathematical knowledge as cohesive or structured rather than a series of isolated" (Baroody et al., 2007) definitions, formulas and procedures and as knowledge that is socially warranted by a community of mathematicians.

Over the semester, students eventually appreciated the utility of maps as a means of depicting networks of conceptual interconnections within topics and highlighting connections between concepts, definitions and formulas. However, attaining this more conceptual understanding of mathematics was hard work and required much reflection, social negotiations and individual research on their part.

As Novak and Cañas (2006) found, learners struggle to create good maps as they are engaged in a creative process of "creating new meanings and new ways to represent these meanings" which can be challenging, especially to learners who have spent most of their life learning by rote (p. 9).

The findings suggested that with more time and practice students can become proficient and adept at constructing maps whilst simultaneously deepening and expanding their theoretical knowledge of the structure of mathematics. Challenges faced by the students included (1) the importance of getting quality feedback from their peers, (2) the need to sustain students' motivation to seek more meaningful propositions by (a) revising inappropriate nodes, (b) selecting more rich and meaningful linking words, and (c) reorganising concept hierarchies, and (3) the need for them to further develop their self-confidence and fluency in presenting mathematical justifications and counter-arguments during social critiques. Furthermore, students experienced difficulties reconciling the type of linear learning style they are used to (e.g., by memorizing definitions) to the hierarchical nature and organization required for concept mapping. The progressive quality of students' maps over the semester confirmed that students' deepening understanding of their topics were very much influenced by the scaffolding and social interactions during critiques, the newly established socio-mathematical norms of the classroom setting, and the socially-validated structure of mathematics. That is, the deliberate alteration of the classroom environment and practices particularly in terms of the individual/social orientation of the learning, and the form and extent of the focus of assessment (Pratt & Kelly, 2005) resulted in a more conceptual and theoretical understanding of their topics. Findings also extended the literature on the impact of social negotiations of meanings, interactions and critiques on the development of students' conceptual understanding of topics, which in this study, was greatly facilitated with the visual mapping of students' progressive conceptions on hierarchical maps over time. Finally, using the meta-cognitive tools promoted a higher level of self-reflection and lateral thinking which, generally motivated students to critically analyse their perceptions of mathematics knowledge and, specifically encouraged deeper, conceptual understandings of topics.

Implications

Findings from the study imply that the concurrent use of concept mapping and social critiques as part of the culture and practices within mathematics classrooms has the potential to promote the development of mathematical thinking, reasoning, and effective communication which are most desirable skills to be successful in mathematics learning. Doing so as early as primary level would be an area worthy of further investigation.

References

Afamasaga-Fuata'i, K. (1998). *Learning to solve mathematics problems through concept mapping & Vee mapping*. Samoa: National University of Samoa.

Afamasaga-Fuata'i, K. (1999). Teaching mathematics and science using the strategies of concept mapping and Vee mapping. *Problems, Research, and Issues in Science, Mathematics, Computing and Statistics, 2*(1), 1–53. Journal of the Science Faculty at the National University of Samoa.

Afamasaga-Fuata'i, K. (2001). New challenges to mathematics education in Samoa. *Measina A Samoa 2000, 1*, 90–97. Institute of Samoan Studies, National University of Samoa.

Afamasaga-Fuata'i, K. (2002). A Samoan perspective on Pacific mathematics education. In B. Barton, K. C. Irwin, M. Pfannkuch, & M. O. J. Thomas (Eds.), *Mathematics education in the South Pacific*. Proceedings of the 25th annual conference of the Mathematics Education Research Group of Australasia (MERGA-25) (Vol. 1, pp. 1–13). Auckland, New Zealand: University of Auckland.

Afamasaga-Fuata'i, K. (2004). Concept maps and Vee diagrams as tools for learning new mathematics topics. In A. J. Cañas, J. D. Novak, & F. M. Gonázales (Eds.), *Concept maps: Theory, methodology, technology*. Proceedings of the First International Conference on Concept Mapping (Vol. 1, pp. 13–20). Navarra, Spain: Dirección de Publicaciones de la Universidad Pública de Navarra.

Afamasaga-Fuata'i, K. (2005a). Mathematics education in Samoa: From past and current situations to future directions. *Journal of Samoan Studies, 1*, 125–140.

Afamasaga-Fuata'i, K. (2005b). Students' conceptual understanding and critical thinking? A case for concept maps and Vee diagrams in mathematics problem solving. In M. Coupland, J. Anderson, & T. Spencer (Eds.), *Making mathematics vital*. Proceedings of the Twentieth Biennial Conference of the Australian Association of Mathematics Teachers (AAMT) (Vol. 1, pp. 43–52). Sydney, Australia: University of Technology.

Afamasaga-Fuata'i, K. (2006a). Innovatively developing a teaching sequence using concept maps. In A. Cañas & J. Novak (Eds.), *Concept maps: Theory, methodology, technology*. Proceedings of the Second International Conference on Concept Mapping (Vol. 1, pp. 272–279). San Jose, Costa Rica: Universidad de Costa Rica.

Afamasaga-Fuata'i, K. (2006b). Developing a more conceptual understanding of matrices and systems of linear equations through concept mapping. *Focus on Learning Problems in Mathematics, 28*(3 & 4), 58–89.

Afamasaga-Fuata'i, K. (2007). Communicating Students' Understanding of undergraduate mathematics using concept maps. In J. Watson, & K. Beswick, (Eds.), *Mathematics: Essential research, essential practice. Proceedings of the 30th annual conference of the Mathematics Education Research Group of Australasia*, (Vol. 1, pp. 73–82). University of Tasmania, Australia, MERGA.

Afamasaga-Fuata'i, K. (2007a). Using concept maps and Vee diagrams to interpret "area" syllabus outcomes and problems. In K. Milton, H. Reeves, & T. Spencer (Eds.), *Mathematics essential for learning, essential for life*. Proceedings of the 21st biennial conference of the Australian Association of Mathematics Teachers, Inc. (pp. 102–111). University of Tasmania, Australia: AAMT.

Afamasaga-Fuata'i, K. (2007b). Communicating students' understanding of undergraduate mathematics using concept maps. In J. Watson & K. Beswick (Eds.), *Mathematics: Essential research, essential practice*. Proceedings of the 30th Annual Conference of the Mathematics Education Research Group of Australasia (Vol. 1, pp. 73–82). University of Tasmania, Australia: MERGA.

Afamasaga-Fuata'i, K., Meyer, P., & Falo, N. (2007). Primary students' diagnosed mathematical competence in semester one of their studies. In J. Watson & K. Beswick (Eds.), *Mathematics: Essential research, essential practice*. Proceedings of the 30th Annual Conference of the Mathematics Education Research Group of Australasia (Vol. 1, pp. 83–92). University of Tasmania, Australia: MERGA.

Afamasaga-Fuata'i, K., Meyer, P., Falo, N., & Sufia, P. (2007). Future teachers' developing numeracy and mathematical competence as assessed by two diagnostic tests. Published on AARE's website. http://www.aare.edu.au/06pap/afa06011.pdf

Australian Association of Mathematics Teachers (AAMT) (2007). The AAMT standards for excellence in teaching mathematics in Australian schools. Retrieved from http://www.aamt.edu.au/standards, on July 18, 2007.

Ausubel, D. P. (2000). *The acquisition and retention of knowledge: A cognitive view*. Dordrecht, Boston: Kluwer Academic Publishers.

Baroody, A. J., & Bartel, B. H. (2000). Using concept maps to link mathematical ideas. *Mathematics Teaching in Middle Schools, 5*(9), 604–609.

Baroody, A. J., Feil, Y., & Johnson, A. R. (2007). An alternative reconceptualization of procedural and conceptual knowledge. *Journal of Research in Mathematics Education, 38*(2), 115–131.

Brahier, D. J. (2005). *Teaching secondary and middle school mathematics* (2nd ed.). New York: Pearson Education, Inc.

Hansson, O. (2005). Preservice teachers' view on $y=x+5$ and $y=\pi r^2$ expressed through the utilization of concept maps: A study of the concept of function. In H. Chick & J. L. Vincent

(Eds.), *Proceedings of the 29th conference of the international group for the pscychology of mathematics education* (Vol. 3, pp. 97–104). Melbourne: PME.

Hatano, G. (2003). Foreword. In A. J. Baroody & A. Dowker (Eds.), *The development of arithmetic concepts and skills: Constructing adaptive expertise* (pp. 11–13). Mahwah, NJ: Erlbaum.

Hiebert, J., & Carpenter, T. P. (1992). Learning and teaching with understanding. In D. G Grouws (Ed.), *Handbook of research in mathematics teaching and learning* (pp. 65–97). New York: Macmillan.

Liyanage, S., & Thomas, M. (2002). Characterising secondary school mathematics lessons using teachers' pedagogical concept maps. In B. Barton, K. C. Irwin, M. Pfannkuch, & M. O. J. Thomas (Eds.), *Mathematics education in the South Pacific.*Proceedings of the 25th annual conference of the Mathematics Education Research Group of Australasia (MERGA-25), July 7–10, 2002 (pp. 425–432). New Zealand: University of Auckland.

Marsh, G. E., & Ketterer, J. K. (2005). Situating the zone of proximal development. *Online Journal of Distance Learning Administration, Volume VIII, Number II Summer, 2005, University of West Georgia, Distance Education Center.* Retrieved on August 20, 2007 from http://www.westga.edu/~distance/ojdla/summer82/marsh82.htm

Mintzes, J. J., Wandersee, J. H., & Novak, J. D. (Eds.). (2000). *Assessing science understanding: A human constructivist view.* San Diego, CA, London: Academic.

National Council of Teachers of Mathematics (NCTM) (2007). *2000 Principles and Standards.* Retrieved on July 28, 2007 from http://my.nctm.org/standards/document/index.htm

Novak, J. D. (2002). Meaningful learning: The essential factor for conceptual change in limited or appropriate propositional hierarchies (LIPHs) leading to empowerment of learners. *Science Education, 86*(4), 548–571.

Novak, J. D., & Cañas, A. J. (2006). The theory underlying concept maps and how to construct them. Technical Report IHMC Map Tools 2006-01, Florida Institute for Human and Machine Cognition, 2006, available at: http://cmap.ihmc.us/publications/ResearchPapers/ TheoryUnderlyingConceptMaps.pdf

Novak, J. D., & Gowin, D. B. (1984). *Learning how to learn.* Cambridge: Cambridge University Press.

Piaget, J. (1969). *Science of education and the psychology of the child.* New York: Grossman Publishers.

Pratt, N., & Kelly, P. (2005). Mapping mathematical communities: Classrooms, research communities and master classes. Retrieved on August 20, 2007 from http://orgs.man.ac. uk/projects/include/experiment/communities.htm

Richards, J. (1991). Mathematical discussions. In E. von Glaserfeld (Ed.), *Radical constructivism in mathematics education* (pp. 13–51). London: Kluwer Academic Publishers.

Ruiz-Primo, M. (2004). *Examining concept maps as an assessment tool.* In A. J. Canãs, J. D. Novak, & F. M. Gonãzales (Eds.), *Concept maps: Theory, methodology, technology.* Proceedings of the First International Conference on Concept Mapping September 14–17, 2004 (pp. 555–562). Navarra: Dirección de Publicaciones de la Universidad Pública de Navarra, Spain.

Schoenfeld, A. H. (1996). In fostering communities of inquiry, must it matter that the teacher knows "the answer." *For the Learning of Mathematics, 16*(3), 569–600.

Steffe, L. P., & D'Ambrosio, B. S. (1996). Using teaching experiments to enhance understanding of students' mathematics. In D. F. Treagust, R. Duit, & B. F. Fraser (Eds.), *Improving teaching and learning in science and mathematics* (pp. 65–76). New York: Teachers College Press, Columbia University.

Sternberg, R. J., & Williams, W. (Eds.). (1998). *Intelligence, instruction, and assessment theory into practice.* Mahwah, NJ, London: LEA Lawrence Erlbaum Associates.

Swarthout, M. B. (2001). The impact of the instructional use of concept maps on the mathematical achievement, confidence levels, beliefs, and attitudes of preservice elementary teachers. Dissertation Abstracts International, AADAA-I3039531, The Ohio State University, Ohio.

Vygotsky, L. S. (1978). *Mind in society.* Cambridge, MA: Harvard University Press.

Williams, C. G. (1998). Using concept maps to access conceptual knowledge of function. *Journal for Research in Mathematics Education, 29*(4), 414–421.

Part V
Future Directions

Chapter 16
Implications and Future Research Directions

Karoline Afamasaga-Fuata'i

Empirical evidence presented in the preceding chapters demonstrated that there is much to be gained educationally in general, and mathematically in particular, for learners and teachers of mathematics at all levels, through the development and application of an innovative approach to teaching mathematics by utilising the meta-cognitive strategy of concept mapping, available computer software and internet resources and recent developments in mathematics education to improve the processes involved in the planning and teaching of mathematics by teachers and in the learning and assessment of students' problem solving skills and most important, the constructive development of students' conceptual understanding of mathematics. Whilst various authors from different parts of the world, investigated and presented the work of one, a few, or many teachers/students, subsequent findings nonetheless suggest potentially viable approaches that could be usefully adopted and adapted in mathematics classrooms anywhere in the world to address the recurring problems experienced by many students as they struggle to make sense of mathematics problems, concepts and processes. These findings have implications for learners of mathematics both in schools and various university programs.

For example, Novak & Cañas proposed a new model of education that involved the use of expert concept maps to scaffold learning and the collaborative construction of concept maps or knowledge models using CmapTools and readily available Internet resources. Having students explore real problems, develop knowledge models and communicate their mapped understanding or knowledge models, as oral, written or video reports are key aspects of the proposed model. Whilst work at the Institute has focused mainly in science education, much remains to be done in mathematics education.

In addition to concept maps, some authors examined vee diagrams, as an additional meta-cognitive tool, to determine the extent of students' analysis and understanding of the *integrated linkages* between methods of solving problems and the conceptual structure of mathematics. Afamasaga-Fuata'i, in the second chapter,

K. Afamasaga-Fuata'i (✉)
University of New England, Armidale, Australia
e-mail: kafamasa@une.edu.au

K. Afamasaga-Fuata'i (ed.), *Concept Mapping in Mathematics,*
DOI 10.1007/978-0-387-89194-1_16, © Springer Science+Business Media, LLC 2009

provided evidence from a case study of a primary preservice teacher's analysis of the measurement content strand in primary mathematics, which demonstrated the richness of interconnections between antecedent concepts of measurement such as length, distance, area, volume and capacity, and developmental trends across the different stages of the primary level. Most important were the visual displays of networks of propositions on concept maps and theoretical and procedural entries on vee diagrams, which effectively encapsulated the interconnections between the curriculum's Knowledge and Skills and Working Mathematically Syllabus Outcomes. The student's reflections provided further evidence of the value of the meta-cognitive strategies in enabling her to become more pedagogically aware of developmental trends and interconnections most relevant to make mathematics learning and problem solving more meaningful and conceptual for her future students. Examining the impact of this kind of preparatory work on teaching practice and students' meaningful learning of mathematics is worthy of further investigation.

In their chapter, Schmittau & Vagliardo provided evidence from a case study of a primary preservice teacher, who constructed a concept map which reflected the results of her in-depth exploration of the meanings associated with and underlying the concept of positional system, its antecedent concepts and the complexity of interrelationships within a conceptual system. The student's concept map addressed both a serious deficiency in current elementary mathematics programs in the United States and provides a reliable direction for future mathematics curriculum development. Findings underscore the importance of conducting similar analyses to include other fundamental concepts of mathematics to promote the view of mathematics as a conceptual system.

Afamasaga-Fuata'i, in Chapter 4, provided additional evidence from another case study, which demonstrated the usefulness of maps/diagrams in making explicit the conceptual structure of the fraction strand of a primary mathematics curriculum. Through a series of progressive concept maps, the post-graduate student demonstrated his evolving conceptual and hence pedagogical understanding of the development of the fraction concept across the primary level. The conceptual labels, and propositional links apparent in the final overview map were substantively more enriched than the initial maps whilst through the construction of vee diagrams, the student engaged in the processes of critical analysis and synthesis, thinking and reasoning, justifying and explaining his knowledge and understanding to highlight the interconnections between mathematics principles, concepts and procedures. The findings imply that the routine construction of concept maps and vee diagrams and regularly expecting students to provide justifications for their mapped interrelationships and methods of solutions can promote the careful development of solutions that make sense and can therefore be justified and rationalized (cf. Ellis, 2007, p. 196), and inevitably over a long period of time, could lead to learning mathematics that is more meaningful and more conceptual in regular mathematics classrooms. Such an undertaking and anticipated outcome requires further research.

In Chapter 5, Afamasaga-Fuata'i & McPhan provided data from a project, which introduced concept maps as an innovative learning and assessment tool in two Australian schools. The evidence provided suggested that implementing innovation in schools and incorporating it into classrooms is dependent on teacher change; a long-term process with the most significant changes in teacher attitudes and beliefs occurring after teachers began implementing the innovation successfully and observed changes in learning (Guskey, 1985). Student maps provided snapshots of their understanding of a topic at a particular time. Assessing the displayed inter-connections revealed the extent to which meaningful learning took place, leading to further reflection and discourse between student and teacher and/or amongst students. The teachers indicated that they would continue with concept mapping as a strategy to brainstorm ideas about new topics and as an assessment tool after a unit. Peer tutoring was highly recommended as a means of working collaboratively with new students or new schools. These suggestions provide viable directions for further research with more schools in anticipation of a broader impact of concept mapping on student learning and assessment.

Pozueta & Gonzàlez, in their chapter provided evidence of meaningful learning of proportionality in a Spanish second grade secondary classroom by comparing students' pre-instruction and post-instruction concept maps to an expert concept map. Indicators of meaningful learning included increase in numbers of concepts used, reduction in errors or inaccurate propositions, increased clarity in hierarchical levels and coherence with the inclusivity of the concepts, and increase in progressive differentiation reconciliated integratively. Findings revealed concept mapping as a useful tool to enable the design of an innovative and conceptually more transparent instructional module on proportionality and also to check the pupils' prior knowledge and monitor their learning process. Potential directions for future research include the investigation of the relationship between concept mapping, some personality aspects and academic performance as a means of predicting students' learning capacity.

Schmittau in Chapter 7 provided evidence from a case study of two secondary preservice teachers who, after receiving the same instruction on the concept of multiplication in a graduate mathematics education course constructed vastly different concept maps, which revealed their differential understanding of the nature of mathematics as a conceptual system, the conceptual content of mathematical procedures and requisite pedagogical content knowledge to mediate such understandings to future learners. One student demonstrated that she had internalized the concept in its systemic interconnections while the other continued to see it through formalistic lens. Findings suggest the need for teachers to not only examine the conceptual connections of superordinate concepts as revealed in curriculum documents and textbooks but that they should also analyse the concepts' historical conceptual development to fully grasp the nature of mathematics as a conceptual system. Further research is necessary to conduct similar analyses with other fundamental concepts in mathematics.

Afamasaga-Fuata'i, in Chapter 8, provided evidence that demonstrated the use of concept maps by an Australian secondary preservice teacher to develop a teaching sequence on derivatives and a lesson plan to introduce the formal definition. Of his own volition, he constructed additional concept maps of the 2-year senior mathematics curriculum first to conceptually and pedagogically situate the topic at the macro-level before a microanalysis of the conceptual interconnections of derivatives to prior knowledge and subsequent learning. Findings imply that concept mapping has the potential to explicate preservice teachers' pedagogical content knowledge and understanding of the relevant syllabus in more conceptually-based and interconnected ways, for further discussion and subsequent assessment of their developing pedagogical competence to mediate meaning of mathematical concepts and processes.

In Chapter 9, Vagliardo presented evidence of the value of concept mapping, when combined with historical research, as an important epistemological tool that can make transparent the conceptual essence of a mathematical idea such as logarithms. Other direct uses such as exposing the operating understanding of important mathematics concepts held by both teachers and students provided the means to identify substantive focus for curriculum design and provide pedagogical direction for positive student learning in mathematics. Findings imply that concept mapping can instruct and guide mathematics educators toward a pedagogy of significant cognitive consequence. Further research is necessary to examine the classroom application and implementation of these ideas, and its subsequent impact on student learning.

Ramirez, Aspèe, Sanabria & Neyra, in their chapter, provided useful guidelines for constructing concept maps and vee diagrams. Using various examples of physical phenomena, they elaborated a number of strategies to guide students' constructions to explicate their understanding of various modeling functions visually on concept maps and subsequently to illustrate, on vee diagrams, how these could be appropriately applied in the context of modeling physical phenomena. The practical suggestions provided reinforced the simplicity of the strategies of concept mapping and vee diagramming and encouraged both students and teachers to consider using them. The challenge therefore for further research is to investigate various models of incorporating concept mapping and vee diagrams in classrooms to further enhance students' learning outcomes in both physics and mathematics.

Caldwell, in Chapter 11, provided evidence from a project that used a concept mapping approach to course planning and lesson planning, development of student learning and assessment of student learning of Algebra I in middle school. Caldwell not only elaborated the processes of planning and constructing comprehensive concept maps for a topic which involved groups of teachers but also provided an assessment model to evaluate teachers' and students' concept maps. Further research into the long-term use of this concept mapping approach in classrooms and its impact on mathematics achievement would be most worthwhile.

Afamasaga-Fuata'i, in Chapter 12, using data from a group of Samoan undergraduate mathematics students, demonstrated that the concurrent use of concept mapping and vee diagrams, as means of supporting students' learning and development of their understanding of new topics, contributed significantly in highlighting the close correspondence between a topic's conceptual structure and its methods and the realization that constructing maps/diagrams requires and demands deeper understanding of interconnections than simply knowing what the main concepts and formulas are. Findings imply that undergraduate mathematics learning can be a more meaningful and conceptually enriching experience through the concurrent use of these metacognitive tools. Further research into the routine use of concept mapping and vee diagrams in undergraduate mathematics and its impact on students' creativity and flexibility in problem solving can explicitly inform the improvement of teaching practices in undergraduate mathematics.

Pérez, in Chapter 13, provided evidence of the use of concept maps by an instructor to guide his teaching and the impact of this approach on students' comprehension and learning of calculus. Findings showed the improvement of students' skills in numerical reasoning, abstract reasoning and spatial relationships as a result of integrating concept maps in the teaching-learning process. Further research is necessary to examine the long-term impact of using concept maps in both teaching and learning calculus.

In Chapter 14, Afamasaga-Fuata'i provided a case study that demonstrated the impact of the concurrent use of concept mapping and vee diagrams on a Samoan student's understanding of, and proficiency in solving problems on, differential equations. Newly established socio-mathematical norms within the classroom promoted mathematical dialogues, critiques and meaningful discussions of each other's ideas. Findings underscore the value of using maps/diagrams as means of making transparent one's developing mathematical understanding for peer critiques. Further research would be worthwhile in identifying potential barriers to incorporation of these metacognitive strategies in undergraduate mathematics classrooms.

Afamasaga-Fuata'i, in Chapter 15, presented evidence from a group of Samoan undergraduate students' work with concept maps, which demonstrated that, initially, students experienced difficulties reconciling the type of linear learning style they were used to (e.g., by memorizing definitions) to the hierarchical nature and organization required for concept mapping. The progressive quality of students' maps over the semester confirmed that students' deepening understanding of their topics were very much influenced by the scaffolding and social interactions during critiques, the newly established socio-mathematical norms of the classroom setting, and the socially-validated structure of mathematics. That is, the deliberate alteration of the classroom environment and practices particularly in terms of the individual/social orientation of the learning, and the form and extent of the focus of assessment (Pratt & Kelly, 2005) resulted in a more conceptual and theoretical understanding of their topics. Findings imply that the simultaneous use of concept mapping and social critiques in class has the potential to promote the development of mathematical thinking, reasoning and effective communication in mathematics classrooms.

Premised on the research findings presented in this book, further research is desirable to fully investigate the potential value, advantages and limitations of using concept maps as a metacognitive tool for planning, teaching, learning and assessment for, and of, learning in whole school settings and in the preparation of prospective teachers of mathematics. How we can effectively introduce concept mapping into mainstream mathematics education is an issue that is worthy of further research. Moreover, collaborative construction of concept maps in mathematics classrooms at different levels by using an appropriate software or CmapTools and its potential impact on student learning is another fertile area for further research. Worthy also of further investigation is the identification of the types of mathematics courses or topics that students are likely to find most conducive for concept mapping to optimize the development of their conceptual understanding of the relevant mathematical concepts, processes and procedures particularly those often encountered in routine problem solving in anticipation of subsequently solving novel and/or more challenging problems. Besides the applications of concept mapping already presented in this book, further research is necessary to explore other innovative pedagogical uses of concept maps and concept mapping in mathematics.

Whilst the chapters in this book reported on research of a minute part of the plethora of problems faced by mathematics students in educational institutions, the authors hope that, through the empirical evidence generated from working with real students, teachers and educators, the findings herein will provoke further research and discussions in the mathematics education community about how this metacognitive strategy can contribute more efficiently and effectively in enhancing the meaningful learning of mathematics and proficient solving of more challenging mathematics problems.

References

Ellis, A. B. (2007). Connections between generalizing and justifying: Students' reasoning with linear relationships. *Journal for Research in Mathematics Education*, *38*(3), 194–229.

Guskey, T. (1985). Staff development and teacher change. *Educational Leadership*, *42*, 57–60.

Pratt, N., & Kelly, (2005). Mapping mathematical communities: Classrooms, research communities and master classes. Retrieved on August 20, 2007 from http://orgs.man.ac.uk/projects/include/experiment/communities.htm.

Index

Printed in the United States of America